Human Rights and Human Nature

IUS GENTIUM

COMPARATIVE PERSPECTIVES ON LAW AND JUSTICE

VOLUME 35

For further volumes:
http://www.springer.com/series/7888

Marion Albers • Thomas Hoffmann • Jörn Reinhardt
Editors

Human Rights and Human Nature

Springer

Editors
Marion Albers
Faculty of Law
University of Hamburg
Hamburg, Germany

Thomas Hoffmann
Philosophy Department
Otto-von-Guericke University Magdeburg
Magdeburg, Germany

Jörn Reinhardt
Faculty of Law
University of Hamburg
Hamburg, Germany

ISSN 1534-6781 ISSN 2214-9902 (electronic)
ISBN 978-94-017-7987-6 ISBN 978-94-017-8672-0 (eBook)
DOI 10.1007/978-94-017-8672-0
Springer Dordrecht Heidelberg New York London

Printed on acid-free paper

Springer is part of Springer Science+Business Media (www.springer.com)

Contents

Chapter 1
Human Rights and Human Nature: Introduction

Marion Albers, Thomas Hoffmann, and Jörn Reinhardt

Recourse to the concept of human nature has always played a prominent role in the justification and defense of human rights. The idea has its roots in ancient philosophy and in religious traditions and also extends to early modern natural rights thinking. "Human nature", as a concept, is still widely used in contemporary philosophical and juridical human rights discourse as a way of justifying the universal and egalitarian validity of the claim of human rights. And in spite of historical changes in the use of the concept of human nature and its ontological implications, the idea that basic rights belong to all humans in the same way is re-affirmed in central contemporary human rights documents.

The specific nature of the human species seems to provide answers to some of the most controversial questions concerning human rights: questions about their scope and content, their universality and their basis of justification. Arguments from human nature are taken up to substantiate claims for human rights, and sometimes even to limit the growing list of rights in international human rights documents. In any case, human nature is taken as an argument that makes a difference.

But the attempt to reconsider the relevance of human nature for human rights is exposed to several difficulties. In the first place, it requires a reevaluation of the almost infinite and complex lines of "naturalistic" arguments which exist in the context of fundamental rights. The claim that basic human rights can be justified with recourse to arguments from human nature is a thesis about the justification of the universal validity of basic rights for human beings. But it is obviously not an ontological thesis about the existence or the origin of the rights in question. Since

M. Albers (✉) • J. Reinhardt
Faculty of Law, University of Hamburg, Rothenbaumchaussee 33, 20148 Hamburg, Germany
e-mail: marion.albers@uni-hamburg.de; joern.reinhardt@uni-hamburg.de

T. Hoffmann
Institute for Philosophy, Otto-von-Guericke-University Magdeburg, Universitätsplatz 2, 39106, Magdeburg, Germany
e-mail: Thomas.Hoffmann@ovgu.de

M. Albers et al. (eds.), *Human Rights and Human Nature*, Ius Gentium: Comparative Perspectives on Law and Justice 35, DOI 10.1007/978-94-017-8672-0_1,

human beings do not have rights in the same way as they have hands or legs, rights cannot be understood as a "natural" kind. The function of referring to human nature in the context of human rights is a different one, namely an argumentative function.

Of the various directions naturalistic arguments can take, one can distinguish, roughly speaking, between two main approaches, an externalist and an internalist one. Externalist approaches try to justify elementary moral and legal principles from "outside" our normative or evaluative vocabularies. The reductive ambition is to trace a normative concept back to certain non-normative descriptions of human beings as formulated in the vocabularies of the natural sciences. The result of these descriptions is supposed to make plausible why every human being, simply as a member of the species, has basic rights – regardless of whether the existing laws of particular states actually give their citizens such a right or not. In contrast, internalist approaches try to justify the universal validity of human rights by referring to a concept of human nature that is normative from the beginning. A prominent example is that of the various versions of "neo-Aristotelian naturalism" that are used in some contemporary moral theory approaches but may also be used to justify human rights and their claim to universality. The objections that are almost reflexively made against the various naturalisms (the Is-Ought relation and the objection of a naturalistic fallacy) affect externalist approaches more than non-reductive approaches. Nevertheless, major problems remain for naturalistic argumentations of any kind.

Presenting universal human rights as somehow being natural rights is not the only idea that looks back on a long tradition; indeed, the critique of this idea is as old. Most objections against the very idea of human rights still echo in one way or another Edmund Burke's critique or Jeremy Bentham's famous polemic of "nonsense upon stilts". Despite their different convictions and political commitments, both authors sharply criticized the natural law-justification of modern human rights documents. Their main objection centered on the argument that substantial rights can only exist within the framework of an existing legal system; this is required not just for conceptual reasons but also for effective enforcement. According to this critique, the idea of the existence of pre-constitutional rights, whether they be natural or even moral, is highly misleading, for the concept of a pre-constitutional right is as plausible as the concept of wooden iron.

Further difficulties concern the content of human nature. A list of essential features of human nature runs the risk of being only a particular and culturally bound expression. The more materially rich and normatively full of content this list gets, the more likely it will lose its generality. On the other hand, if the list is kept extremely formal and abstract in order to guarantee its generality, it will hardly have sufficient content to justify substantial rights like human rights. But how can "human nature" be understood appropriately in the context of human rights?

This leads to the second difficulty attempts to reconsider the relevance of human nature for human rights are exposed to. The notion of human nature itself must be further clarified. However, natural sciences have not only changed our understanding of the human being. Medical and biotechnical interferences as well as the developments in the life- and neuro-sciences make it possible to change essential features of the human species. From a naturalistic point of view, human nature itself is subject to transformation and transgression in an unprecedented manner. However, change even of fundamental characteristics or abilities might also be identified as part of human nature. At least, the vanishing line between the natural and the artificial challenges the idea that basic human rights could be "grounded" in certain features of our common human nature. Beyond that, biotechnical developments and especially the possibilities of biotechnically induced human enhancement represent various challenges to particular human rights positions.

Both discussions refer to one another. Without sufficient understanding of the possible status of human nature for human rights and of the role the concept of human nature plays, it must remain unclear what kind of argumentation is appropriate within the human rights discourse. At the same time, a clarification of the idea of human nature is necessary. This must involve the specific discussions on how to describe the human being, on corporality or on biotechnical alterations (to name only a few), because these discussions influence the understanding of human nature as well as the idea of human rights and its possible foundations.

Against this background, the essays in this volume explore the significance of our understanding of human nature for our understanding of human rights. The contributions in the first part of the book – The Role of Nature in Human Rights Discourse: Foundations and Limitations – explore the possibilities and limits of arguments from nature in the context of human rights and focus on the relation between the concept of human nature, normative principles and anthropological arguments. Is it possible and plausible to justify the universal and egalitarian validity claim of human rights by referring to the concept of human nature? Can the concept of human nature provide a basis for the understanding of human rights? Or does the idea of human rights in its modern, post-1945 manifestation essentially go beyond human nature?

Corinna Mieth argues in her chapter "The Double Foundation of Human Rights in Human Nature" that there are two aspects of human dignity that are simultaneously two aspects of human nature. One aspect concerns the normative, moral status of persons that is connected with their ability to act morally. The other aspect concerns the empirical status of persons that is connected with their neediness and vulnerability. It is this second aspect that leads us to determine the substances of human rights: There are some goods that are indispensable for a decent life – and these goods should be protected by human rights. However, the second aspect, Mieth claims, is connected to the first. Hence, human rights are founded in two aspects of human dignity. Mieth develops this idea with a view to the example of extreme poverty as a violation of human dignity and as a violation of human rights.

In his chapter "Human Rights and Human Animals", Bernd Ladwig assumes that whoever seeks to provide justifying reasons for human rights seems to be, in some way or another, committed to universal anthropological claims. Nonetheless, as Ladwig argues, it is clear that anthropology alone cannot provide sufficient grounds for human rights. In addition to an anthropological footing, we also need recourse to at least one valid moral principle. This anthropological approach provides the argumentative framework within which we can derive, according to Ladwig, some but not all of the specific contents of human rights.

Thomas Hoffmann argues in his contribution "Human Rights, Human Dignity, and the Human Life Form" that it is possible to formulate a justification of the idea of human rights by referring in a first step to the term of human dignity because statements including 'human dignity' refer anaphorically to generic sentences that articulate the concept of human nature. The concept of human nature is neither a set of elements (i.e. extensions) nor an empirical (e.g. biological) predicate. Rather, the concept of human nature is the form of its exemplars, as Hoffmann says (with reference, inter alia, to Aristotle): It is a life form. But a life form is also the norm that determines what is in general "naturally good" (Ph. Foot) for the individuals which exemplify this life form. If one understands the concept of human nature this way, then one is able to justify human rights by referring to human nature, since now human rights can be characterized as the attempt to secure the possibility of the natural flourishing of human beings in general.

A neo-Aristotelian path is also taken by Harun Tepe in his chapter "Rethinking Human Nature as a Basis for Human Rights". Tepe discusses the well known "capability or capabilities approach" (M. Nussbaum and A. Sen) and what he calls an "ontological anthropological approach" by the Turkish philosophers Takiyettin Mengüşoğlu and Ioanna Kuçuradi. According to Tepe, the systematic knowledge of potentialities and conditions for the actualization of these potentialities of human beings offered by the ontological anthropological approach is the prerequisite for the justification of human rights.

In the chapter "The Relationship Between Human Nature and Human Rights. The Confucian Example" Mateusz Stępień elaborates on the idea of rights as grounded in human nature from within the Confucian tradition. From the observation of a lack of any concept of rights in Confucianism, Stępień analyzes the impact of the particular vision of human nature developed by Mengzi on the "discursive space of Confucianism" and the elements of Mencian theory that determine, first, the lack of recognition of rights, and second, even the existence of the discourse on rights within Confucian philosophical tradition. Just as classic concepts of natural law in the Western tradition do not necessarily develop a concept of (individual and subjective) rights, the Confucian (Mencian) idea of inborn goodness was rather an obstacle for developing a concept of rights in the Middle Kingdom, not its stimulus.

Frederik von Harbou offers a naturalistic account of human rights that refers to scientific facts about human beings but does not imply reductionism. The aim of his chapter, "The Natural Faculty of Empathy as a Basis for Human Rights" is to show why human rights are both conceptually and empirically based on the natural human faculty of empathy. Von Harbou understands human rights as expressing a certain minimal standard of morality that implies an altruistic motivation. Both analytical arguments and neuropsychological findings suggest that original altruistic behaviour may only be explained by compassion, which ultimately requires empathy. Empathy is a natural and cross-culturally developed human faculty.

In his chapter, "Human Rights and Basic Needs" Peter Schaber, on the other hand, argues that core human rights are based in the person's basic entitlement to exercise normative authority. According to Schaber, human rights are not grounded in the basic needs of human beings. Rather, both civil liberties and social rights are intended to protect the exercise of this normative authority. Schaber does not claim that the idea of normative authority can ground all or even most human rights, but in his view it is clear that it grounds more human rights than the idea of basic needs.

The objection against the conceptual connection between 'human right' and 'human nature' in Arnd Pollmann's chapter "Human Rights beyond Naturalism" is much stronger. He claims that a plausible human rights approach can and should abstain from any substantial references to human nature. Although philosophical accounts of human rights have to be based on the presupposition of membership in the human species, the idea that human rights are 'natural rights', Pollmann argues, is highly misleading. Pollmann interprets human rights as 'constitutional rights' from the conceptual outset. The subject of human rights is not the pre-political or natural human being but the somehow anticipated, democratically transformed, and also revolutionary subject of his or her own future.

Jörn Reinhardt's contribution, "Human Rights, Human Nature, and the Feasibility Issue" concerns the problem of feasibility. Arguments from nature were not only used to justify fundamental rights. For natural rights thinking, human nature was an instance of critical evaluation as to whether an idea of rights is feasible or not. The different ambitions that are connected with the very idea of rights, especially in the modern natural right tradition, could often be traced back to divergent anthropological assumptions on human dispositions and conduct. In his contribution, Reinhardt's aim is to show how the feasibility requirement applies to fundamental rights. Even though the idea of a "natural right" (in its many variations) and post-1945 human rights are two highly distinct phenomena, a central premise in both discourses is that a concept of rights must be realistic (or rather realizable). Reinhardt explores to what extent arguments from human nature are helpful to deal with the feasibility issue.

The second part of this volume – Species manipulation and the transformation of human nature: Challenges to human rights – addresses the challenges to human rights that result from the transformation of human nature and species manipulations. How do the transformations of human nature change our understanding of human rights and particular basic rights? Which of the possible manipulations of the human body are ethically and legally justifiable? What are the arguments to protect human nature against manipulation? Or should there, in fact, be a human right to attain enhancement and manipulation?

In his chapter, "How to Protect 'Human Nature' – By Human Dignity, Human Rights or with 'Species-Ethics' Argumentations?", Georg Lohmann turns his attention to the current developments of medical genetic engineering that change the methodical attitude towards human nature and also express a change in the evaluation of human nature. Lohmann asks if we could or should protect human nature by human dignity and human rights, or if we need other normative standards. To find an answer he probes the hypothesis that human nature can be protected by human rights and the "untouchability" (Unantastbarkeit) of human dignity. After having discussed arguments brought forward by Jürgen Habermas, Lohman examines three normative arguments for the undisposability (Unverfügbarkeit) of human nature and compares Habermas' species-ethical argument for the undisposability of human nature to similar "post-metaphysical" challenges like an intercultural understanding of human dignity as a justified principle of human rights.

Markus Rothhaar's chapter "Species, Potentiality and Their Manipulation" concerns arguments of potentiality and species membership in the ongoing debate regarding the moral status of human embryos. According to Rothhaar, both of these arguments directly point to the relationship between nature and normativity. As such, they are only convincing if one assumes that species membership and developmental potential merely provide criteria rather than the foundational reasons for moral status. In light of this, the chapter examines proposals made by Kant, Fichte, and Spaemann on how to understand the relationship between the reasons for and the criteria of moral status and contrasts these with recent developments in genetics that have made both species membership and embryo potential, to some extent, manipulable.

Ingrid Schneider's chapter, "The Body, the Law, and the Market: Public Policy Implications in a Liberal State", explores the interaction of the law, the market, and public policy in the governance of body parts which are used for medical ends. Schneider lays out the legal and philosophical groundwork by exploring the universal norm of non-commercialization of the human body, as enshrined in legal documents, declarations and professional codes of ethics. She also reconstructs major philosophical arguments regarding the relationship of humans to their bodies, in particular vis-a-vis Kant's philosophy of morals, and analyzes how continental European law deals with the duality of the human body as person and "material thing". After these analyses, she puts these legal and ethical norms, deontological principles and consequentialist reasoning to the test by scrutinizing models and arguments brought forth to justify financial incentives for organ procurement.

In her chapter "Collection of Human Tissue Samples in Biobanks: Challenges to Human Rights and Human Nature" Bianka Dörr turns the attention to the issue of biobanks. In recent years, biobanks have become major strategic and powerful tools for undertaking medical-scientific research. However, the use of human tissue and donor-related data for research and biomedical applications raises important legal and ethical questions. Dörr focuses on the concept of informed consent and discusses it within utilitarian and human rights approaches while exploring its implications for human nature. She argues that a clear commitment to a human rights approach should be adopted, one that values and respects the individual as the sample donor and asks for his/her informed consent in cases where his/her bodily material will be used for current and future research.

Tetsu Sakurai discusses the consequences of a step that is controversial on its own terms: "Should Society Guarantee Individuals the Right to Maintain 'Normal Functioning'?: A Genetic Minimalist Approach in a Globalized World". His chapter focuses on the idea of a "genetic decent minimum", which requires public policy to provide all members of a society with a certain genetic endowment that will enable them to participate in all spheres of life as normal competitors and collaborators. Sakurai considers the moral implications of the development of genetic technology in the laboratories of affluent societies in light of the fundamental inequalities that exist on a global scale. He takes up Audrey Chapman's criticism of biocentrism from her egalitarian point of view and examines her prediction that, if expensive genetic enhancement technology is used in rich societies, it will inevitably aggravate the economic gap between the rich and poor countries by creating "doubly-strong competitors."

In the final contribution "Enhancement, Human Nature, and Human Rights", Marion Albers centers on human enhancement and analyzes the consequences enhancement possibilities and activities have on the understanding of human nature and human rights. After analyzing and clarifying the concept of human enhancement by presenting fields and visions as well as conceptual dualities such as "therapy/enhancement", "normality/supranormality" or "naturalness/artificiality", she concludes that enhancement is a complex, inherently reflexive concept. A closer analysis of the enhancement debates shows that human nature, human dignity, identity, autonomy or equality are the origin of oppositional arguments. The problem of enhancement reveals how varied and in need of contextualization these concepts are. The relationship between the concept of human nature on the one hand and human rights on the other has always been complex, and in the present-day human rights discourse "human nature" might be assigned a particular role just because of its ambiguity and rich implications. The enhancement problem can be seen as a catalyst for reflexivity because it sets off new discussions on fundamental questions. It enriches the discourse on human nature and human rights and, in turn, benefits from being part of such a discourse.

This volume began to take shape at a workshop organized at the XXV. IVR World Congress of Philosophy of Law and Social Philosophy in August 2011. The workshop was followed by a Colloquium at Magdeburg University in the conference centre in Schloss Wendgräben in November 2011. During the ensuing period, the contributions of this volume were worked out, discussed and refined. Especially, we would like to thank Erik Kravets, M.A. and Audrey Kravets, J.D. for their help and constructive comments during the editing process.

Part I
The Role of Nature in Human Rights Discourse – Foundations and Limitations

Chapter 2
The Double Foundation of Human Rights in Human Nature

Corinna Mieth

> [. . .] recognition of the inherent dignity and of the equal and inalienable rights of all members of the human family is the foundation of freedom, justice and peace in the world. (UDHR, Preamble)

> All human beings are born free and equal in dignity and rights. They are endowed with reason and conscience and should act towards one another in a spirit of brotherhood. (UDHR, Art. 1)

In many international documents[1] as well as philosophical theories (see Griffin 2008; Ashford 2013; Nussbaum 2006, 2008), human dignity is considered to be a source of human rights. In this paper I want to have a closer look at the connections between human rights, human dignity and human nature. I will hold that there are two aspects of human dignity that can be considered as two aspects of human nature. One aspect concerns the normative, moral status of persons that is connected with their ability to act morally. The other aspect concerns the empirical status of persons

[1]In the Preamble of International Covenant on Economic, Social and Cultural Rights (General Assembly resolution 2200A (XXI)) of 16 December 1966, entry into force 3 January 1976, in accordance with article 27 it says: "Considering that, in accordance with the principle proclaimed in the Charter of the United Nations, recognition of the inherent dignity and of the equal and inalienable rights of all members of the human family is the foundation of freedom, justice and peace in the world, recognizing that these rights derive from the inherent dignity of the human person" and in the International Covenant on Civil and Political Rights adopted by the General Assembly of the United nations in 1966, it says also "Considering that, in accordance with the principles proclaimed in the Charter of the United Nations, recognition of the inherent dignity and of the equal and inalienable rights of all members of the human family is the foundation of freedom, justice and peace in the world, recognizing that these rights derive from the inherent dignity of the human person".

C. Mieth (✉)
Institute for Philosophy I, Ruhr-Universität Bochum, Universitätstr. 150, 44780 Bochum, Germany
e-mail: corinna.mieth@rub.de

M. Albers et al. (eds.), *Human Rights and Human Nature*, Ius Gentium: Comparative Perspectives on Law and Justice 35, DOI 10.1007/978-94-017-8672-0_2,

that is connected with their neediness and vulnerability. It is this second aspect that leads us to a determination of the substances of human rights. There are some goods that are indispensible for a decent life. These goods should be protected by human rights.[2] But the second aspect is connected to the first: The recognition of all other human beings as equal moral persons (of their equal moral status, of their dignity) is the foundation of the recognition of their human rights.

In the following, I will develop the idea that human rights are founded in two aspects of human dignity by using the example of extreme poverty as a violation of human dignity and as a violation of human rights. I will hold that there are indecent standards of living that do not constitute violations of dignity and there are violations of dignity that do not constitute violations of human rights. In the first part of the paper, I will ask what it means to lead a life in dignity. I will hold that this question can be properly understood as a question about decency, about a decent standard of living that is adequate with regard to our nature as physical human beings. In the second part, I will show that the qualification of the standard of living of certain persons as a violation of dignity requires not only an inadequate standard of living but also the non-consideration of the moral status of the persons in question. This goes back to the foundation of dignity in our moral nature. In the third part, I will spell out under which conditions severe poverty can be considered to violate human dignity. In the fourth part, I will hold that, although human dignity can be considered to be a foundation for human rights, not all violations of human dignity violate human rights.

2.1 A Decent Life as a Life According to Human Nature

> [E]xtreme poverty is a violation of human dignity and might, in some situations, constitute a threat to the right to life. (UN General Assembly Resolution 134, 18 December 1992)

This formulation might puzzle us: Here the UN General Assembly does not qualify extreme poverty as a human rights violation. It qualifies it instead as a violation of human dignity and – under some conditions – as a threat to, not a violation of, the right to life. There is a tension between the evaluation of human dignity and the right to life. One might get the impression that the addressed violation of dignity is weaker than the threat of a human rights violation of the right to life. This might puzzle us further since, in prominent cases that might come to mind when violations of human dignity are concerned, we might think of severe forms of torture, rape and

[2] I will not focus on what exactly these goods are in this paper. Of course, it is disputable as to what counts as a necessary or indispensible good. For my purpose here, it might suffice to determine these goods in the most general and formal way: that they are necessary for a decent life. For a position that proposes goods that are necessary for an autonomous life cf. Gewirth (1980). For a position that draws basic rights back to basic goods, the substances of these rights, cf. Shue (1996, for a critical comment see Mieth 2008).

degradation. In these cases, a violation of human dignity seems to be even more serious than a violation of the right to life. Although the German judicial system provides a defence to people who kill in self-defence or kill those who constitute a threat to innocent persons; however, torturing in place of killing in these scenarios violates human dignity and is therefore illegal. For Germany, a violation of human dignity seems even more severe than a violation of the right to life.

But for the context of extreme poverty this does not seem to be true: On the one hand, extreme poverty is often seen as incompatible with human dignity or even as a violation of human dignity, but this does not seem to imply that there is a right to some goods that are necessary for subsistence. Society accepts that actively harming someone is forbidden, but whether aiding in poverty relief is compulsory or not is controversial. Not to help people who suffer from extreme poverty is often considered a mere violation of a weak and imperfect positive duty that is normatively weaker than a violation of negative duties not to harm others. In the case of poverty, a violation of dignity seems to be less serious than a violation of or a threat to the right to life.

I think that if extreme poverty is addressed as a violation of human dignity, we should differentiate between two aspects of human dignity that go back to the two aspects of human nature mentioned above. The first aspect refers to the unconditional normative status of persons. The second aspect refers to the empirically conditioned possibility to lead a decent life. I will hold that both aspects must be concerned in order to qualify a state of affairs as a violation of human dignity. In the last part of the paper I will show that not all forms of violations of human dignity are considered human rights violations.

Thesis 1: Human dignity has two aspects. The first is the equal moral status of persons. The second is the necessary conditions for a decent life.

In this section I will have a closer look at the second aspect and what it means to lead a decent life. I think that this aspect is addressed by the UN General Assembly as well as in the often quoted report from former World Bank President Robert McNamara:

> Poverty at the absolute level – which is what literally hundreds of millions of men, women, and most particularly children are suffering from in these countries – is life at the very margin of physical existence. The absolute poor are severely deprived human beings struggling to survive in a set of squalid and degraded circumstances almost beyond the power of our sophisticated imaginations and privileged circumstances to conceive. (World Bank 1976 annual meetings of the Boards of Governors: summary proceedings, Washington D.C, 14)

This statement focuses on the fact that poverty means a "life at the very margin of physical existence". This idea is also addressed by the UN Assembly when they qualify poverty as a threat to the right to life. They posit that there is "a set of squalid and degraded circumstances" that extreme poverty is combined with. The questions for us are: (1) What does it mean to lead a decent life? (2) Under which circumstances can poverty be described as a violation of dignity?

Avishai Margalit considers dignity as another word for self-respect. The connection between dignity and self-respect is also central for Peter Schaber's position. We can differentiate between three dimensions of self-respect.

First, there is self-determination in the sense of the possibility to select between acceptable alternatives (Schaber 2010, 52).

Secondly, there is self-evaluation as Rawls points out: "it includes a person's sense of his own value, his secure conviction that his conception of his good, his plan of life, is worth carrying out." (Rawls 1971, 440) Furthermore, "self-respect implies a co ability, so far as it is within one's power, to fulfill one's intentions." (Ibid.)

Both dimensions include the wish to be acknowledged by others and the necessity to be positively evaluated by others. According to Rawls, self-respect is the most important basic good for a good life.[3] Beyond these respects, which refer to a good human life, self-respect has a moral dimension. Kantians would speak of moral autonomy here. For a Kantian, self-respect means to acknowledge oneself and others as beings that are able to act morally: that are able to put themselves under moral laws. In this respect, Kantians consider autonomy as the ground of dignity.

For what I will describe here as a violation of human dignity, Margalit uses the term humiliation. The important observation in our context is that there are two senses in which the term humiliation, and hence violation of dignity, can be used. The first that I will adopt here goes back to the differentiation between misfortune and injustice (Cf. Shklar 1990, 51ff.). We can only speak of an injustice, a violation of someone's moral or legal rights or claims, if other human beings are involved who have violated their duties, or if unjust institutions are in place which systematically deny someone equal respect, equal rights and life-prospects. The same applies to violations of dignity. The term humiliation that Margalit refers to is ambiguous since "not only behavior is liable to humiliate people. Conditions of life are also capable of providing sound reasons for feeling humiliated. Conditions are humiliating, however, only if they are the result of actions or omissions by human beings. Conditions ascribed to nature cannot be considered humiliating on my view" (Margalit 1999, 9). So Margalit uses the term humiliation synonymous with what I refer to as a violation of human dignity. The latter seems to be more precise since Margalit affirms that there is also a common "secondary, metaphorical sense in which people see the very conditions of human existence, such as old age, handicaps, or ugliness, as reasons for feeling humiliated [. . .]. [T]he secondary sense involves humiliation as a result of natural life conditions" (Ibid., 10). In this second sense, humiliation does not constitute a violation of human dignity.

---

[3] And self-respect is, to some degree, dependent on the basic structure of a society for Rawls. Only if the basic structure is legitimized with regard to the equal moral status of the individuals whose life-prospects are shaped by the basic structure, will self-respect be fully developable for most members of society.

By contrast, Margalit characterizes humiliation in the first sense "as rejection from humanity" (ibid., 135). He distinguishes between "three concepts of humiliation – treating humans as nonhuman, rejection, and acts intended to lead to lack of control or to highlight one's lack of control" (ibid., 146). He emphasizes, these are three "different senses rather than different meanings. [. . .] Humiliation in all its senses is especially closely linked to the negative justification for respecting humans, which involves the prohibition of humiliation as a type of cruelty that can be directed only at human beings. Taking away a creature's control by tying or locking it up is clearly also a manifestation of cruelty to animals, but what is unique to loss of control as a way of humiliating humans is not merely the cruelty of physical confinement but the symbolic element, which expresses the victim's subordination" (Ibid., 147).

Thesis 2: Loss of control can be humiliating, but it is not necessarily a violation of dignity. This is only the case if it,

- *is due to the acts of others (torture, exploitation, slavery, unnatural dependency)*
- *is leading to the loss of acknowledgement by others (e.g. caused by dependency)*
- *is constituting a violation of moral autonomy (non-respect of the equal moral status of other persons).*

So, humiliation as a violation of dignity has two aspects: the physical one that can manifest itself as a loss of control over one's body or one's life-conditions. This loss must be combined with the loss of being acknowledged by others in order to constitute a humiliation or a violation of dignity. For Margalit, this is understood as the denial of the "ability of freedom", that constitutes the deciding feature of being human (ibid.). In the interpretation proposed here, this means not to be respected as a moral being (a being with the ability to act morally). It means a violation of the moral status of the person.

Thesis 3: A decisive aspect of a violation of dignity consists in the non-consideration of the equal moral status of persons.

So again: at what point is poverty humiliating? We remember the well-known description that Robert McNamara has given of severe poverty as "a life at the very margin of physical existence. [. . .] By any objective standard, absolute poverty is an anachronistic tragedy in our century. A tragedy because it is a condition of life beneath the level of human decency and anachronistic because there are now at hand the economic and technological means to end it." (World Bank 1976 annual meetings of the Boards of Governors: summary proceedings, Washington D.C., 37)

Using the above-mentioned definition alone, severe poverty may be classified as neither humiliation nor a violation of dignity. If the severe poverty in question were somehow "natural", not dependent on other human being's actions or omissions, it would not constitute a violation of human dignity.

Thesis 4: The description of severe poverty as a condition beyond decency is not sufficient to qualify it as a violation of dignity.

2.2 What Does It Mean to Not Respect the Moral Status of Persons?

Many philosophers hold that the decisive feature of the non-consideration of the equal moral status of other persons consists in their instrumentalization.

> Rape violates the bodily, mental, and emotional life of a woman, affecting all her opportunities for development and functioning. Rape, we might say, does not remove or even damage dignity, but it violates it, being a type of treatment that inhibits the characteristic functioning of the dignified human being. It is inappropriate to use a human being as a mere tool in that way, because a human being should not be used as a mere tool: respect for human dignity prevents that. (Nussbaum 2008, 359)

Others hold that, in the case of torture, the person who is tortured is treated as a mere means in order to get information. But is this what makes torture or rape such a moral evil? That a person is treated as a mere means would also be true if there were no physical harm done (e.g., falsely promising to release the person if they give up the information, administering a truth serum).

False promises are instrumentalizations, but they are not necessarily humiliating. They constitute violations of the other person as a moral being whose ends deserve respect within the consideration of what ends we can legitimately follow, insofar as they violate the first aspect of dignity. But they do not necessarily involve the violation of the second aspect – the ability to lead a decent life.

Thesis 5: The prohibition of instrumentalization is too broad to determine relevant violations of human dignity.

On the other hand, the prohibition of instrumentalization seems too narrow to determine relevant cases of human-dignity-violations. Extreme poverty is not necessarily a form of or a result of instrumentalization. Here, the second aspect of dignity is affected. But the first aspect, the violation of respect for the equal moral status of other persons, is not solely violable by instrumentalization.

Nevertheless, slavery and exploitation that often accompany severe poverty do constitute extreme forms of instrumentalization. Elizabeth Ashford (2013) and James Griffin (2008) have recently pointed out that people who are suffering from severe poverty are extremely vulnerable to such forms of instrumentalization. We should differentiate here between 1. instrumentalization as a cause of poverty and 2. vulnerability to instrumentalization as a result of poverty. Thomas Pogge (2008) argues that we, the citizens of the wealthy and influential states, contribute to and profit from world poverty by maintaining an unjust economic order that does not prohibit exploitation e.g. by child labour. But the moral problem of extreme poverty is not solely explainable through reference to instrumentalization or the danger of or vulnerability to instrumentalization. If firm X offers child labour in Bangladesh, then, per Ashford's account, the poor are forced to waive their human rights if there is no other opportunity to survive. They are forced to not insist on having their human rights, rights that should protect them against such treatment. Still, it is not necessarily a solution to simply forbid firm X from offering child labour in

Bangladesh. Forbidding child labour would be a good step toward protecting human rights because it may ensure that the children are not exploited. From a moral point of view, if the poor are not instrumentalized by firm X any more, this would also be an advantage. But in the end, the poor would still suffer from severe poverty.

So if extreme poverty does not go back to instrumentalization or makes the poor vulnerable to the threat of instrumentalization, is there still a violation of dignity? Severe poverty in the very early days of mankind seems to not go back to instrumentalization. As such, it is questionable as to whether it is possible to describe this cases as a violation of human dignity. The same holds true for severe poverty under circumstances that are not influenceable by other human beings. Peter Schaber uses the example of a man who is shipwrecked. He is alone on a deserted island where he can find nothing to eat. His options are extremely restricted. He has no choice; he will, sooner or later, die of starvation or illness. Still, since there are no other human beings around, there is no violation of human dignity in this situation. And, according to Margalit's definition, there is also no humiliation.

Severe poverty, as such, does not constitute a violation of human dignity. Severe poverty means a life beyond human decency in the sense that people are lacking essential goods that would be necessary to lead a decent life. But so far we can only speak of a violation here if other human beings brought these circumstances about or instrumentalize the poor who are vulnerable to instrumentalization because they do not have much choice. What about cases in which other human beings are involved, not by instrumentalizing the poor, but by not helping them? Not helping someone does not seem to infringe upon the prohibition of instrumentalization. Is it possible that it constitutes a violation of human dignity nevertheless?

Astonishingly, we find steps in that direction by Kant. The Categorical Imperative does not solely entail a prohibition of instrumentalization. A look at the decisive passage from the Groundwork confirms this.

> So act that you use humanity, whether in your own person or in the person of any other, always at the same time as an end, never as a means. (Kant 1785, 4: 429)

We can distinguish two infringements against this imperative.

1. To use a person as a mere means (this excludes treating her as an end)
2. To not use a person as a mere means, but also not to treat her as an end

If the moral status of persons plays a decisive role in the qualification of an act or an omission as a violation of human dignity, and if it is possible to disregard the moral status of a person not solely by using her as a mere means but also by not treating her as an end, then there exist violations of human dignity that cannot be explained as infringements against the prohibition of instrumentalization.[4]

---

[4]What I leave aside here is the possibility of self-instrumentalization and self-degradation which would occur as violating duties towards oneself in Kantian philosophy. For Kant and Schaber the assumption of duties towards oneself is crucial. In Kant, the possibility to disregard oneself comes from the idea that the humanity in us belongs to our "real" moral nature of the homo noumenon. By contrast, my point is to claim that there should be equal respect for the physical conditions

Thesis 6: The prohibition of instrumentalization is too narrow to cover all relevant violations of human dignity.

The crucial requirement we can find in the formula of humanity of the categorical imperative is to treat others as ends, since this includes not using them as a means as well as a prohibition of not treating them as ends. What does it mean to treat another person as an end in and of herself? To treat others as ends means that we have to restrict our purposes if they are incompatible, in a morally relevant way, with those of other persons. From this, we can derive negative and positive duties. Dignity refers to the moral status of other persons. From this, it follows that we must respect others because of their ability to act morally and that we must not subordinate their purposes to ours without reflecting their moral weight. But these are the purposes of our empirical selves; in Kantian language, of the homo phenomenon.

Now we have the means to describe failure to help as a violation of dignity. Whosoever does not help someone who is in dire need, although help was easily possible, does indeed not instrumentalize this person, but neither does he treat her as an end in herself. To treat her as an end would mean to respect that her purpose to survive – which is an objective need – has severe moral significance. From this respect towards the other person and her purposes would follow that I must act accordingly. If my purpose is to get home as fast as I can and the other person's purpose is to survive, her purpose restricts mine: I have to help her even if I have to give my purpose up, would restrict my purpose not to go through any trouble in a morally relevant way. But this approach is in danger of being still too broad to qualify as relevant violations of dignity.

First, there can be many purposes of other moral persons which I can promote but do not have to promote. The problem with the Kantian approach is that Kant has no theory of goods that helps us to differentiate between purposes, the realization of which we must not prohibit, purposes we must promote, and purposes with regard to which this is not the case. Kant conceptualizes a very general duty of virtue whose content is to promote others' happiness in the Doctrine of Virtue. But this is not plausible since it, too, is unqualified. Kant does not differentiate between necessary empirical conditions of a decent life and further purposes that others may promote as part of a duty of virtue but which need not be promoted as part of a duty of justice. That is because Kant tends to emphasize the first aspect of dignity at the cost of the second. Instead, I propose that here the empirical aspect of dignity must come into play. I hold that the material conditions for a decent life (i.e. a life in dignity), which means a life of self-respect and self-determination, are precisely what others must respect, protect or promote. These conditions are also the substances of human rights.

---

necessary to lead a decent life and, insofar for the homo phaenomenon and his vulnerability and neediness, that is equally important for our human nature. On my account, the first aspect of dignity, the equal moral status of persons, must not be stressed at the cost of the second.

Thesis 7: The requirement to treat others as ends is also too broad. It must instead refer to justified moral demands that refer to our double human nature and the conditions of a decent life. The content of the justified moral demands results from what we consider the conditions for a decent life.

2.3 Does Poverty Constitute a Violation of Human Dignity?

Let us remember the two aspects of human nature that constitute the two aspects of dignity and must be violated in order to qualify an act or omission as a violation of human dignity. The empirical aspect refers to the conditions of a life in dignity. This refers to a life in self-determination in which people can select between acceptable options (Schaber 2012). This does not only refer to the options, but also to the conditions of the selection that Nussbaum has in mind. The protection of and the providence for these conditions are the purposes of human rights. These conditions can be considered the substances of human rights. Human-rights-corresponding duties refer to the protection of and the providing of these goods. Here, human dignity is violable. It is violable on the one hand by violations of human rights, i.e. the right to freedom that is violated by exploitation. This is the point that is emphasized by Griffin and Ashford: extreme poverty makes people vulnerable to infringements of their rights that they put up with in order to survive. Here, extreme poverty can constitute a reason for self-instrumentalization. But it would be cynical to qualify this as a violation of duties towards oneself under these circumstances. Rather, these circumstances go back to a life of extreme poverty where others were able to provide for the goods in question and do not do so. The normative aspect of dignity is affected in both cases, too. The human ability to act morally constitutes the normative aspect of dignity that is non-volatile but, nevertheless, violable. It is violable by not respecting someone as an equal moral subject.

How do these two aspects refer to each other if we want to qualify an act or omission as a violation of human dignity? First, it is necessary that the equal moral status of another person is not respected. Second, by this respectless act or omission that lacks respect, the material conditions for a life in dignity must be affected by actively depriving someone of an objective good or by not providing for this good although one could easily do so. Both conditions together are sufficient to qualify an act or omission as a violation of human dignity.

Thesis 8: Upon closer inspection, even in the case of torture, the violation of human dignity consists of the destruction of the possibility of self-determination. This is what the violation of human dignity by torture has in common with the violation of human dignity by severe poverty. The tortured person is actively deprived of self-determination. People in severe poverty who are lacking the objective goods needed for a life in self-determination are, by destruction of these goods or by not providing these goods, deprived of self-determination by others.

Under these conditions, severe poverty can be classified as a violation of human dignity. But what about relative poverty that also can, like severe poverty, go back to the non-respecting of others as equal moral persons? Adam Smith distinguishes between necessary goods and luxuries. What qualifies a good as a necessary good depends on the standard of a society. Necessary goods are not only necessary for physical, but also for social, survival. For example, he describes a necessary good in nineteenth century England as owning shoes and a linen shirt without which one would be "ashamed to appear in public".[5] So he considers necessaries to be "not only those things which nature, but those things which the established rules of decency have rendered necessary to the lowest rank of people" (ibid.). We often refer to this "rules of decency" that are not met when we refer to poverty. That people cannot afford what is considered a decent standard of living is true not only for severe poverty but also for relative poverty (cf. Neuhäuser 2010).

And it is relative poverty that Rawls and Margalit, whose theories I have referred to above, have in focus. For both of them, a life in self-respect depends on the acknowledgement of one's own life-plan by others. Rawls' solution to the problem of relative poverty is that differences concerning income and wealth are only legitimate if they have the best possible effect on the least advantaged people within a society. And this is founded in his idea that institutions are only legitimate if they are shown to express the equal moral value of persons whose life-prospects they shape (cf. Rawls 1971, § 40 and Mieth 2009). Also in Margalit's view, even more demanding, a society can solely refer to itself as decent if it grants to its citizens extensive social rights that go far beyond the minimum of existence.

We have two alternatives here: Either we restrict violations of human dignity to the violation and destruction of the equal moral status of a person or non-providence for necessary goods as condition for a life in decency. As a result, severe poverty

[5]"By necessaries I understand not only the commodities which are indispensably necessary for the support of life, but whatever the custom of the country renders it indecent for creditable people, even of the lowest order, to be without. A linen shirt, for example, is, strictly speaking, not a necessary of life. The Greeks and Romans lived, I suppose, very comfortably though they had no linen. But in the present times, through the greater part of Europe, a creditable day-labourer would be ashamed to appear in public without a linen shirt, the want of which would be supposed to denote that disgraceful degree of poverty which, it is presumed, nobody can well fall into without extreme bad conduct. Custom, in the same manner, has rendered leather shoes a necessary of life in England. The poorest creditable person of either sex would be ashamed to appear in public without them. In Scotland, custom has rendered them a necessary of life to the lowest order of men; but not to the same order of women, who may, without any discredit, walk about barefooted. In France they are necessaries neither to men nor to women, the lowest rank of both sexes appearing there publicly, without any discredit, sometimes in wooden shoes, and sometimes barefooted. Under necessaries, therefore, I comprehend not only those things which nature, but those things which the established rules of decency have rendered necessary to the lowest rank of people. All other things I call luxuries, without meaning by this appellation to throw the smallest degree of reproach upon the temperate use of them. Beer and ale, for example, in Great Britain, and wine, even in the wine countries, I call luxuries. A man of any rank may, without any reproach, abstain totally from tasting such liquors. Nature does not render them necessary for the support of life, and custom nowhere renders it indecent to live without them" (Smith 1776, Book 5, chapter 2, part 2, article 4).

as a violation of human dignity would have the same extension as the violation of social and economic rights. (That is a common position represented e.g. by Griffin 2008, Schaber 2010/2012, Gewirth 1980) Or we stretch the conception in order to encompass relative poverty.

Another problematic point is the fact that we do ascribe an indecent life to poor people. When this ascription leads to a self-description of a life beyond dignity in the first place, the ascription can lead to a violation of the self-respect of poor people. Furthermore, in this case, the ascription is self-fulfilling. On the other hand, with regard to a self-elected and self-determined life in poverty as with a mendicant or an ascetic, the ascription is wrong. Here again, it becomes evident that the empirical aspect is not sufficient to detect a violation of human dignity.

Thesis 9: Poverty is a violation of human dignity if poverty is so severe that the material conditions of self-determination are not given and if, thereby, the moral status of the poor is violated by others. This can be the case with exploitation as well as by not aiding the poor by providing necessary goods.

2.4 Human Rights and Human Dignity

So far we qualified violations of human dignity as violations of the equal moral status of other persons (normative aspect) plus reference to a basic good that is necessary in order to lead a decent life (empirical aspect). When it comes to human rights, both aspects are equally important. The first aspect is important because human rights are rights that express the equal moral status of persons since human beings do hold them simply because they are human, independent of social status, race, gender, etc. The second aspect concerns the substance – the content – of human rights. It is exactly because human beings are vulnerable physical beings that they need human rights in order to be protected against what Henry Shue (1996) called "standard threats" to a decent life; they need their liberty and subsistence secured against standard threats. But only other human beings can threaten and violate, as well as protect, the substances of human rights in a morally and legally relevant way. If we consider human rights only as legal rights, there may be violations of human dignity in the above described sense that do not qualify as violations of human rights. This is especially problematic with regard to extreme poverty. The underlying sources of legal human rights are moral human rights. Their normative foundation lies in human dignity, referring to the equal moral status of persons as well as referring to a decent human life. As I have shown above, both aspects must be considered in order to qualify an act as a violation of human dignity. With regard to qualify an act as a violation of moral human rights, we might see the matter analogously. Either violations of dignity in a narrow sense are the same as violations of moral human rights, or there are violations of human dignity that do not constitute violations of moral human rights. This would still be incompatible with treating others according to their equal moral status, but not with regard to basic goods that are necessary to lead a decent life. The answer to these questions – which acts or omissions do count as violations of dignity in a full or narrow sense and to what extent are they violations of moral human rights – will depend on the empirical analysis of what the necessary conditions of a decent life are exactly.

References

Ashford, E. 2013. Severe poverty as a systemic human rights violation. In *Cosmopolitanism versus non-cosmopolitanism: Critiques, defenses, reconceptualizations*, ed. G. Brock, 129–155. Oxford: Oxford University Press.

Gewirth, A. 1980. *Reason and morality*. Chicago: The University of Chicago Press.

Griffin, J. 2008. *On human rights*. Oxford: Oxford University Press.

Kant, I. 1785. Groundwork of the metaphysics of morals. In *Practical philosophy* (The Cambridge edition of the works of Immanuel Kant), ed. M. J. Gregor. Cambridge: Cambridge University Press, 1996.

Margalit, A. 1999. *Politik der Würde. Über Achtung und Verachtung*. Frankfurt a. M. : Fischer.

Mieth, C. 2008. World poverty as a problem of justice? A critical comparison of three approaches. *Ethical Theory and Moral Practice* 11(1): 15–36.

Mieth, C. 2009. Menschenwürde und soziale Gerechtigkeit: Überlegungen zu Kant, Rawls und Margalit. In *Der Wert der Menschenwürde*, ed. Ch. Thies, 133–148. Paderborn: Schöningh.

Neuhäuser, C. 2010. Zwei Formen der Entwürdigung: Absolute und relative Armut. *ARSP* 96(4): 541–556.

Nussbaum, M.C. 2006. *Frontiers of justice: Disability, nationality, species membership*. Cambridge, MA: Harvard University Press.

Nussbaum, M.C. 2008. Human dignity and political entitlements. In *Human dignity and bioethics: Essays commissioned by the President's Council on Bioethics*, ed. A. Schulman. Washington, DC: The President's Council.

Pogge, T. 2008. *World poverty and human rights. Cosmopolitan responsibilities and reforms*, 2nd ed. Cambridge: Polity Press.

Rawls, J. 1971. *A theory of justice*. Cambridge, MA: Harvard University Press.

Schaber, P. 2010. *Instrumentalisierung und Würde*. Paderborn: Mentis.

Schaber, P. 2012. *Menschenwürde*. Stuttgart: Reclam.

Shklar, J. 1990. *The faces of injustice*. New Haven: Yale University Press.

Shue, H. 1996. *Basic rights*. Princeton: Princeton University Press.

Smith, A. 1776. *An inquiry into the nature and causes of the wealth of the nations*, ed. K. Sutherland. Oxford: Oxford University Press, 1993.

World Bank. 1976. *1976 annual meetings of the Boards of Governors: Summary proceedings*. Washington, DC: World Bank. URL: http://documents.worldbank.org/curated/en/1976/10/12609933/1976-annual-meetings-boards-governors-summary-proceedings. Last accessed 7 Jan 2013.

Chapter 3
Human Rights and Human Animals

Bernd Ladwig

3.1 Introduction

In order to provide a foundation for human rights, it seems inescapable to use contributions from anthropology for the two-fold reason that, first, only human beings can adopt a moral standpoint from which their rights may be grounded and respected and that, second, human rights are owed to human beings simply as such. Now, the object domain of anthropology is precisely what can be said about humans as such. Whoever seeks to provide justifying reasons for human rights therefore seems to be, in some way or another, committed to universal anthropological claims.

Nonetheless, it is clear that anthropology alone cannot provide sufficient grounds for human rights. In addition to an anthropological footing, we also need recourse to at least one valid moral principle—and thus to a source of reasons that cannot be derived solely from empirical claims about human nature. This gives me cause to adopt a rather modest argumentative goal in this article. All I try to show is that some anthropological arguments are necessary to justify some of the specific contents of human rights. To that end, I will portray humans as a special sort of animal endowed with two natures (Sect. 3.2). This anthropological approach provides the argumentative framework within which we can derive some, but not all, of the specific contents of human rights (Sect. 3.3).

B. Ladwig (✉)
Political Sciences, Free University Berlin, Ihnestr. 22, 14195 Berlin, Germany
e-mail: berndladwig@hotmail.com

M. Albers et al. (eds.), *Human Rights and Human Nature*, Ius Gentium: Comparative Perspectives on Law and Justice 35, DOI 10.1007/978-94-017-8672-0_3,

3.2 How Should We Conceptualize Anthropology?

3.2.1 *Our Dual Nature*

Anthropology, as the term is used in this article, is defined by its concern with human nature. This, however, is a rather abstract statement that requires at least four demarcations. First, we should concern ourselves exclusively with *substantial* claims about human nature. To be sure, political philosophy is never ultimately free of anthropological baggage, at least in a purely formal sense. Even Friedrich Nietzsche, who once proclaimed that humans are more "uncertain, changeable, indeterminate than any other animal" (Nietzsche 1887/1967, 13) thereby made a bold statement about humans as such. What he was saying about human beings concretely, however, is that any attempt to determine human nature in a contentful way must necessarily be in vain.

Second, the word "nature" is intended to express something typical or essential for human beings. All successful candidates for such features must fulfill two conditions: (1) they must be present in normal human cases, and (2) they must have sufficient import for us (Birnbacher 2009, 190). Distinguishing traits like the use of tools or of language, reflexivity, morality, a sense of the beautiful and the enjoyment of music evidently fulfill both conditions. Although not all individual human beings partake of these characteristics, they indeed play a more-than-marginal role in the lives of the vast majority of members in all human societies.

Third, claims about human nature must be empirically verifiable. They cannot be held to be true a priori, and thus solely by virtue of their meaning. On the other hand, they also cannot rely on essential characteristics or entities like an immortal soul or the will of God, which lie outside the range of any imaginable methodological verification.

Fourth and lastly, "nature" is not to be entirely equated with a "second nature" that is acquired through education, shaped by culture, structured by norms, embodied in institutions, and internalized through habit (see McDowell 1996). Humans also have a "first nature", resulting from biological evolution, without which we would lack the capabilities that make any second nature possible (Tomasello 1999). Cultural phenomena are relevant to anthropology only insofar as their relationship with our first nature comes into view. This relationship, however, should be determined in a non-reductionist fashion—in both directions.

Many neo-Darwinian biologists, on the one hand, see humans as animals amongst animals, and the meta-theoretical attitude that they apply to us is really not fundamentally different from their attitude towards other animals. They hold second nature to be epiphenomenal to first nature, expressing, for example, the law of genetic optimization.[1] Cultural anthropology, on the other hand, is primarily

[1]For an example, see Wilson (2002); against the reductionist tendencies in naturalistic approaches such as Wilson's see Illies (2006).

interested in the second nature that we find on display in the variety of cultural forms of life. Many of the representatives of cultural anthropology urge us to see even first nature itself as a cultural construction. If this were true, it would effectively turn biologism on its head: the alleged biological basis of culture, rather than the vast variety of cultures encountered all over the globe, would prove to be epiphenomenal.

I read both perspectives—the biologistic and the culturalistic—as one-sided. The only anthropological approach that strikes me as genuinely fruitful for my purposes is one that empirically and theoretically demonstrates how our first and second nature are bound up with one another. Both biology and the study of culture are indispensable in achieving an unabbreviated understanding of human beings, and neither have a derivative status vis-à-vis the other.

The nexus between the first and second nature of human beings was already of decisive concern for two representatives of classical philosophical anthropology, Arnold Gehlen (1940/1986) and Helmuth Plessner (1927/1981). Both responded to the Darwinian challenge; both sought to lend intelligibility to the spirit and culture of human beings on the basis of their bodily comportment; and both research programs were anti-dualistic. Correspondingly, both chose an entry point for their approaches that was psycho-physically neutral: the key-concept for Gehlen in that regard was "action" whereas Plessner ascribed the same role to "positionality". The finer details of their programs, however, are not important for me here. All I want to emphasize is the effort of both to think of biology and culture as inseparable but to do so without collapsing one pole into the other. This is the basic intention that I follow here, in that I wish to avoid any dualistic Menschenbild as well as the one-sidedness of either biologism or culturalism.

To begin with the biological side: humans surely are products of evolution. As a consequence, it would be a genuine wonder if we could not ascribe a great number of human features to natural history, features that we can recognize trans-culturally, from the shape and functions of the body to mating patterns right up to our susceptibility to intoxicating substances (see Diamond 1992). But some characteristics that are evolutionarily acquired also serve as our springboard into a history of a much different sort: humans have always steered their development via the historical channels of cultural traditions, and this explains why such development can proceed so quickly and branch out in the way that it does. There is simply no sequence of genetic changes that could have brought forth such a diversity of forms of life in such a short time (Tomasello 1999, 4).

3.2.2 Humans as *Position-Taking* Animals

In defiance of the fashionable neo-Darwinian analogy between genes and "memes" (see Blackmore 1999), we need to fundamentally distinguish the mechanisms of cultural change from the mechanisms of genetic transformation.[2] Cultural developments partly proceed according to a logic of justification. Their progression largely depends on what strikes humans as convincing; and that which they find convincing depends to a significant degree on what really *is* convincing. The logic of cultural development cannot be sufficiently understood without taking its normative features into account.[3]

Humans are condemned neither to blithely trust the impressions of their senses, nor to simply follow their immediate impulses. They can obtain a critical distance from both through reflection. A person taking a reflexive stance seeks to uncover the considerations counting for or against a particular belief or action. To that end, she may be ready to disregard even the strongest impressions of her senses and to resist even her most forceful impulses, at least for the time being. It is this type of freedom that marks her as a normatively responsible person.

An important medium that enables us to scrutinize opinions and actions is propositional language. This is the medium par excellence that allows us to achieve distance from both first impressions and first-order volitions. We can use singular terms to uncouple objects from all kinds of particular situations in order to subsume those objects under judgments from a wide variety of perspectives, incorporated into concepts. We can save the results and apply them to further situations and persons. We can pass them on to future generations by connecting them to a stockpile of knowledge that expands in a cumulative fashion. We can construct conditional, counterfactual propositions and depict possible worlds that deviate from our own. Last, but not least, we can create completely new types of facts: institutional facts, like legal claims or exchange rates, which are based on constitutive rules (see Searle 1995). And because all of these human-specific capacities are both a result of biological evolution and are normatively structured, it is wisest to explore them by means of a conceptual division of labor.

Human biologists tend to pose the question in empirical terms, seeking to unravel the natural historical occurrences that have enabled humans to enter into the accelerating and differentiating channels of cultural development. Why, for instance, can humans identify with others in such a way that creates shared intentionality that may, in turn, issue in dialogical reflection and argumentative dispute? (see Tomasello 1999). Only insofar as we can intersubjectively coordinate the way in

[2]There are, to be sure, phenomena related to cultural fluctuations that do seem to correspond to the model of "memetic" transmission through imitation, for example, *fashion*. But it is precisely because any particular prevailing fashion is hardly a rationally reconstructable phenomenon that its abrupt disappearance is hardly surprising. The arbitrariness of particular fashions corresponds directly to their unsteadiness.

[3]What I mean, in general, here is that any human form of life is confronted with the task of avoiding mistakes, and no form of life could have emerged without a vast number of actions that were based on true, well-grounded beliefs. Moral problems are only a special case of the generalized normative imperative of discovering the right and avoiding the wrong.

which we relate to things and events are we able to attain propositional knowledge. We thus produce three-way-relationships between first person, second person, and reference object, or, to use Donald Davidson's term, we triangulate (Davidson 1982, 317–327). Due to its normative structure, this sort of conduct makes knowledge possible. But insofar as we are biologically geared towards triangulation, it also constitutes a proper object for the natural-scientific investigation of our species.

Philosophers are normally most concerned with reflexively interrogating the universal structures of being-in-the-world, such as "question, reflection, rationality, responsibility" (Tugendhat 2007, 10). If we label these structures "anthropological", as Ernst Tugendhat does, the end result is a *Menschenbild* that is tailored to our second nature but not reducible to it. Humans have a second nature because they are position-taking animals. This is a fundamental feature we can recognize reflexively simply because we ourselves take up a position in the very act of recognizing it. Philosophical anthropology is thus position-taking in a second-order sense, as it attempts to reconstruct those structures common to all human position-taking.

3.2.3 Humans as Position-Taking *Animals*

But is this really all that can be said about human beings as such? I think that there are two reasons why we can and should say a great deal more. First, our reflective capacities are made up of cognitive skills that we share with other primates and perhaps also with many animal species beyond that; the normative bearing of our orientation does not spring up ex nihilo. Such a bearing presupposes that we find our way by means of differentiation in a world structured by significances (point A). Second, an elementary aspect of our status as position-taking animals is bound up with problems that confront us because we are the kind of creature that we are. Our lives are bound to our bodies, as are the lives of all other animals. Furthermore, we experience our lives, we can suffer, we have intentions, and we require and seek companionship, just like members of many other species. But we are unique amongst animals due to our status as creatures that can reflect upon all of this, questioning the veracity of our beliefs and the rightness of our actions (point B).

Concerning point A, the anthropologist Michael Tomasello—surely someone who would not deny the existence of human-specific achievements—points us toward some of the basic capabilities shared by all primates:

> Non-human primates definitely do have an understanding of all kinds of complex physical and social events, they possess and use many kinds of concepts and cognitive representations, they clearly differentiate between animate and inanimate objects, and they employ in their interactions with their environments many complex and insightful problem-solving-strategies. (Tomasello 1999, 19)

According to this statement, we are clearly not the only subjects that have intentions and beliefs. A chimpanzee is not acting without reason when she steers clear of certain members of her group or when she uses sticks to fish down bananas from trees. She draws distinctions and chooses courses of action in accordance with the significances carried for her by things, events, states of affairs, and relations.[4] This is a necessary yet not sufficient condition for normativity.

What we find additionally in humans and only in humans is the capacity to assess theoretical and practical beliefs independent of situations and to test them argumentatively. That makes us aware that they may be true or false as beliefs. Amongst all the animals that have concepts and beliefs, only we also have a concept of the beliefs that we have. Normativity, thus, involves both living in a world of significances and having a position-taking stance vis-à-vis that world. The significances of the world provide subjects with reasons that we, in distinction to other animals, can judge as valid or invalid. As occupants of the normative space of reasons, we therefore confront problems that other living creatures do not (see McDowell 1994). We can do something wrong and we can be aware of the fact of our fallibility. Notwithstanding this fundamental difference to all other animals, however, the appropriateness of our answers is in part a function of the fundamental features that define us *as animals of a particular type*. Such features restrict from the beginning what might reasonably count as a problem for us, as well as what might be a proper way to solve it.

This brings us to point B: on what basis could we determine whether or not a foreign form of life is human a foreign form of life is human? Certainly not solely by detecting that the normal members of such a form of life are engaged in reflection and argumentation. We could imagine creatures that engage in argumentation like we do, but do not confront many of the same challenges that we have to face: such creatures might have no sensitivity to pain or no sense of honor and dignity at all; they might live forever or not result from fertilization; instead of being birthed by female members of the species, they might fall readily from the trees. These hypothetical creatures might nourish themselves simply by breathing, and they might never sing or laugh, indeed, they might never feel any compulsion whatsoever to do so.

To be sure, this would mean that these creatures would precisely *not* engage in argumentation the way that we do. They would not care to know, for example, what happens to them after death, whether or not a specific pain must be endured, or whether or not a particular behavior would dishonor their parents or stifle the development of their children. Their conversations would revolve around neither the procurement of sustenance, nor around distinguishing healthy sexuality from sexual perversion, nor around the question of whether play, song, and laughter are basically pleasing or, on the contrary, signify dangerous distractions. And they would have no religion that promised to give comprehensive answers to all those and other 'deep questions'.

What separates us from these dreamed-up creatures are the distinguishing features of our human condition. The anthropologist Donald Brown has compiled

[4]"Drawing distinctions" and "choosing" are of course not meant in the sense of propositionally reflected modes of interacting with the world, but rather in the sense of *knowing how*, which we can ascribe to the chimpanzee without also ascribing *knowing that*.

more than 200 universal human characteristics, including practices like poetry, music, and dance, the burial of dead, values like hospitality, generosity, reciprocity and sexual restraint, concepts like good and evil, right and wrong, parents and children (Brown 1991). Brown's results might be controversial in their particulars. Nevertheless, to toss them all together strikes me as not only false, but rather as senseless, for the reason just indicated: any hypothetical foreign life form would be recognizable to us *as human* in the first place only if we were able to attribute substantial features to it that we and the foreigners have in common.

And even if a foreign life form obviously deviates from our own, it might be worthwhile to search for underlying commonalities. The religions are an example. Their contribution to the variegated character of human culture does not need to be belabored today. But it is perfectly sensible and only apparently paradoxical to trace the diversity of *answers* provided by religion back to the universality of certain fundamental *problems* we all share as members of the same species. Religions provide answers to the full spectrum of problems and experiences that served as my reason for distinguishing the fictive life forms that I just presented from our own human condition: the ability to suffer, moral vulnerability, finitude, sexuality, dependence on nourishment, joy in play, music, and humor.

The arbitrariness of many religious answers arouses the impression that the questions are arbitrary as well. But it is possible that a question is unavoidable, even if the answer lacks all rational foundation. The denigrations of women and of homosexuality, for instance, are contingent—and clearly morally untenable—answers to the general problem of how to control our drives. There are two sources of the contingency of the answers: the creativity and imagination of human beings and the diversity of their living conditions.

Problems are not mutely "there", waiting to be discovered. They must be interpreted and weighed. Their interpretation and evaluation depend, first, upon how humans understand their form of life as a whole. They also depend upon the conditions—themselves in need of interpretation —within which a form of life unfolds: nomads in the desert will (or perhaps must) express care for the weak and respect for elders differently from the residents of a wealthy large city. The holistic nature of interpretation makes it difficult to isolate problems and to achieve intercultural unity concerning better or worse solutions.

One could argue that the need to identify problems by means of *interpretation* ultimately counts in favor of the culturalistic perspective: it is only within symbolically structured forms of life that human beings can achieve a sense of their challenges, and forms of life are holistically structured. But we do not need to attain an Archimedean point of view to be able to say something of substance about humans as such (similarly Nussbaum 1993, 242–269). We can also approach the issue interpretively. From the perspective of the participant, interpretations are there in order for us to deal with a reality in which we often fail and that we may misconceive. And any attempt to understand foreign forms of life implicates us, at least virtually, in precisely that set of problems whose identification the "foreigners" are seeking to achieve (see Habermas 1981, 167ff). We could obtain no proper understanding of Chinese medicine, for example, if we did not presuppose that some

Chinese practitioners struggle with the problems of frailness and finitude that we basically also confront. Taken more generally, the sense for relevance and aptitude must be sharable if it is to be possible to understand foreign realities at all.

3.3 What Anthropology Can and Cannot Contribute to Human Rights

3.3.1 Four Problems

A skeptic might respond: that's all well and good, but what is to be gained from this discussion for our understanding of human rights? On Brown's list of human universals we find, for instance, on one hand things as far-flung as conflict regulation, empathy and moral sentiments, the prohibition of killing, and the positive valuation of generosity, and on the other ethnocentrism, the dominance of men in the 'public' sphere, and even rape. Even this small set of examples suggests four objections to the use of anthropological claims in grounding human rights.

First, some universals could simply turn out to be dependent upon changeable conditions. They might be based on unfortunate but also modifiable circumstances confronted by all previous forms of life. According to Karl Marx, we remain stuck in the pre-history of mankind characterized by scarcity and the resulting struggle for existence. The achievement of communism would mean that we would no longer be plagued by a whole host of problems. Things like justice, for example, would then no longer be of concern to us. It requires a bit less fantasy to depict a world free of the political dominance of men over women. In that sense, Richard Rorty recommends that we stop searching for presumed cultural invariants (Rorty 1993, 111–134). Instead, we should cultivate our creativity in a way that enables us to produce the necessary conditions for the changes that we desire. More comfortable living conditions, enhanced security, and general cultural sensitization would perhaps make universal progress possible, as many kind children of prosperity already demonstrate today.

Second, all of the items on Brown's list are the object of empirical-universal claims, but there is nothing normative that follows from such claims in an unmediated way. It simply does not follow from the claim that "humans condemn the unnecessary or selfish killing of their own kind in all forms of life" that we should universally condemn all such killing. The condemnation of murder and manslaughter stand on the same footing of factuality with their no-less-universal prevalence. In order to proceed from an "is" to an "ought," we would at least need some premise with normative content, which anthropology, as an empirical discipline, cannot provide.

Third, lists like Brown's provide no sorting principle that would enable us to separate human rights-relevant characteristics from those that are either irrelevant or even impediments to our pursuit. We might be happy to find out that helpfulness is valued throughout the world and that random killing is condemned everywhere. But all the same, we experience a great deal of unpleasantness when confronted with the prevalence of group-related prejudice and male privilege. One possible response would be to normatively determine *first* what we want, and *then* look and see whether or not what we want corresponds to human universals. But that could not be an approach that would give anthropology a foundational role. The burden of providing foundations would entirely rest on normative pre-suppositions, and to claim otherwise would be to argue in circles (on this objection, see Jörke 2005, 58).

Fourth, even where the results of anthropology comport with our moral intuitions, they are far too general; they lack the specific traits that turn an ethics into a conception of human rights. That concerns, first of all, the concept of rights itself: human rights require that any human being may expect consideration, for their own sake, with regard to interests that are so important that they provide sufficient ground for external duties. To be sure, no durable human form of life is possible if it shows no moral consideration of fundamental interests at all. Human beings would require a few certainties be afforded to any hypothetical form of life: they must not be harmed or insulted without sufficiently strong reasons, they must be able to rely on partners in cooperation, they must not be deprived of important goods arbitrarily, they must not be imposed with arbitrary burdens, they must receive help in emergency situations, and they must experience some care and affection, at least during periods of one-sided dependency as in early childhood. But we can easily imagine such certainties being in place without the idea that their presumptive beneficiaries have a valid claim to them. They might instead be part of a purely duty-based morality (see Feinberg 1970, 243–257).

Further characteristics of human rights morality are its universalism, its egalitarianism, and its individualism: all individual human beings have valid claims to equal respect and consideration. But that is a historically late interpretation of morality rather than an essential feature of all morally-regulated forms of life. It is the consequence of complex learning processes that have broadened our knowledge of empirical commonalities throughout human forms of life as well as modified our conception of morally relevant reasons for exclusion and unequal treatment. And last but not least, this development also corresponds to the modern-occidental critique of metaphysics, whose results and moral consequences have still not been accepted everywhere, e.g., with regard to the rights of religious dissenters.

Besides these *formal* features of human rights-based morality there is also a particular feature of its *content* which is likewise not common to any human morality. A crucial part of the morality of human rights is the esteem for personal autonomy. Proponents of human rights are convinced that all human beings need ample scope to find out how to lead their life in accord with reasons appropriate for them. Regarding their content, human rights are thus specifically tailored to secure free space for self-responsible reflection and action. If someone believed, for instance, that he knew the will of God, and that everything hinges upon acting in accordance with this knowledge, then he would have grave problems with a morality that allowed 'non-believers' to commit grave mistakes. In this regard as well, the critique of metaphysics might well have supported the idea of human rights. After all, we owe tribute to the Enlightenment for the insight that our existential judgments are fundamentally fallible and that we therefore carry responsibility for them that no one else can carry in our place. In this article, however, I do not wish to enter into an elaboration of how to justify the essential characteristics of a human rights-based morality. The intention behind my allusions was simply to point out that more belongs to this species of moral thinking than moral universals alone.

All four of these objections carry a certain weight. Together, they significantly restrict the range of relevance for anthropological arguments in human rights-based morality. But their scope should also not be overestimated. The first objection, concerning the preliminary character of some of our problems, gives us reason to invest a great deal of methodological care in identifying typical traits that we can use to recognize human forms of life as such, even apart from the changeability of the framework conditions. Inherent in this objection is the warning not to underestimate a truly typical trait of human beings: their creativity. But this creativity might nevertheless have to stand the test of certain fundamental problems inherent in all human forms of life, and it remains possible that human rights are a universally desirable answer to some problems of this sort. The theoretical possibility of leading human understanding into an indeterminate gray zone by means of science fiction scenarios does not speak against this. Our question is not one of what *Übermenschen* or creatures that were a mixture of human and machine would long for. We wish to know what is needed by creatures that have a sufficient amount in common such that they all count as human beyond any reasonable doubt. The protective function of human rights must be tailored only to such creatures and to no others.

The second objection is rather popular among philosophers but not too difficult to disarm. In order to avoid an is-ought fallacy, one must simply be careful to ensure that at least one of the premises in an argument with a normative conclusion is itself a normative one. To be sure, the assertion that something is in the interest of a hypothetical human being has no normative content in and by itself. But within a rights-based approach, some interests play a role as content-laden reasons for determining basic value claims. And the selection of interests is itself tailored to this normative purpose. Whatever one may make of that, an is-ought fallacy is not the problem here.

Beyond that, I cannot comprehend how one could hope to characterize human interests without using some kind of ‘thick’ concepts that contain valuations inseparable from descriptions (see Putnam 1990, 1993, 143–157). To give an example: to say of a human being that he or she would *suffer* if someone took away his or her children implies that it would be *bad* for this person if someone took away their children. This is not a normative proposition by itself, but how would a moral conception look that was wholly indifferent to human suffering and well-being? And someone who insisted on avoiding all thick concepts like “suffering” would scarcely use a language in which we would continue to recognize our human interests at all.

The third objection, concerning the lack of normative selectivity, deserves a bold response: the sorting principle behind the selection of anthropological characteristics *is*, in fact, a normative one, but this does *not* mean that we can dispense with the empirical contribution of anthropology. Both follow directly from questions concerning those universal interests that could be capable of lending content to human rights. After all, the assertion that there are interests that are both universal and of import for human rights is a fallible hypothesis. And our normative interest in finding a positive answer also does not force us to neglect the problematic sides of human nature.[5] It is precisely due to our dangerous proclivities that we claim human rights in the first place. We need human rights exactly because we are possible rights-violators as well (Höffe 1998, 29–47).

The fourth objection held that the anthropological features of any morality are much too general to lead us to a rights-based approach. If we accepted this objection, we would once more have to concede that anthropological arguments are not *sufficient* to ground human rights. But this objection does not forbid us from holding that such arguments are *necessary* all the same. After all, human rights need content, and the principle of personal autonomy cannot exhaust this content by itself. However central the aspect of personal autonomy might be, those who absolutize this one value, like many liberals, thereby ignore other pre-conditions and fundamental components of human flourishing. They one-sidedly stress that humans are *position-taking* animals, but underemphasize that we are also position-taking *animals*. If we emphasize self-determination monolithically, then we fall short in recognizing the creaturely side of human beings.

3.3.2 A Four-Dimensional Approach of Interests Relevant for Human Rights

In order to paint a more comprehensive picture, we can separate out the possible contents of human rights into four dimensions: (a) existence, (b) well-being, (c) self-determination, and (d) moral status. Self-conscious subjects of human rights wish to lead truly human lives (existence), and they want to flourish in an active (self-determination), as well as a passive (well-being), respect. Human flourishing also has a reflexive dimension consisting of the self-respect of persons who have internalized their normative standing as subjects of human rights (moral status). Consequently, one genuine way of harming human beings is by denying them the status of free and equal bearers of rights.

Now, from the fourth objection we just looked at we can infer that anthropology, as such, does not provide access to the dimensions of self-determination and of moral status. As already shown, to esteem personal autonomy and to recognize each human as having rights simply by virtue of their humanity are not features of *any* human morality. They are rather late achievements, indebted to conflict-mediated processes of enlightenment. In the dimensions of existence and well-being, however, anthropological arguments are indispensable. With the term “existence”, I am

[5] For a critical appraisal, see Jörke (2005, 98f).

referring to a truly human way of being-in-the-world as a position-taking animal. This definition extends beyond a mere biological conception of existence. Merely being alive in a biological sense does not provide us with any morally significant content. Human beings can be alive in a minimal sense without being currently or even potentially conscious—imagine someone living for years in a coma from which they will never wake, or think about infants born without a functioning cerebrum (anencephalics).

For this reason, I will enumerate all those interests in the *existence* dimension whose fulfillment is a necessary condition for leading one's life as a position-taking animal. These interests ground the most elementary demands: to life, to a minimal threshold of bodily and mental health, including freedom from unbearable pain, to free room for perception and action, to some degree of affection and care, as well as to a minimal education and an introduction into the normative space of reasons. All of these elements are described as minimal because not every illness, sensory deprivation, spatial confinement, lack of human contact, or lack of certain concepts or valuable bits of knowledge is sufficient to deprive us of the possibility of consciously and actively participating in a human form of life.

All assertions about interests in the elementary regard of existence are assertions about anthropological universals. The same is fundamentally true of the human rights that are bound up with such interests: we need anthropological knowledge, for example, in order to be able to say universally what a right to subsistence would have to include, either as part of or as an extension of the right to life, such as access to nutrition, clothing, shelter, drinking water, sanitary facilities, life-preserving medical care, and the like. If these all sound a bit trivial, that wouldn't be such a bad thing from the standpoint of human rights, for it would show that it is relatively easy to reach agreement concerning the basic pre-conditions for a successful human life (see Hampshire 1983, 142f; Kekes 1994, 44–60; Seel 1995, 83–87).

Reaching consent with respect to the interest dimension of *well-being* is not nearly as easy. To this dimension belong human rights-relevant interests on the passive side of our existence, including pleasant emotions and positive experiences. To be sure, in typical cases, this side can only be analytically separated from the active side of the exercise of certain faculties. The feeling of arousal that sexual intercourse produces, for instance, cannot be achieved without the act itself—or at least without the substitutive power of fantasy. And even deep suffering is often due to the suppression or denial of possible action. Human beings, in general, also do not wish to enjoy their feelings in a way that is independent of certain practices. Otherwise, they could regularly provide themselves with good feelings by means of an "experience machine".[6]

Nevertheless, pleasure and similar types of favorable feelings are intrinsically good while intense pain and similar types of unpleasant feelings are intrinsically bad, and not only because they have a conducive or adverse effect on our freedoms.

---

[6]That is the famous critical example from Nozick (1974, 42–45).

Whoever reflects upon their own well-being, and does so freely, does not thereby reduce all welfare-relevant goods to components or pre-conditions of this very same freedom. Even beyond the dimension of existence, well-being in the narrower sense of pleasant feelings remains a basic, irreducible component of a successful life. Furthermore, individual experiences of joy or suffering are not simply isolated with respect to one another. They are connected to a whole condition of being that manifests itself in our moods and determines our memories and expectations (see Wolf 1986, 242–265).

In the dimension of well-being, fundamental components of a successful life and not merely its preconditions are at stake. Nevertheless, anthropological assertions are possible in that regard too, if only on a very high level of generality. Human beings are not fundamentally distinguishable from one another in the elementary senses of bodily suffering, and the sources of suffering are also fundamentally the same: beating, stabbing, intense pressure, pounding, burning, deprivation of oxygen, and the like. But even joy and pleasure have certain universal causes and forms of expression in common.

Sexual intercourse, for instance, is a very important source of especially intense feelings for many people, feelings that they seek out again and again. And it plays this role in all known forms of life; irrespective of how they end up regulating sexual behavior within their respective societies. A second source of pleasurable feelings, which provides joy to people in all forms of life we are familiar with, is music. Whoever presumes that these two things are irrelevant to human rights should think of the Afghan Taliban: they have succeeded in brutally suppressing sexuality, above all female sexuality,[7] and have cracked down on music, dance, and related forms of entertainment with equal vigor. Anthropological findings like those of Donald Brown justify the conclusion that the Taliban have thus done violence to human nature.

3.3.3 A Look at a Related View: Martha Nussbaum's Capabilities Approach

Martha Nussbaum has engaged in similar reflections for quite a long time now. She, too, does not limit herself to listing what humans need in order to lead specifically human lives, but rather is bold enough to make assertions about basic components of the well-being of any fellow humans as well. With regard to her diverse lists of fundamental human goods and capabilities, she has often received harsh and fundamental criticism. Therefore, I would like to at least point out how my own suggestions relate to hers and to discuss the extent to which such criticism is justified in one case or the other. In this way, I will at the same time lend added specificity to what I said in the last section. I will use the discussion of Nussbaum to

[7] See Physicians for Human Rights (1998).

more precisely determine the scope of, and proper boundaries for, anthropological argumentation in a moral theory of human rights.

One could immediately object, noting that a list of wished-for things, from the use of the five senses to contact with plants and animals, does not lead us directly to a conception of human rights. For that purpose we would at least have to show that all of the items on the list are sufficiently pressing. Second, it must be shown that—and how—specified duties of third parties can be related to all points on the list. But who is supposed to guarantee, for instance, that all humans are able to have contact with animals if they so wish? Third, we expect respect for and realization of human rights first and foremost from political actors and with a view to the basic structure of society (see Pogge 2002, 47). But who would want to make the provision of contact with animals a central question of politics and to a structural characteristic of all minimally acceptable social orders?

To be sure, that is an objection not against Nussbaum's anthropological approach as a whole, but rather against certain items on her list. I'm now inclined to defend her even with respect to her arguments about the capability of contact with non-human forms of life; why should we not undergird reasons for environmental and species protection with the humans rights-based argument that access to other species is of central significance for the flourishing of great numbers of people in all cultures that are known to us? This argument strikes me as an empirically plausible one. I also do not think that it is an exaggerated expectation of politics that it at least does not willingly curtail possibilities of enjoying nature, and that it perhaps even actively promotes those possibilities, within limits. But I wish to concentrate on more central objections to Nussbaum's suggestions, particularly those that strike at the core of her anthropological claims.

First, there is the objection that her list extends far beyond universals that apply to all of humanity. In this sense, it is perceived as at least partially an expression of specifically Western values, which is quite obvious with regard to the central item on her list, "practical reason." Human beings do not everywhere esteem—and do not do so to the extent that Nussbaum would prefer —the capacity "to form a conception of the good and to engage in critical reflection about the planning of one's life" (Nussbaum 2006, 77). This objection, however, is not directed toward the positive valuation of personal self-determination as such, but rather toward its confusion with capacities for which an anthropological foundation can perhaps be found. And as was shown above, I think that we should simply grant this objection. Whoever wishes to convince relativists of the universal preferability of a self-determined way of living will not go far enough with anthropological arguments. But it does not follow from this that such arguments are useless in general.

Nussbaum's emphasis on the capacity for autonomy has important consequences for the structure and the status of her substantive proposal as a whole. She ascribes to the capability of practical reason—as well as to those of sociability—an "architectonic position" within her conception. This fits well with her claim made clear in recent writings that the capabilities approach is one species of a human rights approach. Additionally, she declares her list to be open, abstract, and vague (Nussbaum 2006, 76–79). This seems to me to be an asset, not least with respect to a general objection that discourse theorists of human rights bring to the table against outcome-oriented views with anthropological content. According to that objection, agreement concerning the concrete contents of human rights can only be the result of a deliberation open to all those affected. Philosophy should restrict itself to making the moral point of view explicit and to forming universal propositions concerning the indispensable framing conditions for practical discourses—which, however, might itself bring some minimal content to the table in order to provide an effective and inclusive structure of justification (Cf. Forst 2007, 711–740).

But why should this prevent philosophy from seeking to further enrich the contents of human rights discourses by way of founded proposals? Such proposals could help actors to better recognize what basic interests they share and to better assert themselves against the forces of prejudice, ideology, and the false modesty of those who are particularly disadvantaged (see Ladwig 2007a). We need anthropological arguments anyway when we begin to think about the demands of future human beings, for their needs can play a role in our current discourses only by means of advocacy. Nevertheless, philosophers should see themselves *as participants* in an advisory capacity, and their proposals should be vague enough that they leave free space for concretization, which can indeed only be carried out by those affected. But at the same time, their proposals should be substantial enough to limit the playing field to what can be plausibly asserted in human rights discourses, and to draw these limits not only with respect to constitutive rules and pre-conditions of discourse itself.

Beyond this, Nussbaum has stressed from the beginning that her list is intended to include capabilities and not functionings as such. This is another aspect of her approach that is indebted to the architectonic role of self-determination in her approach. An example she repeatedly gives is that we should not force anyone to eat enough, but that if one is to be able to fast, one must also have *the possibility* of having enough to eat in the first place. Nussbaum therefore demands effective freedoms rather than particular "beings" such as being well-nourished and "doings" such as practicing a certain religion. As attractive as that seems to be, however, the capability approach raises two questions.

First, in certain regards it simply makes no sense to concentrate on capabilities to achieve functionings instead of the functionings directly. This is certainly true with regard to some of the assets that are fundamentally relevant to human rights. Let us take the need to speak a language as an example. Someone who can speak can also be silent, and being silent makes sense only for someone who is able to speak—but in order to do so, such a person must already be in possession of a language. Being in command of a language as such, however, is a functioning, not a capability. The same is true for the possession of self-respect: how could one be capable of displaying self-respect when she does not have it in the first place? Or let us take a devastating disease like malaria. If it were to die out once and for all, then no one would ever again have the possibility of suffering from it, and surely no reasonable person would regret this. Being free of malaria would then be a functioning, but it would no longer be a capability. Capabilities, if we use the term in a clear-cut manner, imply *options*: one could potentially leave them unutilized. For this very reason capabilities contribute to personal autonomy. But certain elementary goods that are of fundamental import to human rights must be secured directly in the form of functionings and not of capabilities.

Second, and even more fundamental, it is unclear why we should be interested in capabilities as such. The direct answer seems to be that capabilities provide us access to intrinsic goods. But then the value of the capabilities is entirely dependent upon the value of the goods in question. Only insofar as the latter are relevant to human rights is the same true for having access to them. In any case, this applies to the *essential dimension* of such goods: no particular type of music might be of significance for human rights, but the enjoyment of music in general surely is. The possibility of such enjoyment should be secured from a human rights standpoint because music is a good from which sufficient numbers of people in all human forms of life actually derive pleasure.

3.3.4 Three Types of Interests

Against this last example one might object that some people could well be happy without music, just as there are asexual persons to whom sexual excitement is entirely foreign. The capabilities approach acknowledges such possibilities. How can an anthropological approach preserve this central asset of Nussbaum's approach while simultaneously avoiding its weaknesses? In order to answer this question, I will suggest distinguishing between three types of interest that are of significant import for human rights and to which some anthropological universals do belong: inalienable interests, prima facie interests, and probable interests. This distinction, however, should not be equated with relating the relevant universals to the interest dimensions of existence and well-being, as distinguished in the previous section (more on this a bit later). The three types of interests can be grasped as follows:

I call interests "*inalienable*" when they are indispensable pre-conditions for the capacity of humans to orient themselves in the world and to lead their lives as self-conscious persons. Such interests must be fulfilled if humans are to exist as position-taking animals in the first place. The capacity for speech, for instance, belongs in this category. The same is true of the experience of care and affection, especially in the case of small children. When it comes to such interests, mere capabilities simply do not suffice—children directly demand certain functionings, such as being unconditionally loved. All of those interests, of course, fall into the existence dimension by definition. However, it would be far too hasty to also suppose the opposite: not all interests that fall into the existence dimension are therefore also inalienable. For at least one interest that will be mentioned in just a moment, that is clearly not the case.

Prima facie interests are interests that a person has without him or her necessarily desiring their fulfillment under all circumstances. Declarations of value concerning prima facie interests are open to the possibility of overriding considerations. Even some interests that are anthropologically universal are ones that we have only as prima facie interests. As Bernard Gert has shown (Gert 1998, 83f), human beings can sacrifice a great deal without being irrational, but they can sacrifice certain fundamental goods only if they see pressing reasons to do so. Examples include basic health, the absence of persistent strong pains, and even the continuation of one's life.

From a human rights standpoint, life itself is a good that undoubtedly falls in the existence dimension of morally relevant interests. Nonetheless, most proponents of human rights think that one should be allowed to give it up under certain conditions. Reasonable persons can come to the conclusion that they should cease living as soon as possible. Perhaps one prefers a shorter life free of grave suffering to one that is longer but riddled with agony. In the latter case, life is no longer a good for the person in the usual sense. To be sure, life is not only a good of extraordinary weight, as indicated by the passion with which almost all people cling to it. It is also the condition of the possibility of access to any experimental goods whatsoever. From this it follows, on one hand, that only someone who has access to any intrinsically valuable goods also has reason to see her own life as a good for her. This is why it need not be unreasonable for an individual to long for their own death. On the other hand, it follows that whoever has reason to hope for the further enjoyment of any intrinsic good has at least one—self-regarding—reason to want to go on living, even when this reason is not necessarily a decisive one (see Ladwig 2007b, 17–39).

Consequently, someone who throws their life away on a whim strikes us as being crudely irrational. We are inclined to see them as pathological cases. Martyrs cannot be taken as counterexamples, first, because they hold the mission for which they give up their life as dignified, and second, because they typically want others to perceive their death as an ultimate sacrifice. But one can only see something as a sacrifice which one takes to be valuable. Thus, the death of a martyr is intelligible only when one accepts a prima facie interest in the continuation of life.

As this last example already makes clear, an agreement on prima facie interests does not exclude the possibility of deep and even deadly moral conflicts. But in contrast to irreconcilable basic values (see Brandt 1967, 75–78), there are possible meeting points for understanding. Defenders of human rights can thus attempt to unmask the reasons that their opponents give for abandoning prima facie interests as merely ideological: for who would want to sacrifice their life in a flight of fancy? And because some interests that bear strong weight for human rights are prima facie interests, this justifies fulfilling them in the form of capabilities rather than in the form of functionings. The justification lies in the particular nature of such goods themselves, connected with the value of personal autonomy.

For the third category, no better expression comes to my mind than "*probable interests*". What is meant here are interests that not all, but rather the vast majority of humans have in common. Such interests are anthropologically universal in all hypothetical forms of life, and this makes them *typical* human interests. A standard example is once again the interest in sexual satisfaction. The relatively marginal number of asexuals counsels us against defining this interest as one of the strictly universal characteristics of our species. And indeed, we would certainly not want to exclude asexuals from the human race by definition. Nonetheless, it is highly likely that, in general, humans everywhere in the world and at any time in history experience *some* sexual needs at *some* considerable level of urgency.

As a consequence, to presuppose some relevance of sexuality allows us to identify problems typical for any human forms of life, although not for any and all individuals participating in such forms of life. This, of course, does not justify the further assumption that any forms of life need to be as sexualized as many contemporary societies in 'the West'. In any event, this leads to a normatively significant conclusion: whoever drastically curtails sexual freedoms in a given form of life or turns sexual intercourse into torture—as, for instance, with women via genital mutilation—drastically violates the human rights of a significant percentage of the members of that form of life.

Nonetheless, in the particular nature of such a good lies a compelling reason for guaranteeing it only in the form of a capability. The continuing prevalence of sexual interests does not give us a right to any detrimental *normalizing* encroachment into the lives of asexuals. Moreover, it would be bizarre to imagine a state that would be concerned with producing a projected level of sexual satisfaction amongst its citizens. A corresponding duty could not be fulfilled in a way that is tolerable from a human rights standpoint, and, therefore, could not be understood as a human rights-related duty. Finally, sexuality plays a role as a prima facie good only for those who experience some sexual desires. If it were otherwise, then people who were committed to total sexual abstinence would be pathological cases, which, however, is not necessarily true.

3.3.5 And for Those Who Suffer in Silence?

The heading "normalization" also points us to a possible objection of a fundamental sort, which we should therefore tackle now, at the end of this article. I have conducted my discussion by conceptualizing humans as position-taking animals: we do not have one nature but two, and the task of philosophical anthropology is to explicate the basic characteristics of their interrelationship in a non-reductionist manner. Consequently, the contents relevant for human rights that anthropology sheds light upon also appear to be valid only for position-taking animals. But, unfortunately, not all human beings are position-taking animals. Indeed, not even all humans who are neither unborn nor brain dead have this status, although they surely all count as subjects of human rights. If the Menschenbild is socially exclusive, then the corresponding spectrum of goods seems to be as well. From a human rights standpoint, however, a spectrum of goods would have to be socially inclusive. This, in short, is the objection.

But the jump from a demanding Menschenbild to an exclusionary concept of human rights is overly hasty. As position-taking animals, we also discover some interests that we possess not only by virtue of our capacity to take position. We share such interests with individuals who are not normatively responsible subjects of judgments and actions. And we, too, are not just accountable persons—we are also creatures who can suffer and who experience intense and long-lasting pains, for instance, as intrinsically bad. But if a particular interest gives *us* grounds for moral consideration amongst all responsible persons, then this also commands us to extend that same consideration to *all of those* who are sufficiently similar to us in the relevant regards (see Ladwig 2007a; following Seel 1995, 319). Excluding the immature members of our species with respect to something like bodily suffering would clearly be morally arbitrary.

Our capacity for judgment certainly *modifies* even those interests that we have not only because we are capable of judgment. Accountable persons suffer differently from other animals—human and non-human—because they are aware of their suffering and can judge it to be meaningful or meaningless. They do not value freedom of action alone. Instead, they value the freedom to act for reasons they can autonomously judge as valid. They are not merely mortal, they also can reflect upon their mortality; they don't need physical and emotional proximity to others alone, they need recognition from others as their equals. But even when such problems are interpreted and conceptualized, they remain akin to those of other creatures that do not possess the same kind of awareness that we do but nonetheless still *feel*. All of us were once creatures of this kind. And who can tell if or when we will once again belong to this category?

References

Birnbacher, D. 2009. Was kann die Anthropologie zur Politik und ihrer theoretischen Fundierung beitragen? In *Politische Anthropologie. Geschichte—Gegenwart—Möglichkeiten*, ed. D. Jörke and B. Ladwig, 179–194. Baden-Baden: Nomos.

Blackmore, S. 1999. *The meme machine*. Oxford: Oxford University Press.

Brandt, R.B. 1967. Ethical relativism. In *Encyclopedia of philosophy*, vol. 3, ed. P. Edwards, 75–78. New York: Macmillan.

Brown, D. 1991. *Human universals*. New York: McGraw-Hill.

Davidson, D. 1982. Rational animals. *Dialectica* 36(3): 317–327.

Diamond, J. 1992. *The third chimpanzee: The evolution and future of the human animal*. New York: Harper Perennial.

Feinberg, J. 1970. The nature and value of rights. *Journal of Value Inquiry* 4: 243–257.

Forst, R. 2007. The justification of human rights and the basic right to justification: A reflexive approach. *Ethics* 120(4): 711–740.

Gehlen, A. 1940/1986. *Der Mensch. Seine Natur und seine Stellung in der Welt*, Gesamtausgabe Bd. 3. Frankfurt a. M: V. Klostermann.

Gert, B. 1998. *Morality: Its nature and justification*. New York: Oxford University Press.

Habermas, J. 1981. *Theorie des kommunikativen Handelns. Bd. 1, Handlungsrationalität und gesellschaftliche Rationalisierung*. Frankfurt a. M: Suhrkamp.

Hampshire, S. 1983. *Morality and conflict*. Cambridge, MA: Harvard University Press.

Höffe, O. 1998. Transzendentaler Tausch. Eine Legitimationsfigur für Menschenrechte? In *Philosophie der Menschenrechte*, ed. St. Gosepath and G. Lohmann, 29–47. Frankfurt a. M: Suhrkamp.

Illies, C. 2006. *Philosophische Anthropologie im biologischen Zeitalter*, Zur Konvergenz von Moral und Natur. Frankfurt a. M: Suhrkamp.

Jörke, D. 2005. *Politische Anthropologie. Eine Einführung*. Wiesbaden: VS Verlag für Sozialwissenschaften.
Kekes, J. 1994. Pluralism and the value of life. In *Cultural pluralism and moral knowledge*, ed. E. Fraenkel Paul, F.D. Miller, and J. Paul, 44–60. Cambridge: Cambridge University Press.
Ladwig, B. 2007a. Menschenrechte und menschliche Natur. Bausteine zu einer kritischen Theorie der Menschenrechte. *Leviathan—Berliner Zeitschrift für Sozialwissenschaften* 35: 85–106.
Ladwig, B. 2007b. Das Recht auf Leben—nicht nur für Personen. *Deutsche Zeitschrift für Philosophie* 55: 17–39.
McDowell, J. 1994. *Mind and world*. Cambridge, MA: Harvard University Press.
McDowell, J. 1996. Two sorts of naturalism. In *Virtues and reasons. Philippa foot and moral theory*, ed. R. Hursthouse, G. Lawrence, and W. Quinn, 149–179. Oxford: Clarendon Press.
Nietzsche, F. 1887/1967. On the genealogy of morals. In *"On the genealogy of morals"* and *"Ecce Homo"*, Trans. and ed. W. Kaufmann. New York: Random House.
Nozick, R. 1974. *Anarchy, state, and utopia*. New York: Basic Books.
Nussbaum, M.C. 1993. Non-relative virtues: An Aristotelian approach. In *The quality of life*, ed. M.C. Nussbaum and A. Sen, 242–269. Oxford: Clarendon Press.
Nussbaum, M.C. 2006. *Frontiers of justice. Disability. Nationality. Species membership*. Cambridge, MA/London: Belknap Press of Harvard University Press.
Physicians for Human Rights. 1998. *The Taliban's war on women: A health and human rights crisis in Afghanistan*. http://physiciansforhumanrights.org/library/reports/talibans-war-on-women-1998.html. Last download on 6 Feb 2013.
Plessner, H. 1927/1981. *Die Stufen des Organischen und der Mensch. Einleitung in die philosophische Anthropologie*. Berlin/New York: De Gruyter.
Pogge, Th. 2002. *World poverty and human rights. Cosmopolitan principles and reforms*. Cambridge: Polity Press.
Putnam, H. 1990. *Realism with a human face*. Cambridge, MA: Harvard University Press.
Putnam, H. 1993. Objectivity and the science-ethics distinction. In *The quality of life*, ed. M.C. Nussbaum and A. Sen, 143–157. Oxford: Clarendon Press.
Rorty, R. 1993. Human rights, rationality, and sentimentality. In *On human rights: The 1993 Oxford Amnesty lectures*, ed. S. Hurley and St. Shute. New York: Basic Books.
Searle, J. 1995. *The construction of social reality*. New York: The Free Press.
Seel, M. 1995. *Versuch über die Form des Glücks. Studien zur Ethik*. Frankfurt a. M: Suhrkamp.
Tomasello, M. 1999. *The cultural origins of human cognition*. Cambridge: Harvard University Press.
Tugendhat, E. 2007. Anthropologie als 'Erste Philosophie'. *Deutsche Zeitschrift für Philosophie* 55: 5–16.
Wilson, E.O. 2002. *Sociobiology: The new synthesis*. Cambridge, MA: Belknap Press of Harvard University Press.
Wolf, U. 1986. Was es heißt, sein Leben zu leben. *Philosophische Rundschau* 33: 242–265.

Chapter 4
Human Rights, Human Dignity, and the Human Life Form

Thomas Hoffmann

1. When thinking about human rights and human nature, two basic questions come to mind concerning their conceptual relations. First, if there is a relation to begin with and second how this relation, if existing, is to be characterized.

My main concern is with the second question because I answer the first one without hesitation: there must be a relation between the concepts of human rights and human nature. If there was not, we could not understand why human rights are to be the basic rights of humans and *for* humans. If we however could not understand this, we would not be able to frame a justification for human rights being universal and egalitarian rights of all human beings rather than mere historically contingent and culturally relative agreements of subordinated and minor significance. For if we do not understand why human rights are actually basic rights for humans, we are not able to justify that these rights apply equally and in equal manners to all humans as such – irregardless of them manifesting cognitive characteristics of a human *person* or them living in Manila, Mecca, Miami, Manchester or Munich.

If we could not justify such a thing, it would indeed mean that we are not only confronted with the lamentable empirical situation we are constantly confronted with – almost constant violations of human rights. If we do not have a plausible justification at hand for the unrestricted, universal and egalitarian validity of human rights, we are furthermore confronted with the awkward situation of not being able to justify why that, what we mourn, is a mournful violation of human rights at all. Thereby we would have gambled away all moral, political and juridical progress made by legislating human rights. Also, we would not have anything at hand to oppose all the world's villains who violate human rights – not even on a conceptual basis. Having reached this critical point, it does not help to state that we – due to

T. Hoffmann (✉)
Institute for Philosophy, Otto-von-Guericke-University Magdeburg,
Universitätsplatz 2, 39106 Magdeburg, Germany
e-mail: Thomas.Hoffmann@ovgu.de

M. Albers et al. (eds.), *Human Rights and Human Nature*, Ius Gentium: Comparative Perspectives on Law and Justice 35, DOI 10.1007/978-94-017-8672-0_4,

our Christian tradition or the Enlightenment or humanism or the dread of National Socialist atrocities – have a strong positive attitude towards human rights and that we share the preference (since 10 December 1948 or whenever) to honour them.

This statement may all in all be applicable but it is, unfortunately, not an argument which justifies why human rights apply unrestrictedly, universally and egalitarianly to all human beings. For this statement is neither an argument nor does it apply to the validity of human rights! It is rather a *description* of our motivation's genesis to honour human rights. A mere description of our motivation's genesis to honour human rights cannot be a reasonable justification for the validity of human rights. A mere description may, however, provide explaining reasons why we came to honour human rights in the first place. Simply explaining reasons may only let someone realise the reasons why we are motivated to do something. Explaining reasons cannot justify that it is good and right to do exactly what we are motivated to do. A villain who does not care for human rights may therefore listen to our description of our motivation to honour human rights with interest and leniency. And after we are done with our description, he may give us a description of his motivation *not* to honour human rights. Having exchanged our descriptions in such a way, we may state that we will have had a somewhat psychologically, sociologically or historically informative talk. But that's it! For from a normative and evaluative point of view, there is nothing to say of mere descriptions of motivations' genesis – much less to mourn.

I believe such a situation to be as unsatisfactory as it is dangerous. Therefore, it seems important to me to try to formulate not only a psychological, sociological or historical description, but also a justification, making it plausible as to why human rights apply equally and in an equal manner to all human beings *as such*, i.e. *for* human beings *as* human beings. Period!

This is exactly what I attempt to do in the following by initially linking the concept of human rights to the concept of human nature in a certain way and then stating some more basic ideas on the concept of nature itself. These ideas question the popular concept of nature used in the empiristic mainstream of the philosophic modern age and I will at least suggest an alternative concept of nature. Although my ideas on the concept of nature may lead away from common philosophical debates on human rights, they are in my opinion crucial when thinking about the conceptual relation between "human rights" and "human nature." Questioning the popular concept of nature used in the empiristic mainstream of the philosophic modern age and suggesting an alternative concept of nature gives us the opportunity to understand the relation between "human rights" and "human nature" in a way that does not seem implausible.

2. The conceptual relation between "human rights" and "human nature" may first be illuminated by paying attention to the concept of *human dignity*. It is then possible to briefly (but accurately) say that the role of human rights is, ultimately, to preserve the dignity of each human being. The not very spectacular definition I would like to propose is, therefore: human rights serve to protect dignity, i.e. their role is to protect human dignity.

It is exactly this role that morally justifies the validity of human rights – apart from political or juridical justifications. Thereby, moral justification is not only prior to political or juridical justifications, it is the basic form of justification. For what is morally bad cannot turn into something entirely good – even if it is of political use or if it proves itself to be legally plausible because it acts consistently and coherently according to pre-existing law.

Thus, human rights serve to protect human dignity and are morally justified by this protective function – and in a certain way there is little more to add as to how the universal and egalitarian validity of human rights is justified. However, this definition of the justification of human rights is not only a bit formal; it also uses the under-defined and controversial term "human dignity." So nothing has been really clarified by referring to human rights' protective function if nothing is said about the concept of human dignity. If I ended my explanations on this very point, it would be right to argue that I have not met the demand for a justification of human rights but merely shifted the discussion from human rights to human dignity. That is why I am obligated to say something about human dignity that brings us closer to our aim, i.e. to understand the relation between "human rights" and "human dignity."

3. Debates between legal scholars and philosophers, as said, leave the concept of human dignity rather underdefined and the definition remains controversial. Faced with this indefiniteness, some philosophers fear that using the term "human dignity" may be nothing more than "rhetorical gesture" (cf. Birnbacher 1996). And some philosophers even claim that the term "human dignity" is an empty formula, serving as an ideological weapon that can be used for any purpose (cf. e.g. Hoerster 2002, 24). Therefore, these philosophers suggest eliminating the term "human dignity" from our vocabulary once and for all (cf. Hoerster 2002, 24ff.). Other philosophers nevertheless have tried to give a more accurate definition of the term "human dignity." Some assert human dignity as a non-losable natural property of human beings (cf. e.g. Spaemann 1987). Other philosophers decline such a realistic point of view in favour of subjective-reflexive or intersubjective-reciprocal approaches of attributing human dignity where human dignity appears as a non-natural quality of human beings (cf. e.g. Lohmann 2010; Menke and Pollmann 2007; Schaber 2010).

In my opinion, both perspectives lead to a series of conceptual problems. Therefore, I would like to advocate not thinking of human dignity as a certain property of a human being, whether it be natural or non-natural. Hence, I am of the opinion that the expression "... has human dignity" is not a predicate ascribing any property. In this respect, it is accurate to state that the expression "... has human dignity" itself has no content at all, i.e. with this expression no discrete material content is implied representing a certain human property.

But holding this view does not imply to share the opinion that the expression "... has human dignity" should be eliminated entirely from our vocabulary. In my opinion, we should not argue this way at all for such an eliminative position results from a false conclusion. With the expression "... has human dignity" not being a predicate and not ascribing discrete material content, it is misleadingly concluded that this expression is entirely irrelevant for our language-games. This proves to be wrong in two ways. First, it is misleadingly assumed that expressions in the grammatical manner of "... has human dignity" always have to act as logical-semantic predicates, provided that they play a logical-semantic role at all. Second, it is wrongly concluded that an expression in the grammatical form of the expression "... has human dignity", that does not function as a logical-semantic predicate, has to be redundant on a pragmatic-performative level and therefore may be eliminated. This, however, is not accurate.

Although the expression "... has human dignity" does not function as a logical-semantic predicate and does not ascribe discrete material content, it has a crucial function in those situations where we, faced with the inhumane treatment of people, express the *reminder* to honour human dignity. Such a reminder is indeed not an ascription of a property. But its performative-pragmatic significance is to refer to those natural properties *already* ascribed to human beings. In the case of treating humans inhumanely, one reminds that a human being has to be treated according to those properties that generally apply to human beings because they are human beings. In short: one reminds to treat a human being *as* a human being. Or, as Arnd Pollmann once said, summarizing my position in both a witty and amusing way: reminding to honour human dignity means reminding species appropriate husbandry of humans. Speaking of "species appropriate husbandry" precisely describes the point I am trying to make for it is about the nature of those animals that are humans. Human beings are – as implied by the reminder to honour their human dignity – to be treated in a way which does not make it impossible for them to live a life as instances of the human life form which is generally good for instances of the human life form.

If this is the point of a reminder which comprises the expression "... has human dignity", as I claim, it has certain implications for the expression "... has human dignity" on the semantic level. The expression "... has human dignity" is no predicate ascribing a discrete content, but the expression "... has human dignity" functions semantically as a prosentence-forming operator which turns a sentence in which it is included into a prosentence that refers to previous statements – similar to pronouns referring to previous nouns (precisely: nominators or singular terms). Thus, the function of the expression "... has human dignity" is comparable – to a certain point, of course – with the anaphoric referential function of the expression "... is true" as it is described in prosentential theories on truth. (Cf. Grover et al. 1975; Brandom 1994, esp. 299–305.)

4. If it is said that the expression "... has human dignity" functions semantically as a prosentence-forming operator, which turns the sentence it is used in into a prosentence, the question certainly immediately comes to mind to which statements such a prosentence anaphorically refers. If my previous thoughts apply and respecting human dignity means to treat a human being in a way that does not make

it impossible for him to live a life that is good for the instances of the human life form, it seems to me that a prosentence, formed by the prosentence-forming operator "... has human dignity", may only refer to a certain kind of statements, namely judgements that characterise human nature by making the human life form explicit. Making the human life form explicit means signifying the general form of flourishing of humans as instances of their life form. And humans flourishing is what is naturally good for instances of the human life form.

Thereby, prosentences, containing the expression "... has human dignity", not only refer to statements that are making the human life form explicit, but they also refer to norms implied by the explicit statement of the form. By these norms we can judge two things. First, statements on the human life form give us general norms to judge whether a single human as an instance of the human life form is a naturally good or a naturally defective instance of its form. And, second, these general norms are also what we may use to judge whether that which happens to a single human being is either naturally good or bad for its flourishing as an instance of its life form (cf. Foot 2003, 25–80).

Statements on a life form may be named "*natural-historical judgements*", as Michael Thompson puts it (cf. Thompson 1995, 272f., cf. Thompson 2008, 64f.). And this means, with respect to our topic, that the prosentence-forming operator "... has human dignity" generates prosentences that have an anaphoric reference to natural-historical judgements on the human being.

Natural-historical judgements on the human being have a certain logical form, namely the form of generic statements (cf. e.g. Stekeler-Weithofer 2009). Genuine natural-historical judgments on the human being are characterised by a *time-general* (cf. Rödl 2012, ch. 6) relation of the human life form to forms of state and action which instances of the human life form manifest *in general* to flourish *in general* as instances of their life form, i.e. as humans. But this generality is not quantified but a generic generality, i.e. natural-historical judgements on the human being do not express a generality that is *exceptionless*, for they do not consist of sentences that are closed by the universal quantifier and do comprise variables that stand for single individuals or single sets of individuals. So natural-historical judgements do not have the quantified form: *For all x: If x is human, then x is F.*

As generic statements, natural-historical judgements on the human being rather are of the non-quantified logical form: *The human is/does/have F.* Here, "the human" does not represent a single human individual or a set of single human individuals; it does represent a *form:* the human life form. And "F", which is linked in "The human is/does/have F" to "the human" by the time-general copula "is/does/have", also does not represent a certain individual property, empirically assigned or not assigned to a human individual (or several individuals). Rather "F" represents a form of state or action which instances of the human life form manifest in generic generality. If an instance of a life form does not or inadequately manifest single state- or action-forms of its life form, it exemplifies its life form inadequately, i.e. it has natural defect.[1] In general, this is naturally bad for that specific instance for it is an instance of its life form (cf. Foot 2003, 25–37; cf. also Hoffmann 2010).

5. To put it in a perhaps less formal and more vivid way, I will briefly give some examples of natural-historical judgements concerning the human life form. These examples have various levels of abstraction and may concern intellectual as well as physical state- and action-forms of the human.

Some natural-historical judgements concerning the human life form are e.g. "The human being possesses the ability of conceptual spontaneity"; "The human being experiences"; "The human being wants"; "The human being acts"; "The human being talks". Other examples are: "The human being concludes"; "The human being renders a judgement"; "The human being has convictions"; "The human being expresses opinions"; "The human being acts for reasons"; "The human being knows what he does"; "The human being wants to achieve goals"; "The human being intends to achieve what he envisions to be good to achieve". And other natural-historical judgements concerning the human life form are: "The human being has two legs"; "The human being has a heart"; "The human being possesses 32 teeth"; "The human being is chummy"; "The human being has sex"; "The human being resides"; "The human being eats"; "The human being rests".

These natural-historical judgements on the human are linked with each other by the nexus "*in order to*" in natural-teleological judgements (cf. Thompson 1995, 294, cf. Thompson 2008, 77–80). The teleologically structured ensemble of natural-historical judgements on the human, which this way can be made explicit, is, however, what we describe with Michael Thompson as *"natural history of the human being"* (cf. Thompson 2008, 72f.). The time-general natural history of the human is what characterises the human life form as such *in toto*. If we therefore say "human being", we mean what natural history of the human being tells us. And if we say that this or that individual is a human, we say that it is of that life form which is explicated as such and *in toto* in the natural history of human beings. This general form, that the individual is instantiating, is its *nature*.

If we, however, in contrary to logical positivists and their successors, do not equate everything conceptually *eo ipso* and by definition with the mere nominal and everything natural with the empirical, then we may also say, "The form that the individual falls within is its concept" instead of saying "The form that the individual is instantiating is its nature". We therefore may understand the natural history of the human as an explication of the human life form which is both an explication of the human nature and an explication of the concept of the human being. The natural history of the human being then is making the human nature and the concept of the

[1]To say that a single human being has a defect because it is not (or not perfectly) F would be impossible if the quantificational form *For all x: If x is human, then x is F* would be the relevant form. For it follows from "For all x: If x is human, then x is F" and "This x is not F" that "This x is not human."

human being explicit at the same time.[2] And this means in our context: By referring to human nature with the help of the expression "… has human dignity" we refer to the concept and nature of the human being and those natural norms it implies.

Those who appreciate Aristotle may also say: we refer to the real definition of the human. Those who rather vote for Kant, however, may say that we refer to synthetic judgements a priori concerning the human being. Those who prefer Hegel may say we refer to the concept of the human which is the generality in the human itself (cf. esp. Hegel 1831, 26; Hegel 1830, 313). Those who are into the later Wittgenstein may, as well, say: we refer to grammatical sentences on the human – i.e. to remarks telling us how the term "human being" is to be used correctly in practice. And all those who have a soft spot for Heidegger may, in my opinion, also comprise the human life form, which is both its nature and its concept, in another category – in that of *Sein*.[3]

With appropriate "fundamental-ontological" interpretation, even Aristotle's *ousía* or Aquinas' *essentia* might be slightly adjusted to fit Heidegger's *Sein*. Heidegger was good at that, as it is well-known and displayed in his lectures. But I think, instead of reading Aristotle and Aquinas with Heidegger and thereby overly stressing the length of this text, it is also suitable to use Heidegger's argumentation in the first place. If we do so, we may say: the teleologically structured entity of natural-historical judgements on the human life form, as displayed by the natural history of the human, is making human nature and the concept of human being explicit and therefore reveals the *Sein* of human beings. And that means, regarding human dignity, by referring anaphorically to natural-historical judgements on the human life form with the help of the operator "… has human dignity", we refer to human nature and the concept of the human being, which is human *Sein*, which is – to use Hegel once more – the generality in the human itself.

6. Maybe I expect a lot of most readers when it comes to the use of terms. Nature being a concept and a concept being nature might sound rather strange to most recent philosophers' ears. And it may sound no less strange nowadays that the life form is not empirical but a priori and generic. For stating that the *Sein* of the single *beings* is as such not categorised by nomological regularities that are a result of inductive conclusions inferred from the single observations and single descriptions of the single *beings* may appear as a relapse behind modern natural science's achievements. It may easily lead to the assumption that one advocated for an absurd variation on the platonic realism of ideas or a religious idea, which "falls back behind the Enlightment" and "sends us right back to the Middle Age".

---

[2](Hoffmann 2014, Ch. 4.).

[3]The German term »Sein« in the heideggerian use of this word is quite difficult to translate exactly into English, but here and now it is sufficient for us just to think of it as something like *the Being of beings*, that is nearly: the essence of individuals. To get the full meaning of "Sein" I guess one has to read the whole of: Heidegger (1927). Cf. also Heidegger (1957, 166), Heidegger (1950, 33). Cf. also concerning this: Hoffmann (2007, 305–317).

It seems to me, however, that all this sounds dubious only because we take the scientific term of nature and the ontology of the scientistic naturalism for granted nowadays, so much so that we barely reflect on it – much less reflect on it *critically*. The scientific concept of nature and the ontology of scientistic naturalism appear to be so unquestionable that it simply has to appear crazy that concept and nature of the human being might be the same, i.e. the human *Sein* which is the generic generality in the human being itself. And it makes things worse that mainstream logics and philosophy always understand "generality" as a *quantified exceptionlessness* and that this concept of quantified exceptionlessness is not only used to analyse the logical structure of definitions, but also to describe the logical form of nomological sentences in the natural sciences.

I therefore instantly admit that everything I just said may sound unfamiliar to philosophical ears of the present age. But it does not have to be wrong. I believe that it is considerably more accurate than the view of the empiricism-fixated modern age, which intensified to an empiric dogma within the last 300 years, stating that general forms – unless they shall be absurd platonistic ideas or "religious phantasms" – may only be nomological regularities, which emanate ex post and by inductive generalisation from sound natural scientific observation and descriptions of singular incidents.

In opposing natural science's concept of nature as the only acceptable concept of nature and if denying that scientistic naturalism is the only acceptable – or even basic – ontology, one opposes ontological opinions preferred by e.g. Carnap and Quine and also expressed by Wilfrid Sellars in *Empiricism and the Philosophy of Mind*, who wrote that: "[. . .] in the dimension of describing and explaining the world, science is the measure of all things, of what is that it is, and of what is not that it is not" (Sellars 1956, 83). This is the central idea of scientistic naturalism, which dominates our modern view on the world and which – driven by misleading philosophical interpretations of neuroscience's progress – sadly becomes more and more popular concerning our view on ourselves and our we-about-us-discourses.

Although I do not deny that discoveries in natural science may influence our generic judgements on life forms and substance forms of lifeless things and that this may lead to modifications of the nature, of the concept and the *Sein* of the relevant living and lifeless individuals, I still believe that natural science is not the ontological measure of all things and that the scientific concept of nature is not our basic concept of nature. If so, it now raises the question as to where our basic concept of nature has its logical place, if not in the natural sciences. Without being so presumptuous as to believe that I might be able to answer this question ad hoc and entirely, I would now like to denote briefly how a concept of nature had to be characterised if it was characterised as "*hermeneutic naturalism*" – as I put it in contrast to the scientistic naturalism.

If the nature of an individual is understood in the way of hermeneutic naturalism, "nature" then denotes the life form which the individual is instantiating. And this very nature is a specific teleological and logical point within our practice of being-in-the-world, which is conceptual-structured and infinite for us. This infinite totality is given to the single human who is initiated into it beforehand and objectively, i.e. it exists independently of him and is not subject to his subjective judgements or his intersubjectively shared opinions on it. A conceptually structured practice of human being-in-the-world in such a way may be described by Heidegger as *Welterschliessung* ("world-disclosure") (cf. Heidegger 1927, esp. 76, 87). Following John McDowell, we may say that a *Welterschliessung* is infinite – and therefore neither comparable nor relocatable – because we cannot look at it as such and in toto from a *sideways-on perspective* (cf. McDowell 1996, 35f., 81ff.). For every look we take is an instantiation of this *Welterschliessung* in actu (cf. Hoffmann 2007, 380–384).

Due to the human capacity for conceptual spontaneity, he makes the world – along with everything existing in it – accessible with his initiation into a preliminary practise of being-in-the-world. With being initiated into that practice, nature becomes available along with everything existing, because everything has its respective teleological and logical place within the conceptually structured practice of human being-in-the-world. (And it is this place within the conceptually structured practice that does also specify whether a being is in the world as a "brute" given or as a "made" artefact, institution, convention etc.)

7. My explanations so far result in the following: The nature of what exists in the world is given to a human beforehand by his initiation into the conceptually structured practice of being-in-the-world. The immanently existing, and therefore the world, is thus full of sense and meaning. Or maybe one could say, slightly charmed by Hegel, that the natural world is always already the *spirited* world.[4]

Therefore, being born does not mean being thrown into a big bunch of indiscriminatingly given matter at which a human first has to stare and empirically study before going on to sort and categorise it into certain schemes, thereby adding sense to it – with the big bunch of the indiscriminatingly given matter playing the part of nature and the schemes of mind being our concepts. That this is an utterly misleading picture, resulting only from the dogmas of empiricism, has already been clarified by Hegel and Heidegger, as well as by Donald Davidson and John McDowell (cf. Davidson 1974, 1983; McDowell 1995, 1999).

If we do not want to follow this misleading picture, which results from the empiristic dogmas, we have to say – like McDowell does – that the human being, when making experiences, always perceives this or that *as* this or that. And if the human acts, he always knows already that he is just doing this or that (i.e. he has *practical knowledge* of what he is doing – as taught not only by Elizabeth Anscombe). And also by judging this or that, he always believes or claims already *that* this or that is this or that way (and this will not be denied by anyone who has a half-understandable concept of "judgement"), i.e. being-in-the-world means already having comprehended the nature of what can be existing in the world as such. And only after having comprehended this nature one can empirically identify to what extent singly existing individuals may be of their nature in such-and-such way.

[4]Cf. Gadamers' use of *"Welt"* and *"Umwelt"* in: Gadamer (1960, 446f.).

This may be undoubtedly articulated with empirical judgements. In contrast to these judgements, those judgements which make empirical judgements possible by explicating a first and mostly implicit understanding of the nature or the *Sein* of the individuals are no empirical judgements themselves. They are rather generic judgements, which we already spoke of and which we call "natural-historical judgements", as long as they concern life forms and not mere substance-forms of lifeless things.

If we state natural-historical judgements on the human life form we, therefore, explicate aspects of human nature. We then articulate aspects of the human *Sein* as being-in-the-world and this is the teleological and logical place which the human, as such, has within the conceptually structured practice of human being-in-the-world. If we take heed to respect and not to violate human dignity and if we even know in part what we are saying with this, then we are referring to the conceptually structured concept of the human which has already been ascertained in practice, which is its nature and the normative measure of all that of which we take heed.

If we have mastered this concept, then we do not have to empirically explore beforehand whether it is generally good or bad to lie to another human or to betray him, to let him live in immaturity or existence-threatening poverty, to restrain him from expressing his opinion or to live out his sexuality, to let him work until he almost drops dead, to restrain him from his participation in the *polis*, to mutilate, torture or kill him. All this happens to generally be bad for a human being – and one knows that, as long as one has general rules of inference at hand in practice (and theory) which constitute the concept of "human", which is nothing else but human nature.

If my outlined consideration is sound, then there is no use for the idea of something that we can call an "adjunctive image of human dignity". Since perceiving a human as such cannot be understood as a fundamental epistemic ability which takes place alongside any conceptual normative praxis that we have been initiated into. Recognizing a human as a human already implicates all of these norms which we demand to follow, if we call for human dignity to be honoured. This also means that following and not following these norms cannot be rational alike and that – if we claim to be rational – the decision we have to make is not a question of subjective or intersubjective preferences. If we claim practical rationality for our deliberations, intentions and actions there is no additional moral, judicial or political question whether or not to perform an act of recognition and honour independent from the act of perceiving a human being as a human being.

8. Stating everything I have does not, however, imply that natural sciences are obsolete. For, by stating what I have, one does undoubtedly refuse the ontology of scientistic naturalism, but not the natural sciences. One does not refuse them, but in contrast to scientistic naturalism, one does not describe it as the ontological measure of all things. Instead, one does understand that natural-scientific descriptions and explanations are a specific part of our broad practice of being-in-the-world, as a specific (partial) practice of the instrumental handling of things and occurrences in a world made accessible beforehand in a non-natural scientific way.

The natural scientific practices may then – as Heidegger does – be described as a *privation* (cf. Heidegger 1927, 149, 164) of our broad practice of being-in-the-world. For the natural scientific approach subsequently and purposefully "robs" nature of all its conceptual, teleological and normative moments which determine our general handling and daily errands. Having such a perspective toward natural scientific descriptions and explanations does not derogate natural sciences, nor does it make their descriptions and explanations obsolete. For, based on natural scientific questions, natural scientific practices may definitely lead to appropriate answers to natural scientific problems. And these answers – as I already said – may have influence over what we have called forms of substance and life. For natural scientific discoveries may lead us to modify the nature, the concept, the nature and the *Sein* of the respective being one or another way, much like Neurath's mariners may tinker with parts of their vessel out on the open sea (cf. Neurath 1932). But natural scientific descriptions and explanations are neither broad descriptions and explanations of what is "world" in toto, nor are they the fundamental practice of human being-in-the-world. But first and foremost, natural sciences do not formulate the basic, much less the exclusive, concept of nature.

9. As I said before: "The human being possesses the ability of conceptual spontaneity"; "The human being experiences"; "The human being wants"; "The human being acts"; "The human being talks". "The human being concludes"; "The human being renders a judgement"; "The human being has convictions"; "The human being expresses opinions"; "The human being acts for reasons"; "The human being knows what he does"; "The human being wants to achieve goals"; "The human being intends to achieve what he envisions to be good to achieve". "The human being has two legs"; "The human being has a heart"; "The human being possesses 32 teeth"; "The human being is chummy"; "The human being has sex"; "The human being resides"; "The human being eats"; "The human being rests". All of these are natural-historical statements which express what defines nature and the concept of human in general.

If one picked from this list those that are usually labeled as "cognitive", "rational" or "mental" properties, the bundle picked could get tagged as "autonomy". Following Kant, one could believe that human dignity might be a value granted to beings due to their autonomy, i.e. their ability for reasonable deliberation and performance. In doing so, it is apparently impossible to equate "autonomous being" with "human". For from what Kant ascribes to autonomous beings, it is obvious that persons exclusively are autonomous beings. But it is an evident fact that not all human beings meet all cognitive and rational conditions we regard as necessary to describe them justifiably as persons. Consider patients in comas, unborn babies or seriously mentally challenged people.

Does this mean that one can only justifiably remind to honour the dignity of those humans that currently meet the conditions of being a person? According to what I just said, it does not. Indeed, the natural norms concerning the human being implicate all properties and capacities that are a human person's own. If we spell out the concept of the human being in general, we spell out nothing but the concept of a human person. But the concept of the human being, as I said, gains its content from the natural-historical judgements on the human life-form. And these natural-historical judgements are generic judgements, not universally quantified judgements, as I also said before.

Hence, autonomy is one of the characteristic state- or action-forms of the human life form, which we predicate in a generic way when we predicate it by articulating a natural-historical judgement. But even if such a predication proves to be false regarding a specific individual human being, and just because this individual human being does not manifest personhood, this does not mean that it is not an instantiation of the human life form. It is rather an instantiation of the human life form with a natural defect – but no doubt: It is a human being. Therefore, autonomy is indeed a fundamental feature of the human being, but the reminder to honour human dignity does not exclude those human individuals who are obviously not autonomous (in Kant's sense).

If we admonish somebody to honor the human dignity of a single human being, then nothing but the (generic) human life form has to be taken into account if asked whether our reminder is justified. The question as to whether this single human de facto manifests all properties of his life form here and now – e.g. the capacity of autonomy – cannot be the essential one with regard to the warrant of our reminder. Since if a human has to manifest all properties, our reminder would only make sense if it pertained to those humans who are perfect according to the natural norms of the human life form. But assuming this would make our reminder truly redundant just because it would be pragmatically pointless. Because not only are coma-patients or mentally challenged people not perfect, I would even argue that none of us – without intending any offense – is a perfect instantiation of the human life form.

10. Coming to the end, I would like to explain how my proposal is related to the philosophical discussions mentioned at the beginning, which question (i) if it is possible to graduate the ownership of dignity, (ii) if the protection of human dignity is categorical or limited under given circumstances, (iii) how the relation between the concept of human dignity and the concept human rights should be spelled out correctly.

First of all: after what was said before, it should be obvious that we cannot consider these questions as questions for a certain substance or property we call "human dignity", which now has to be specified. But having left behind this misleading perspective, it is now possible to answer the questions (i) to (iii) in a way which should prevent unsolvable conceptual problems and oscillating dilemmas. My account proposes the following answers.

Ad (i): It is empirically obvious that what we refer to if we claim to honour human dignity is fulfilled in different stages by different humans, for there are de facto successful cases and cases of deficit or complete failure when it comes to recognizing a human as human and manifesting practical rationality. This can, nota bene, be the case not only according to other people, but also according to the first person (e.g. because of mental disabilities or massive humiliation by torture and such). Keeping in mind that the normative measure is exemplified by successful cases and that the claim to honour human dignity is a reminder in view of existing deficits, it is obvious that what we refer to by our claim cannot itself be graduated. To graduate it nevertheless is just an expression of practical irrationality or lack of knowledge.

Ad (ii): Taking into consideration what was just said about (i), there is only one possible answer to (ii). For if de facto graduating what we refer to by claiming to honour human dignity is an expression of practical irrationality or lack of knowledge, it would be practically irrational not to try to inhibit these cases and guarantee conditions which makes successful cases possible. Therefore, the special protection of the possibility of successful perceptions of humans and rational actions has to have categorical and unconditional validity. Limiting this protection would therefore also be an expression of practical irrationality or lack of knowledge.

Ad (iii): Following the suggestion to understand the claim to honour human dignity as a reminder that does not have self-contained content but expresses, pragmatically-performatively, that natural properties of human beings and natural norms of the human life form are being ignored, it must be a contradiction to state that the concept of human dignity is equal to or has the same extent as the concept of at least basic human rights. Basic human rights are not what we refer to by claiming to honour human dignity. In fact, basic human rights are the legal attempt to protect the possibility of successful cases of practical rationality concerning humans treating humans as human beings – i.e. treating human beings correctly as instances of the human life form, which is the nature and the concept of the human being. To protect this possibility is their main function. And this function is the normative basis of justifying the universal validity of basic human rights.

References

Birnbacher, D. 1996. Ambiguities in the concept of Menschenwürde. In *Sanctity of life and human dignity*, ed. K. Bayertz, 107–122. Dordrecht: Kluwer.

Brandom, R.B. 1994. *Making it explicit. Reasoning, representing, and discursive commitment*. Cambridge, MA: Harvard University Press.

Davidson, D. 1974. On the very Idea of a conceptual scheme. In *Inquiries into truth and interpretation*, ed. D. Davidson, 183–198. Oxford: Oxford University Press, 1984.

Davidson, D. 1983. A coherence theory of truth and knowledge. In *Subjective, intersubjective, objective*, ed. D. Davidson, 137–153. Oxford: Oxford University Press, 2001.

Foot, P. 2003. *Natural goodness*, Paperback-ed., 25–80. Oxford: Oxford University Press.

Gadamer, H.-G. 1960. *Wahrheit und Methode. Grundzüge einer philosophischen Hermeneutik*, 6th ed. Tübingen: Mohr, 1990.

Grover, D.L., J.L. Camp, and N.D. Belnap. 1975. A prosentential theory of truth. *Philosophical Studies* 27: 73–125.

Hegel, G.W.F. 1830. *Enzyklopädie der philosophischen Wissenschaften im Grundrisse I, Werke*, Bd. 8. Frankfurt a. M.: Suhrkamp, 1986.

Hegel, G.W.F. 1831. *Wissenschaft der Logik I, Werke,* Bd. 5. Frankfurt a. M.: Suhrkamp, 1986.

Heidegger, M. 1927. *Sein und Zeit*, 15th ed. Tübingen: M. Niemeyer, 1979.

Heidegger, M. 1950. Die Sprache. In *Unterwegs zur Sprache*, ed. M. Heidegger, 9–34. Stuttgart: Klett-Cotta, 1959.

Heidegger, M. 1957. Das Wesen der Sprache. In *Unterwegs zur Sprache*, ed. M. Heidegger, 157–216. Stuttgart: Klett-Cotta, 1959.
Hoerster, N. 2002. *Ethik des Embryonenschutzes. Ein rechtsphilosophischer Essay*. Stuttgart: Reclam.
Hoffmann, T. 2007. *Welt in Sicht. Wahrheit, Rechtfertigung, Lebensform*. Weilerswist: Velbrück Wissenschaft.
Hoffmann, T. 2010. Erste Natur, Zweite Natur und das Gute für den Menschen. In *Natürlich gut. Aufsätze zur Philosophie von Philippa Foot*, ed. T. Hoffmann and M. Reuter. Frankfurt a. M: Ontos.
Hoffmann, T. 2014. *Das Gute*. Berlin/NewYork: De Gruyter (forthcoming).
Lohmann, G. 2010. Die rechtsverbürgende Kraft der Menschenwürde. Zum menschenrechtlichen Würdeverständnis nach 1945. *Zeitschrift für Menschenrechte* 1: 46–63.
McDowell, J. 1995. Knowledge and the internal. In *Meaning, knowledge, and reality*, ed. J. McDowell, 395–413. Cambridge, MA: Harvard University Press, 1998.
McDowell, J. 1996. *Mind and world*. Cambridge, MA: Harvard University Press.
McDowell, J. 1999. Scheme-content dualism and empiricism. In *The engaged intellect. Philosophical essays*, ed. J. McDowell, 115–133. Cambridge, MA: Harvard University Press, 2009.
Menke, C., and A. Pollmann. 2007. *Philosophie der Menschenrechte*. Hamburg: Junius.
Neurath, O. 1932. Protokollsätze. *Erkenntnis* 3: 204–214.
Rödl, S. 2012. *Categories of the temporal. An inquiry into the forms of the finite intellect*. Cambridge, MA: Harvard University Press.
Schaber, P. 2010. *Instrumentalisierung und Würde*. Paderborn: Mentis.
Sellars, W. 1956. *Empiricism and the philosophy of mind*. Cambridge, MA: Harvard University Press, 1997.
Spaemann, R. 1987. Über den Begriff der Menschenwürde. In *Das Natürliche und das Vernünftige*, ed. R. Spaemann. München: Piper.
Stekeler-Weithofer, P. 2009. Generisches Wissen in kategorialen Inferenzstrukturen: Zur Metaphysik des Begrifflichen. In *Language and world part two: Signs, minds and actions*, ed. V. Munz, K. Puhl, and J. Wang, 191–215. Proceedings of the 32th international Ludwig Wittgenstein-symposium in Kirchberg am Wechsel, Frankfurt a. M.
Thompson, M. 1995. The representation of life. In *Virtues and reasons. Philippa Foot and moral theory. Essays in honour of Philippa Foot*, ed. R. Hursthouse, G. Lawrence, and W. Quinn, 247–297. Oxford: Oxford University Press.
Thompson, M. 2008. *Life and action*. Cambridge, MA: Harvard University Press.

Chapter 5
Rethinking Human Nature as a Basis for Human Rights

Harun Tepe

Though we hear a great deal about human rights these days, the concept of human rights is considered by many to be a "normative concept" lacking clarity and theoretical foundation. The co-existence of stirring interest and deep conceptual scepticism is remarkable (Sen 2005, 151). Even the questions related to the basic concepts of human rights remain contentious. The clarification of the concept of human rights and justification of human rights norms have been the main theoretical issues for philosophers and social scientists for a long while. Today, the questions dealing with the theoretical and ethical foundations of human rights remain unanswered. In spite of its power, some assert that the idea of human rights is not clarified adequately and not justified objectively. The lists of human rights vary from life and liberty to the long and open-ended catalog of international human rights documents:

> Human rights mean different things to different people. People easily elevate what they want into a human right, while they label what they dislike as a human rights violation. Even those who claim to be using 'objective' criteria find it easier to write about what human rights should be than describe them as they are. This involves entering the 'swampy lowland of messy and confusing problems' and the process of law-making is therefore rarely studied. (Tomaševski 1995, 59)

The problem of ambiguity and subjectivity in human rights partly results from taking the concepts of international human rights instruments for granted. The concept of human rights presented in these instruments, especially in the *Universal Declaration of Human Rights*,[1] has been considered as a criterion in deciding

[1]The most appropriate document to look at to find these intentions is the first drafted one, the *Universal Declaration*, which is an ethical document *par excellence*, up to now probably the best

H. Tepe (✉)
Department of Philosophy and Department of Human Rights, Hacettepe University, Beytepe, 06532 Ankara, Turkey
e-mail: harunt@hacettepe.edu.tr

M. Albers et al. (eds.), *Human Rights and Human Nature*, Ius Gentium: Comparative Perspectives on Law and Justice 35, DOI 10.1007/978-94-017-8672-0_5,

which rights belong to human rights. However, "it is tempting, and apparently uncontroversial, to take explicit, formally adopted international statements on human rights for granted. International human rights instruments are, however, the final stage in a long and protracted process. A formally adopted text may seem clear (although this is not necessarily the case) on what it says, but not on what that means" (Tomaševski 1995, 59).

In these international documents one notices that different conceptions of human rights are applied. Although human rights in the *Universal Declaration* are conceived of mainly as principles of action, concerning how human beings should treat human beings, human rights in the *Covenants* (i.e. *International Covenant on Civil and Political Rights* and *International Covenant on Economic, Social and Cultural Rights*) are considered basic principles for social organization, i.e. as demands concerning the establishment of social relations in a certain manner. The states are supposed to protect human rights by establishing social relations and social organizations according to the norms of human rights (Kuçuradi 1995, 7).

Consisting of two words, "human" and "rights", the idea of "human rights" presupposes a concept of human being (or human nature). Human rights are the moral rights of the individual because he/she is a member of the human species. These rights express some necessities related to each human being. Due to his/her biological nature man is supposed to have some inalienable rights called human rights. Everyone has these basic rights regardless of his/her nationality, race, religion and gender. The supposedly strict connection between the concepts "human" and "right" invites us to concentrate on the concept of human or human nature. However, nowadays a theory of human nature is considered old-fashioned and may be labelled as essentialist and metaphysical.[2] Though one cannot talk about "human rights" without using both the terms of "human" and "right", the assumption is made that there is no need to exhaustively define "human being". Because "we are rather well acquainted with such creatures", we may introduce a core concept, that of a right-holder instead of defining what a human being is. A right-holder is the person who has the right in question (Orend 2002, 15).

The difficulties that arise regarding an adequate definition of human (nature) and the allegedly supposed normativity of the concept human nature dispense with the question of what a human being is or what constitutes human nature. The critics of the idea of human nature mainly focus on a supposedly accepted concept of human

elaborated document of human rights we possess, or, to use a term from present-day philosophy of language, the most carefully performed speech act of the world community Kuçuradi (1995, 5). In spite of it "being the best elaborated document," *Universal Declaration* also seems problematic as the concept of human rights is not sufficiently clarified.

[2]This results from a specific, but very widespread, concept of human nature that there is an essential—biological—human nature which is the same for every human being. This is clear in Rorty's critique of the concept: "To believe that there is an essential human nature is to believe ... that there is one possible description of Man that renders all possible descriptions commensurable" Berry (1986, 127).

nature which defines human being with reference to those characteristics that all human beings share. But is any feature or property shared by all human beings? In spite of our numerous differences, is there something that we can call a common human nature, by which we can recognize each other as human being?

If human nature is adequately defined, not only will a basis for human rights be achieved but also a device for justification of human rights. Nature, in general, has been a justifying device for all moral and legal norms by various doctrines of natural law. But does nature, and in particular human nature, prescribe or proscribe behaviour or does it provide the ultimate criterion of validity for human rights? Those who claim that it does, use the expression "human nature" in a variety of rather vague meanings (Gorecki 1996, 25). One who wants to use human nature as a justification must tackle the challenge of the vagueness of the concept and its alleged normative content.

> The documents of human rights are valid universally only in a theoretical sense. Their validity depends on the theoretical conception of human nature. It is a sort of essentialism that is presupposed for the justification of universal human rights. For instance, the universal equality of mankind is thought to be an implication of the essential human nature. But in practicality, the concept of universal equality is empty.
>
> Human rights may depend on human nature because rights are prescriptive whereas nature is descriptive, and because they both express something human. It is the concept "human" that connects 'right' to 'nature'. But it is also this concept 'human' that causes all sorts of conceptual inflation in the application of human rights documents. (So 1995, 38)

This compels one to focus on the concept of human nature and to grasp different meanings of the concept. The problem with the term "human nature" stems from a certain use of the term, that is, one which reduces being human to a specific character of himself/herself, which supposes that every person has the same traits.

This article will not introduce a new concept of human nature, but instead will provide a different account of human being rather than the accustomed concept of human nature, hoping that these will be guiding ideas that overcome the difficulties with the concept of human nature. Instead of advocating a specific view of human nature, I will examine what is meant with the term "human nature". I will refer to two concepts or accounts of human being developed from the two approaches to human being: one is better known than other, namely the "capability or capabilities approach", whilst the other is the "ontological anthropological approach". Although these two approaches have common features, they differ in many aspects, mainly regarding their ways of dealing with the human being. The capability or capabilities approach considers human being as holders of some capabilities, whereas the ontological-anthropological approach sees man as bearer of some properties (phenomena), some of which constitute the characteristics of human being, but others of which are only the possibilities of him/her. In both of these approaches, human being is understood in a broader sense without reducing him/her to a particular property of human being, like reason, *Geist* or language. If the concept of human nature is understood in this broader sense, the idea of human rights may be justified on the grounds of this new account of human nature or the human being. Using the concept of "human being" instead of "human nature" may help us to avoid accusations directed at the concept of "human nature", and yet it seems inevitable that the concept of human being is to be used in defining and justifying human rights.

5.1 The Idea of Human Rights

We experience a variety of rights, among which we can at first reckon moral and legal rights. Human rights are understood as kinds of universal moral rights that belong equally to all human beings simply by virtue of the fact that they are human beings. Moral rights are experienced as valid claims against others. These claims demand that the others do what they ought to do as right, or refrain from doing what they ought to avoid as wrong. Human rights, experienced as valid claims in this sense, have two further peculiar characteristics: They are ascribed to all humans, and they are directed at political power holders. Wherever positive law implements them, human rights also become legal rights (Gorecki 1996, 4). However, we cannot identify them as legal rights if we are not legal positivist. Though it will be impossible to prevent human rights violation without legal instruments, human rights cannot be identified with laws.

Human rights are rights of the individual who belongs to the human species. They are rights which all persons equally enjoy simply because they are human. What makes a person the holder of these rights is that he/she is a human being. Hence human rights are claimable by everyone and held against everyone, especially against those who run social institutions. They are thought to be shared by all of us no matter our nationality, age, race, gender, language, occupation, religion, income, social standing or sexual orientation. As stated in the *Universal Declaration of Human Rights*, "all members of human family" hold human rights.

Some suggest that prominent human rights documents like the *Universal Declaration of Human Rights* settle the issue of rights-holding definitively. However, the defence of equality and universality are not beyond dispute. While this is a relevant contention, and while we might want to agree with it, what happens, for example, when we step back from the law and reconsider the issue from the point of view of moral reasoning of human rights as general moral rights? Does one really believe that all human beings including serial rapists, mass murders and terrorists enjoy human rights in this sense? Does it even make sense to suggest that people who violate the rights of others nevertheless retain human rights themselves? These kinds of questions that surround the status of being a human rights holder prove that human rights are more about enduring entitlements than about explicitly uttered claims (Orend 2002, 38–39).

An answer to these questions related to human rights holders is given by B. Orend who poses the question, "who holds human rights?". He sought a joint set of necessary and sufficient conditions for status as a human rights holder, conditions that are consistent with the universality and equality residing at the heart of the human rights idea and came to conclusions: to hold human rights, one must be biologically human, one must avoid violating another's human rights and one must have fundamental interests in, or vital needs for, living a life of minimal value. One's having these three characteristics gives everybody else overriding reasons to acknowledge one's status as a fully-fledged holder of human rights (Orend 2002, 65). These three characteristics might be reduced to one: being human or being a member of the human species.

A new concept of human rights that rests on an account of human being is presented by I. Kuçuradi. "Human rights express some necessities, related to each human being. These necessities appear as demands to recognize and observe the value of Man, i.e. they appear as demands to protect individuals for the sole reason that they are human beings" (Kuçuradi 1982, 47). For her the "value of Man" rests on the special place man occupies among other living beings. What gives this place to a human being are his/her characteristics and potentialities as a whole which distinguish him/her from other beings and that these characteristics and potentialities appear as activities proper to human beings and products of these activities. These properties or potentialities constitute "the value of Man" or his dignity. The demands that aim to protect the potentialities proper only to human being constitute the inalienable human rights (Kuçuradi 1982, 47–48). Human rights aim to actualize and develop these potentialities for every individual belonging to the human species. Kuçuradi maintains that "what the *Universal Declaration* intends to word is the conditions of **this** possibility" (Kuçuradi 1995, 81).

Though human rights are interrelated and form an indivisible whole, they also show some remarkable differences: a part of them is directly related to human potentialities, while another to the pre-conditions necessary for their development in general, and still another to the certain (changing) conditions. Kuçuradi restricts the concept of human rights to the first two kinds of demands, that is, the demands concerning the security of the individual and/or the so-called 'fundamental freedoms' and the demands related to the preconditions for the protection of human potentialities in general (rights such as the right to a standard of living adequate for one's health, the right to education, the right to work, etc.).

The first group is directly connected with the exercise of human potentialities. If individuals are not hindered when they exercise activities proper to man only, these rights will be protected by the state or by other individuals. These are the rights which cannot be given by the state, but can only be respected in individuals by individuals or statesman. The so-called "fundamental freedoms" are the legal guarantees of these rights of the individual.

The other group of rights in which all human beings are equal is demands related to the pre-conditions affording each individual the possibility to develop his potentialities: rights such as the right to a standard living, adequate cure for one's health, the right to education, etc. Their fulfilment is different from the first group of rights since they can only be protected indirectly, through other rights given to individuals by a state–through social-economic and (some) political rights–and through public institutions and organizations founded mostly by political decisions (Kuçuradi 1982, 49–50).

5.2 The Concept of Nature

The difficulties with defining human nature arise mainly from the concept of "nature". To clarify what nature is, Aristotle provides a starting point with his analysis of *ousia* that contains different meanings of it as essence, simple bodies, the things composed of bodies and the cause of being and the parts which are presented in such things, limiting them and marking them as individuals.

> We call 'substance' [*ousia*] (1) the simple bodies, i.e. earth and fire and water and everything of the sort, and in general bodies and the things composed of them, both animals and divine beings, and the parts of these. All these are called substance [*ousia*] because they are not predicated of a subject [*hypokeimenon*] but everything else is predicated of them.-(2) That which, being present in such things as are not predicated of a subject, is the cause of their being, as the soul is of the being of an animal.-(3) The parts which are present in such things, limiting them and marking them as individuals, and by whose destruction the whole is destroyed, as the body is by the destruction of the plane, as some say, and the plane by the destruction of the line; and in general number is thought by some to be of this nature; for if it is destroyed, they say, nothing exists, and it limits all things.-(4) The essence, the formula of which is a definition, is also called the substance of each thing.
>
> It follows, then, that 'substance' [*ousia*] has two senses, (A) ultimate substratum [*hypokeimenon*], which is no longer predicated of anything else, and (B) that which, being a 'this', is also separable and of this nature is the shape or form of each thing [*eidos*]. (Aristotle 1966a, 1017b10–25)

The Greek term *ousia*, which bears the meanings of ultimate substratum (*hypokeimenon*), of constitutive and marking part of the individuals, of form (*eidos*), and of essence (*ousia*), in itself has been conveyed in English by the single term "substance". Thus the term "substance" is supposed to bear the different meanings of that of *ousia*, i.e. the ultimate substratum, form of a thing or body, constitutive, limiting and marking part of the things, and the essence. The Latin term '*substantia*' or the English term 'substance" conveys only one of these meanings of the term *ousia* (Kuçuradi 2009, 156–158).

The same problem is encountered with the term "nature" which is supposed to include different meanings of substance. If the "nature" of something is at stake this may mean the ontological specialities of the thing that are present in such things, that is what marks them as individuals or what its essence is. The term "nature" may also mean the totality of "the simple bodies ... and in general bodies and the things composed of them ... and the parts of these" (Aristotle 1966a, 1017b10–15). There are also difficulties at the semantic level as a result of the ambiguity surrounding the concept of nature. As David Hume stressed, the content of the concept of nature changes depending upon the concept for which it is taken to be the opposite. "Nature" means something different when it is used as the opposite of "miracle", to "what is unusual" or to "what is artificial". Further, opposite concepts can easily be found and each time the meaning of "nature" changes (Bayertz 2003, 134).

Aside from its meanings, namely opposite to miracle, unusual and artificial, "nature" conveys both the form of a thing or body, of a constitutive, limiting and marking part of a thing. The question of what the "nature of table" is may actually ask what the form or essence or the constitutive limiting and marking parts of the table is. The question is related to the specificities which mark it as table and differ it from other things as the form and matter of a table. The nature of x is what makes it x and not y: simultaneously, it is its nature that prevents it from being y. It is a matter of definition; to define is to establish a limit or demarcate a boundary. To define in this way is also to identify. Human nature functions as a given, as a constraint, as a limit (Berry 1986, 104).

To conclude, the different meanings of the term nature and the problems caused by these different uses of the term concern also the concept of human nature. The term "human nature" is used by many, but means different things.

5.3 The Concept of Human Nature or Human Being

As with the political basic principles "equality" and "justice", the doctrine of human rights rests upon a certain understanding or conception of "human being" or "human nature". Whereas some cultures may not understand the concept of "human being", or, if they do, may not attach any moral significance to it, the domain of morality may be shaped by a system of social stratification and by the boundaries of the community. However, Donnelly claims that such views are "almost universally rejected in the contemporary world" (Freeman 1994, 492). This does not necessarily warrant the conclusion that the concept of human nature is universal:

> There are, of course, no universally accepted conclusions concerning human nature. Many scholars, especially since John Locke, are of the impression the term 'human nature' is a genuine misnomer, and, in fact, there is no such reality at all. Like other fictional contrivances, 'human nature' is but a reification, having no corresponding reality in the material world .... Most modern theory suggests that man simply is what he does. Some are tame, others wild; some sharing, others hoarding; some monogamous, others polygamous; .... You see, so the argument goes, there is no human nature at all, only human behavior. The possibilities are infinite as to the behavior a given man, or given society, may embark upon. This point of view, which I call the man-as-neutral concept of human nature, is probably the most widely accepted posture concerning human nature among 20th century intellectual community. (Mitchell 1972, 23–24)

Because of the ambiguity of the term "human nature", there have been various challenges: The challenge represented by existentialism attacks the idea of "givenness" and the associated idea that human nature can provide a definitive ideal of human flourishing. The other major challenge, represented by pragmatism, attacks the idea of human nature as a trans-cultural constant which can provide a universal foundation to render the diversity of human experience commensurable (Berry 1986, 122).

However, the challenges to the idea of human nature are not only these two theories of philosophy. Hannah Arendt, for instance, remarks at the beginning of The Human Condition (1958, 10) that nothing entitles us to assume that man has a nature or an essence in the same sense as other things (Berry 1986, 122). Besides, Foucault announces the death of man during the 1960s and sees no other way than to destroy the anthropological ‘quadrilateral’ in its very foundation. Referring the first attempt at this uprooting of Anthropology by F. Nietzsche, Foucault speaks of the death of man as a necessary result of the death of God. Man and God belong to one another, the death of the second is synonymous with the disappearance of the first (Foucault 1994, 341–342).

The thesis on human nature proposed by M. Foucault and by postmodern thinkers has become a paradigm today. It is generally accepted that there is no human nature which will be served as a basis for morality, especially for human rights. This leads us to the question, “if we do not have a concept of human nature, can we speak of rights of man or human rights which has to rest on human nature?” (Heyd 2003, 168).

However, the question of human nature or human being in a broader sense is as old as the history of philosophy. Although philosophers of nature such as Anaxagoras and Heraclitus did not deal with questions related to human beings, but mainly to the question of nature or substance (*arkhe*), they had an implicit concept of human which is easily discernible in their philosophical fragments. Aristotle is generally considered to be the first systematic philosopher who wrote on metaphysics and logic, and also on ethics and politics. The question concerning the characteristics of human being plays a central role in Aristotle’s ethical and political writings because politics, which aims at making people happy (i.e. well), must first know what human nature is or which properties constitute man’s nature. Aristotle believes that the soul consists of two parts, the part which grasps a rule or rational principle, and the irrational part. One division of the irrational element is widely distributed and vegetative in its nature which causes nutrition and growth. There seems to be another irrational element in the soul, namely the appetitive; it is, in general, the desiring element. The vegetative element does not share a rational principle, but the appetitive, and, in general, the desiring element in a sense, has a share of it, in so far as it listens to and obeys it (Aristotle 1966b, 1102 a 25-b 30). Aristotle draws a similar distinction within the part which grasps a rational principle. He assumes that there are two parts of which grasp a rational principle—one by which the individual contemplates kind of things whose originative causes are invariable, and the other by which the individual contemplates variable things; he calls the first of these parts the scientific and the other calculative. Also, he sees three things in the soul which control action and truth: sensation, reason and desire. The origin of an action is choice, and that of choice is desire and reasoning with a view to an end (Aristotle 1966b, 1139 a 5–35).

Aristotle refers to all of these in the context of defining happiness or well-being (*eudaimonia*) and virtue. Since happiness or well-being is seen as an activity of the soul in accordance with perfect virtue, he considers the nature of virtue and then the nature of soul because virtue is related not to the body but to the soul. Hence, “Aristotle believed that political planners need to understand what human beings require for flourishing life. He explicitly said that his ethical writings about the flourishing human life were intended as guides for society’s future politicians, so that they could see what it was they were trying to achieve” (Nussbaum 2011, 125). Many contemporary thinkers hold the opinion that human nature provides objective foundations for ethics and political philosophy. They think that although modern culture has become suspicious of the idea of human nature and the idea of nature has undergone a process of positivistic disenchantment we cannot dispense it. No ethical theory is capable of avoiding a conception of human being which can serve as a guiding principle in the formulation of goals and ideals. The *nature* of human beings is the source of stability, predictability, generality and trajectory of perfection without which no moral conception can be articulated (Heyd 2003, 167–168). Although it seems necessary for ethics and human rights to be based on the idea of human nature, it is not easy to answer the questions of what human being is and which properties are essential to him/her. There are no clear boundaries to the term human nature. We see different concepts of human nature in ethics and politics. Whenever human rights are mentioned, the idea of human nature is assumed most of the time implicitly. As a basic and elusive notion, the notion of human nature is important for both theoretical and practical reasons:

> We can see the political importance of the idea at two levels. Firstly, at the conceptual level, while 'human nature' is not obviously an 'issue' like 'equality' or 'justice' ... it is implicit within them. ...
>
> Secondly, at a popular or informal level, the notion of human nature is politically important because of both the frequency and context of its use in everyday discourse. The frequency with which 'human nature' is invoked is indicated by the almost proverbial use of the expressions 'that's human nature' and 'it's only human nature'. The context in which these invocations are made is one where explanation and justification are called for. ... Accordingly, examination of the concept of human nature is of more than a narrowly academic interest and should, if successful, help throw light on the 'real world' of political experience. (Berry 1986, IX–X)

The concept of human nature has been used as a justifying device for human rights or ethical-legal norms, though it has never been a clear and distinct concept. As it was conceived as the mainspring of any moral norms, the idea of nature generally has a powerful appeal. When we discover that nature compels us to do something, most of us feel strongly pressed and are easily persuaded to follow. That is why, for those who perceive it as the mainspring of any norm, nature constitutes a justification of the norm in the sense accepted here (Gorecki 1996, 25).

It does not seem easy to find what constitutes human nature, although different properties or characteristics of man are given as the basic feature of the human being. Is it human to be rational or social? Is it essentially human to be moral or to have free will? But there are also some other features, like to sin, to compete, to war, to rebel and to destroy. Are these all components of human nature? Unless we find an efficient means to limit the concept of what is essentially human, conceptual inflation of human rights cannot be avoided (So 1995, 39).

In conclusion, the question of human nature may refer to various aspects of human being and thus various answers can be expected:

1. The question of what human nature is may refer to

   > those characteristics and capabilities which only the human being possesses and which therefore distinguish him as a specific creature. ... [It] includes above all human beings' capacity for reason and speech and capacities for intentional action,

> responsibility, etc., which arise from it. It is precisely because the capabilities ..., as a whole, those which define the human being as a specific creature, that they have frequently been identified with 'human nature' in the history of philosophy. ... We are talking here about another new concept of 'human nature', and so as not fall into total conceptual confusion, we should speak here of 'human essence' rather than 'human nature'. (Bayertz 2003, 137–138)

Because of these properties, human being is supposed to have a special place in the cosmos or dignity.

2. The question of what human nature is may refer to some properties which are supposed to be shared by all human beings. To believe in human nature is to believe that humankind possesses some common attributes. These attributes are to be understood not as 'optional extras' but belonging to Man as Man. They are universal in the sense that whenever and wherever humans are encountered these attributes will also be found. This view of universality does not claim that all humans are identical (Berry 1986, 58).
3. Human nature may be understood as the biologically (and psychologically) constitutive properties of members of the human species or as the immutable core of the ideal human being. In the first case, human nature has no normative implications; in the second, it is itself a normative concept and hence cannot serve as the foundation for an ethical system (Heyd 2003, 167).
4. Human needs may define the human nature that gives rise to human rights.

> It is obvious to any reasonable observer that all human beings have vital needs in life and fundamental interests in living a life of at least minimal value to themselves and the world. ... If we want to survive and begin to pursue our personal goals at all, we must have the objects of our vital needs and fundamental interests. It was contended that there are five such objects, claimed by everyone: physical security, material subsistence, personal freedom, elemental equality, and social recognition. (Orend 2002, 95)

In short, then, the question of human nature may refer to different aspects of the human being as observed in the examples cited above and may require still different answers. This leads us to take a closer look at the question of human nature, and see what questions are expected, then make a decision whether or not it is meaningful to answer it. In light of the preceding discussions, it should be clear that we need a new understanding of human nature or human being.

5.4 Justifying Human Rights Through Human Nature (Human Being): Two Approaches

These arguments about human nature prove that a notion of human nature or of the human being seems indispensable, and how it is difficult to clarify and elaborate a concept of human nature and to use it as a tool for the justification of human rights. However, no human rights theory can work without supposing a concept of the human being. A convenient concept of the human being can serve as justification, but it cannot be seen as an "ultimate problem-solver" or ultimate justification of human rights. The following two approaches seem to succeed in attaining a new understanding of human nature for a justification of human rights, though both avoid using the term "human nature".

5.4.1 The Capability or Capabilities Approach (A. Sen and M.C. Nussbaum)

The capability (or capabilities)[3] based approach developed by Amartya Sen concentrated first on means such as income and primary goods, and is regarded as an approach to economic problems, in particular economic injustice and poverty. It was an approach to the assessment quality of life pioneered within economics by A. Sen, and by now is highly influential through the *Human Development Reports of the United Nations Development Programme* (UNDP) (Nussbaum 2000, 70). The capabilities approach can be defined as an approach to comparative quality-of-life assessment and theorizing in basic social justice. The key question to ask is: What is each person able to do and to be? It takes each person as an end; it is not concerned with the total or average well-being, but instead with the opportunities available to each person. It is focused on choice and freedom, holding that the crucial good societies should be promoting for their people a set of opportunities or substantial freedom. As it is pluralist about value, it is concerned with entrenched social injustice and inequality, especially capability failures that are the result of discrimination or marginalization. It ascribes an urgent task to government and public policy—namely, to improve the quality of life for all people, as defined by their capabilities (Nussbaum 2011, 18–19). Sen argues as follows:

> The idea of 'capability' (i.e. the opportunity to achieve valuable combinations of human functionings—what a person is able to do or be) can be very helpful in understanding the opportunity aspect of freedom and human rights. ... Seeing opportunity in terms of capability allows us to distinguish appropriately between (i) whether a person is actually able to do things she would value doing, and (ii) whether she possesses the means or instruments or permissions to pursue what she would like to do (her actual ability to do that pursuing may depend on many contingent circumstances). ... The capability-based approach resists an overconcentration on means (such as incomes and primary goods) that can be found in some theories of justice (e.g. in the Rawlsian Difference Principle). (Sen 2005, 153–154)

Sen focused on capability first in the specific context of evaluating inequality, and after that he tried to explore the possibility of using the capability perspective for analyzing social issues, such as well-being and poverty, liberty and freedom,

---

[3]M. Nussbaum prefers to use the plural form "capabilities" to emphasize that the most important elements of people's quality of life are plural and qualitatively distinct: health, bodily integrity, education and other aspects of individual lives that cannot be reduced to a single metric without distortion. Sen, too, emphasizes this idea of plurality and nonreducibility, which is a key element of the approach (Nussbaum 2011, 18).

living standards and development, gender bias and sexual divisions and justice and ethics. Sen maintains that a truly developed society would enable humans to be or to do, and to live and act, in certain *valuable* ways. He employs general terms "valuable functioning" and "achievements" to cover these intrinsically good "beings and doings". The concepts of functioning and capability have been main concepts of the capability approach and been used for a reconstruction of the social ideals of freedom, rights and justice.

> Perhaps the most primitive notion of in this approach concerns 'functionings'. Functionings represents parts of a person–in particular the various things that he or she manages to do or to be in leading life. The capability of a person reflects the alternative combinations of functionings the person can achieve, and from which he or she can choose one collection. The approach is based on a view of living as a combination of various 'doings and beings', with quality of life to be assessed in terms of the capability to achieve valuable functionings. (Sen 1995, 31)

Sen extends the concept of functioning beyond intentional actions to include any "state of existence of a person". Not only does the choosing initiate action, but also includes the mental state that happens to accompany the activity. Moreover, also included under the concept are states of a person such as their being physically or psychologically fit (Crocker 1995, 154). Martha Nussbaum's concept of functioning is somewhat narrower than that of Sen. Although Sen conceives choosing as a distinguishable functioning, Nussbaum understands choosing as not more than the voluntary or chosen dimension of an intentional human functioning. Sen conceives the mental states of happiness or pleasure (or their opposite) as distinguishable functions that have intrinsic value and can be pursued as such. Although Nussbaum, defining her position as less utilitarian and more Aristotelian, counts "being able to have pleasurable experiences" as one of the valuable human functional capabilities; she refuses to make pleasure a separable functioning (Crocker 1995, 153–155).

Martha Nussbaum developed the new version of this approach, independent of Sen's work, by thinking about Aristotle's ideas of human functioning and Marx's use of them (Nussbaum 1995a, 243–267). This new version is different in several ways from Sen's, both through its emphasis on the philosophical underpinnings of the approach and its readiness to take a stand on what are central capabilities. Sen focused on the role of capabilities in demarcating the space within which quality of life assessments are made, while Nussbaum used the idea in a more exigent way, as a foundation for basic political principles that should underwrite constitutional guarantees (Nussbaum 2000, 70–71). This new version of the capabilities approach is philosophical rather than economical and political, but it has been developed for political reasons and is closely allied to a form of political liberalism. It is also based on the universalist account of central human functions. Nussbaum argues that the best approach to this idea of a basic social minimum—the bare minimum required for respecting human dignity—is provided by an approach that focuses on human capabilities, that is, on what people are actually able to do and to be. She maintains that the capabilities in question should be pursued *for each and every person*, treating each as an end and none as a mere tool of the ends of others. She thus adopts the principle of each person's capability, based on the principle of each person *as end* (Nussbaum 2000, 5).

The capabilities approach has (at least) two versions, in part because it has been used for two purposes: Nussbaum's version puts the approach to work in constructing a theory of basic social justice and adds other notions—as *human dignity*, the *threshold* and *political liberalism*—in the process. Sen's primary concern has been to identify capability as the most pertinent space of comparison for purposes of quality-of-life assessment. Focusing on instances of capability failure that result from gender or racial discriminations, his version of the approach does not propose a definite account of justice. In consequence, Sen does not employ a threshold or a specific list of capabilities, although he thinks some capabilities (e.g. health and education) have a particular centrality, nor does he prioritize theoretical use of the concept of *human dignity*, though he certainly acknowledges its importance (Nussbaum 2011, 19–20).

If we want to see the main concepts of the capability approach in detail, the first question is "what are capabilities?" Capabilities are the answers to the question, "What is this person able to do and to be?". "In other words, they are 'substantial freedoms', a set of (usually interrelated) opportunities to choose and to act. Capability is thus a kind of freedom: substantive freedom to achieve alternative functioning combinations" (Nussbaum 2011, 20).

On the other side of capability is functioning. A functioning is an active realization of one or more *capabilities*. Functionings do not need to be especially active. Enjoying good health is a functioning, as is lying in the grass. Functionings are "beings and doings" that are the outgrowths or realizations of capabilities.

In contrasting capabilities with functionings, we should bear in mind that capability means opportunity to select. The notion of *freedom to choose* is thus built into the notion of capability. To use Sen's example, a person who is starving and a person who is fasting have the same functioning where nutrition is concerned, but they do not have the same capability because the person who fasts is able to not fast, whereas the starving person has no choice (Nussbaum 2011, 25).

Considering the various areas of human life and asking the question "What does a life worthy of human dignity require?", Nussbaum speaks of an ample threshold level of ten central capabilities as a bare minimum. Following a widely shared understanding concerning to the task of a state, namely that a state has the job of enabling people to pursue a dignified and minimally flourishing life, Nussbaum asserts that a decent political life order must be secured for all citizens at the very least at a threshold level of these ten central capabilities.[4] Those ten capabilities are:

1. *Life*: being able to live to the end of a human life of normal length; not dying prematurely, or before one's life is so reduced as to be not worth living.
2. *Bodily health*: Being able to have good health, including reproductive health; to be nourished; to have adequate shelter.

[4]She refers to these central capabilities as "The Shape of the Human Form of Life" (Nussbaum 1995b, 76).

3. *Bodily integrity*: Being able to move freely from place to place; to be secure against violent assault . . . ; having opportunities for sexual satisfaction and for choice in matters of reproduction.
4. *Senses, imagination and thought*: Being able to use the senses, to imagine, think, and reason—and to do these things in a 'truly human' way . . .
5. *Emotions*: Being able to have attachments to things and people outside ourselves; to love those who love and care for us, to grieve at their absence; in general, to love, to grieve, to experience longing, gratitude, and justified anger . . .
6. *Practical Reason*: Being able to form a conception of the good and to engage in critical reflection about planning of one's life . . .
7. *Affiliation*: (A) Being able to live with and toward others, to recognize and show concern for other human beings. . . . . (B) Having the social bases of self-respect and nonhumiliation; being able to be treated as a dignified being whose worth is equal to that of others . . .
8. *Other species*: Being able to live with concern for and in relation to animals, plants, and the world of nature.
9. *Play*. Being able to laugh, to play, to enjoy recreational activities.
10. *Control over one's environment*: (A) Political. Being able to participate effectively in political choices that govern one's life; having the right of political participation, protections of free speech and association. (B) Material. Being able to hold property, and having property rights on an equal basis with others; having the right to seek employment on an equal basis with others; having the freedom from unwarranted search and seizure. (Nussbaum 2011, 33–34)

Sen does not give a list of capabilities, but does find this list of capabilities useful for some minimal rights against deprivation. He also adds that, for other practical purposes, one may need quite a different list and with a given list one cannot avoid the question of valuation. Sen also indicates the problem of dealing with the hierarchy within a given list of capabilities and argues that some of the basic capabilities will figure into every list of the relevant capabilities in every society, but the exact list to be used will have to take note of the purpose of the exercise. For a specific purpose, he maintains, the coverage of capabilities can be narrowed with the use of good sense (Sen 2005, 159).

These given features of human being guide us in driving these ten requirements and needs that constitute the threshold of a good human life. The list of features is composed of two categories: limits and capabilities. A life without both would be lacking and impoverished, if at all human. Obviously then, this would not be a good human life. Therefore, this list of capabilities is a ground-floor or minimal conception of a good life.

Nussbaum describes two distinct thresholds: a *threshold* of capability to function beneath which a life is so impoverished as to not be human life at all. *A somewhat higher threshold*, beneath which those characteristic functions are available in such a reduced way that, although one may judge the form of life a human one, it is not *a good human life*. The concern for human rights and public policy is the latter threshold. For human rights, concern must be "to ask whether all citizens are capable, not just of the bare minimum, but of good life in these areas" (Nussbaum 1995b, 80–81).

Nussbaum characterizes the significance of this list in these terms: "My claim is that a life that lacks any one of these capabilities, no matter what else it has, will fall short of being a good human life" (Nussbaum 1995b, 85). Further, she maintains that the list, and its associated argumentation, ought to be taken seriously by development theorists in the design of development strategies. Public policy must be guided by a conception of the human good that gives the policy maker strong guidance in selecting goals and priorities for the development process. "The basic claim I wish to make . . . is that the central goal of public planning should be the capabilities of citizens to perform various important functions" (Nussbaum 1995b, 87).

Those ten basic human functional capabilities have already been recognized within the scope of human rights. If everyone has the opportunity to use or actualize these functional capabilities in any state or country, then the basic rights or human rights will be protected. As Nussbaum maintains, these needs and capabilities establish the threshold requirement for human being and living humanly. The threshold should be understood as a set of minimum requirements concerning material and institutional arrangements that will prepare the actualization of the potentialities specific to human being (Deveci 2002, 61).

Taking this account of human being[5] as a starting point provides a justification for human rights. If human beings are conceived of as having certain central capabilities, then human rights must secure each member of the human species to actualize these capabilities. The good human life aimed for by every person can be possible only by maintaining human rights. Then the concept of human being or "human nature" in the broader sense gives us grounds to justify human rights. If nobody is allowed to live beneath the threshold of a good life, then human rights and human dignity would be protected. Following Nussbaum one can say that it would be the concern of human rights to ask whether all citizens are capable, not just of the bare minimum, but of a good life in these areas. A life that lacks any of these capabilities, no matter what else it has, will fall short of being a good life, that is, of being a dignified life.

Although Nussbaum and Sen emphasize the close relationship between capabilities and human rights, they do not use the capabilities approach as a device for the justification of human rights; instead, they find this relation more complicated than whether one justifies the other. Nussbaum feels that the capabilities approach can clarify the concept of human rights and inform people about the goals in public policy:

> Moreover, a focus on capabilities, although closely allied with the human rights approach, adds an important clarification to the idea of human rights: for it informs us that our goal is not merely "negative liberty" or absence of interfering state action—one very common understanding of the notion of rights—but, instead, the full ability of people to be and to choose these very important things. Thus all capabilities have an economic aspect: even the freedom of speech requires education, adequate nutrition, etc. (Nussbaum, http://cfs.unipv.it/sen/papers/Nussbaum.pdf, 17)

---

[5]"The Capabilities approach is not a theory of what human nature is, and it does not read off from innate human nature. Instead, it asks, among the many things that human beings might develop the capacity to do, which ones are the really valuable ones, which are the ones that a minimally just society will endeavor to nurture and support?" (Nussbaum 2011, 28).

Sen shares the same view as Nussbaum that the capabilities approach may help to explain human rights, but he thinks that though these two concepts—human rights and capabilities—go well with each other, all kinds of human rights cannot be analysed within this approach.

> There are many human rights for which the capability perspective has much to offer. However, human rights to important process freedoms cannot be adequately analysed within the capability approach. (Sen 2005, 163)

Nussbaum, on the other hand, dwells on the common ground of the capabilities and human rights approaches. "The common ground between the Capabilities approach and human rights approach lies in the idea that all people have core entitlements just by virtue of humanity, and it is a basic duty of society to respect and to support these entitlements. . . . The capabilities on my list overlap substantively with the human rights recognized in the Universal Declaration and other human rights instruments" (Nussbaum 2011, 62). The relationship that Nussbaum and Sen draw implies that capabilities may serve to define and ground human rights, as I have argued above.

5.4.2 The Ontological-Anthropological Approach (T. Mengüşoğlu and I. Kuçuradi)

Takiyettin Mengüşoğlu is one of the eminent thinkers of Turkish philosophy, not only introducing a new philosophical approach in Turkey but also a new theory of philosophical anthropology. He introduced the ontological approach originating from works of Aristotle and Nicolai Hartmann in Turkey mainly through his philosophical works on ontology, epistemology and philosophical anthropology. Mengüşoğlu applied the ontological approach to the problems of philosophical anthropology and tried to define human being as a whole, instead of reducing human being to any peculiar characteristics or properties of man like reason, conscience or *Geist*. He labelled this new approach "ontological-anthropological" and claimed that it enables us to see human being as a whole and that any epistemological approach which tries to grasp human phenomena merely as objects and neglects to connect them with life of the person who attempts to know these phenomena has to fail.

Mengüşoğlu begins with Kant's approach to human being and develops a new theory which considers human being as a concrete totality. The ontological basis of human being is his bio-psychic concrete totality by means of which all activities are realized. Thus, Mengüşoğlu suspends all dualist views of human being and puts forward a view of philosophical anthropology that handles the human being as a living creature who possesses some human phenomena and achievements that constitute his/her "nature". Some of these phenomena are peculiarities of human being, while some are products or achievements of human being. The question, what is human being and what is the difference between an animal and human being, can be answered by referring to the phenomena and achievements of the human species.

Human being is considered to be a being which knows, loves and dedicates oneself to something, which is active, has free will, can feel values and is the creator of art and technology (Mengüşoğlu 1988, 13–17). This is an open-ended list. Mengüşoğlu calls these phenomena "conditions of existence of the human being" because we meet them actually or potentially wherever and whenever there is a human being. Those are closely interrelated and one cannot be realized without the others. Those phenomena are neither only psychic nor biological. We face concrete human being as a whole in his simple works of daily life and also in his most complicated activities like science, technology and art. In fact, the concrete human being as a whole can be seen in his activities and his achievements that are realized by these activities. Human beings can be conceived in those phenomena and achievements (Mengüşoğlu 1988, 49).

Mengüşoğlu divides these phenomena and achievements into two groups, some of which are the capabilities found in human nature. The language, natural and genetic capabilities, namely bio-psychic nature and disharmonic nature as a whole, are his structural specificities and capacities. All of them can be observed in every human being; one can develop them or can keep them primitive, but he/she cannot abolish them. Whereas phenomena such as knowledge, art and technology, religion and state are the achievements of the human species, they are not achievements of every single individual of the human species. Each member shares them, but only in capacity or potentiality.

Mengüşoğlu considers philosophical anthropology to be the basic discipline of philosophy because every view provided by political philosophy or ethics presupposes a view of man and asserts that ethical problems can only be solved from the perspective of philosophical anthropology. The epistemological point of view dominant in philosophy for a long time is faced with obstacles in conceiving the ontological nature of ethical and anthropological problems.

Ioanna Kuçuradi, who is a student of Mengüşoğlu, has developed her own views of value and human rights on the basis of this philosophical anthropology. Taking the ontological anthropological approach to human being as a starting point, she introduced a new theory of value and ethics in addition to her new conception of human rights. The basis for all these views was Mengüşoğlu's approach which seemed, by Kuçuradi, to be indispensable for ethics and theory of value, but which she thinks requires elaboration.

For Kuçuradi, the originality or the most significant peculiarity of this view of anthropology lies in the will to understand man by means of the activities special to him/her. This perspective or point of view seems to be the most important contribution of Mengüşoğlu to the contemporary philosophical anthropology. In these analyses of Mengüşoğlu we see that the characteristics of human being are presented along with his potentialities, as Kuçuradi points out. Kuçuradi, however, differentiates between human phenomena or characteristics and human potentialities which were not clearly distinguished by Mengüşoğlu (Kuçuradi 1997, 84). This stance marks the starting point of her new concept of human rights.

The idea of human rights is also justified by means of understanding human dignity or the value of man. Thus, human rights which point out certain necessities related to each human being appear as demands to recognize and observe the value of man. In other words, they aim to protect each person for the simple fact that he/she is a human being. The "value of man" or "dignity of man" means the special place of man among other living beings. What provides this special place to man is all his/her characteristics and potentialities which distinguish him/her from other living beings. They appear as activities proper to men and as products of these activities. These are properties of human beings only, besides those that human beings possess in common with other living beings; what constitutes the "value of man" or his dignity are these properties or potentialities. Thus, protecting these ontological properties of man in individuals or developing man's potentialities in individuals becomes the right of every single member of the species called human. Hence, a due respect for man's potentialities becomes the right of every individual. Human rights aim to provide every individual with the chance to develop these potentialities (Kuçuradi 1982, 47). Specifying the potentialities of human being, namely to build up a concept of human being, appears as a prerequisite for identifying which rights are human rights.

5.5 Conclusion

The set of properties which constitute the essence of human being is signified by a concept of human nature or of human being. Many thinkers of our time have held the opinion that human nature provides objective foundations for ethics and human rights. In this article, I argue that the concept of human rights presupposes—or is grounded in—a concept of the human being or human nature. However, the concept of human nature sounds generally metaphysical because of the accustomed understanding of it. The problem with the concept of human nature stems from the fact that human being is thought to embody certain characteristics which are expected to be found in every member of the human species. It is assumed that there is a human nature which is innate and the same in every human being.

Then we may expect that the problem will be solved if we have a new understanding of human nature, one that is open-ended and does not ignore man's potentialities alongside actual properties. The two views presented above, namely the capabilities approach and the ontological anthropological approach, seem to be offered to satisfy this requirement; both views refrain from making use of the concept of "human nature" as they conceive of man as a being of possibilities or capabilities besides some actual characteristics.

> The capabilities approach is not a theory of what human nature is, and it does not read norms off from innate nature. Instead, it is evaluative and ethical from the start: it asks, among the many things that human beings might develop the capacity to do, which ones are the really valuable ones, which are the ones that a minimally just society will endeavor to nurture and support? An account of human nature tells us what resources and possibilities we have and what our difficulties may be. It does not tell us what to value. (Nussbaum 2011, 28)

Although Nussbaum differentiates the capabilities approach from an account of human nature, it is very clear that she provides us with a concept of human being. "Human beings come into the world with the equipment for many 'doing and being' (. . .) and we have to ask ourselves which ones are worth developing into mature capabilities" (Nussbaum 2011, 28). In addition to this introduction of a concept of human being, the capabilities approach tries to specify which capabilities are worth developing. As such, it includes a dimension of value which is not necessarily inherent in every conception of human being.

However, the capabilities approach provides us with a basis to determine and justify human rights by means of its implicit concept of human being. Human rights aim to secure the condition for every human being to realize his/her "doings or beings" or to develop the capabilities worth developing. Nussbaum and Sen emphasize the close relationship between the capabilities approach and human rights, indicating that they have common ground, namely the view that all people have certain core entitlements just by virtue of their humanity. Nussbaum asserts that the capabilities on her list overlap substantially with those human rights recognized in the Universal Declaration of Human Rights. And they play a similar role, providing a basis for both cross-cultural comparisons and for constitutional guarantees. The capabilities approach supplements the standard human rights approaches by grounding rights claims in bare human birth and minimal agency, by articulating the relationship between human rights and human dignity more clearly than most standard rights account, by more clearly setting out the relationship between human entitlements and those of other species and finally by spelling out the relationship between human rights and duties (Nussbaum 2011, 63). All these supplementations to human rights demonstrate how efficient the capabilities approach is in grounding human rights:

> Capabilities, like human rights, supply a moral and humanly rich set of goals for development . . . . In effect, capabilities cover the terrain occupied by both the so-called first-generation rights (political and civic liberties) and the so-called second-generation rights (economic and social rights). And they play a similar role, providing an account of extremely important fundamental entitlements that can be used as a basis both for constitutional thought within a nation and for thinking about international justice. (Nussbaum 2006, 284)

Similarly, the ontological anthropological approach developed by Mengüşoğlu and Kuçuradi provides another convenient basis for grounding or justifying human rights by means of the human phenomena as properties and potentialities. The activities proper to man and products of these activities, as properties and potentialities, constitute the "nature" or the dignity of man. Human rights are norms generated by the knowledge of the value of these characteristics and potentialities. These norms state the conditions considered necessary for the human kind to fulfil the potentials special only to the human species (Kuçuradi 2007, 66). The systematic knowledge of potentialities and conditions for the actualization of these potentialities of human being offered by the ontological anthropological approach seems to be the prerequisite of the criteria and justification of human rights. If human rights are considered demands, then a part of these demands is connected directly with the exercise of human potentialities. What they demand is that the individuals should not be hindered while exercising activities proper to man only. Another part of these demands is related to the conditions necessary for development in general, while another part is to certain conditions. Although there are various differences among human rights, the concept of human rights requires a concept of human being. In conclusion, human rights cannot dispense with the concept of human nature or human being, but the concept of human nature has to be taken in a broader sense, including not only human properties but also human potentialities. In spite of our infinite differences, we also have something in common that we can call "human nature" or "human being" by which we can differentiate human beings from other creatures; it is by that means that we can justify human rights.

References

Arendt, H. 1958. *The human condition*. Chicago: University of Chicago Press.

Aristotle. 1966a. *The works of Aristotle*, vol. VIII, Metaphysica. Trans. W.D. Ross. Oxford: Clarendon Press.

Aristotle. 1966b. *The works of Aristotle*, vol. IX, Ethica Nicomachea. Trans. W.D. Ross. Oxford: Clarendon Press.

Bayertz, K. 2003. Human nature: How normative might it be? *Journal of Medicine and Philosophy* 28(2): 131–150.

Berry, C. 1986. *Human nature*. London: Macmillan Education Ltd.

Crocker, A.D. 1995. Functioning and capability: The foundations of Sen's and Nussbaum's development ethic, Part 2. In *Women, culture and development: A study of human capabilities*, ed. M.C. Nussbaum and J. Glover. Oxford: Clarendon Press.

Deveci, C. 2002. Human rights with Aristotelian universalism. In *Human rights in Turkey and the world in the light of fifty-year experience*, ed. I. Kuçuradi. Ankara: Hacettepe University.

Foucault, M. 1994. *The order of things. An archaeology of the human sciences*. New York: Vintage Books.

Freeman, M. 1994. The philosophical foundations of human rights. *Human Rights Quarterly* 16(3): 491–514.

Gorecki, J. 1996. *Justifying ethics. Human rights and human nature*. New Brunswick/London: Transaction Publishers.

Heyd, D. 2003. Human nature: An oxymoron? *Journal of Medicine and Philosophy* 28(2): 151–169.

Kuçuradi, I. 1982. Philosophy and human rights. In *Philosophical foundation of human rights*, ed. I. Kuçuradi, 47–52. Ankara: Hacettepe University.

Kuçuradi, I. 1995. Human rights instruments questioned in the light of the idea of human rights. In *The idea and documents of human rights*, ed. I. Kuçuradi, 3–12. Ankara: Philosophical Society of Turkey.

Kuçuradi, I. 1997. 20. Yüzyıl Antropolojisinde Takiyettin Mengüşoğlu'nun Yeri. In *Yüzyılımızda İnsan Felsefesi*, ed. İ. Kuçuradi, 75–85. Ankara: Philosophical Society of Turkey.

Kuçuradi, I. 2007. Etik İlkeler ve Hukukun Temel Öncülleri Olarak İnsan Hakları. In *İnsan Hakları: Kavramlar ve Sorunlar*, ed. İ. Kuçuradi, 55–68. Ankara: Türkiye Felsefe Kurumu.

Kuçuradi, İ. 2009. Aristoteles'in Ousia'sı ve Substans Kavramı. In *Çağın Olayları Arasında*, İ. Kuçuradi, 149–162. Ankara: Türkiye Felsefe Kurumu Yayınları.

Mengüşoğlu, T. 1988. *İnsan Felsefesi*, İnsanın Varlık Yapısı ve Nitelikleri, İnsan ve Hayvan Dünya ve Çevre. İstanbul: Remzi Kitabevi.
Mitchell, J. 1972. Why study human nature. In *Human nature: Theories, conjectures, and descriptions*, ed. J. Mitchell, 23–32. Metuchen: The Scarecrow Press Inc.
Nussbaum, C.M. 1995a. Non-relative virtues: An Aristotelian approach. In *The quality of life*, ed. M. Nussbaum and A. Sen, 242–268. Oxford: Clarendon Press.
Nussbaum, C.M. 1995b. Human capabilities, female human beings. In *Women, culture and development. A study of human capabilities*, ed. M. Nussbaum and J. Glover, 61–104. Oxford/New York: Oxford University Press.
Nussbaum, C.M. 2000. *Women and human development, the capabilities approach*. Cambridge: Cambridge University Press.
Nussbaum, C.M. 2006. *Frontiers of justice, disability, nationality, species membership*. Cambridge/London: The Belknap Press of Harvard University Press.
Nussbaum, C.M. 2011. *Creating capabilities, the human development approach*. Cambridge/London: The Belknap Press of Harvard University Press.
Nussbaum, C.M. *Beyond the social contract: Capabilities and global justice*. http://cfs.unipv.it/sen/papers/Nussbaum.pdf. Accessed 11 Nov 2011.
Orend, B. 2002. *Human rights concept and context*. Peterborough/Lancashire: Broadview Press.
Sen, A. 1995. Capability and well-being. In *The quality of life*, ed. M.C. Nussbaum and A. Sen. Oxford: Clarendon Press.
Sen, A. 2005. Human rights and capabilities. *Journal of Human Development* 6(2): 151–166.
So, H.-Y. 1995. Negativistic conception of human rights. In *The idea and documents of human rights*, ed. I. Kuçuradi, 37–43. Ankara: Philosophical Society of Turkey.
Tomasevski, K. 1995. The right to define what human rights are: International human rights law making. In *The idea and the documents of human rights*, ed. I. Kuçuradi. Ankara: Philosophical Society of Turkey.

Chapter 6
The Relationship Between Human Nature and Human Rights. The Confucian Example

Mateusz Stępień

6.1 Introduction

There are many difficult and persistent questions about human rights ("HR") on which interdisciplinary inquiry may shed some light (e.g. Panikkar 1982; Angle 2002; Sim 2004). One of them is the relation between the visions of human nature and HR. For a long time, within the dominant Western discourse, the connections between "nature" or "human nature" and rights had been worked out in various ways. Aquinas, Grotius, Pufendorf, de Vattel, Rousseau tried to derive certain rights from variously understood "nature", "natural order" or "human nature". This long-standing approach, called by Ch. Beitz the naturalistic one (2009, 51), was strongly influenced, especially beginning in circa the mid-sixteenth century, by the concept of human nature developed in the tradition of natural law on the grounds of jurisprudence and theology. Of course, there is still a question about the possible relations between jurisprudential and teleological findings and the rise of thinking in the categories of "laws of nature" in the so called "natural philosophy" of those days (Daston and Stolleis 2008). Regardless of that problem, HR discourse, from its beginning, has been basically essentialist, anti-sociological, and has presupposed that HR are something more than mere contractualism and not an artificial cultural product. Even nowadays, in influential doctrinal works, a distant echo of such a way of thinking may be noticed. Despite the fact that in definitions presented by A. Gewirth ("We may assume, as true by definition, that human rights are rights that all persons have simply insofar as they are human"), A. J. Simmons ("rights possessed by all human beings/at all times and at all places/simply in virtue of their humanity"), J. Griffin ("a human right is a right that we have simply in virtue of

M. Stępień (✉)
Department of Sociology of Law, Faculty of Law and Administration, Jagiellonian University, ul. Golebia 24, 31-007 Krakow, Poland
e-mail: mateusz.stepien@uj.edu.pl

M. Albers et al. (eds.), *Human Rights and Human Nature*, Ius Gentium: Comparative Perspectives on Law and Justice 35, DOI 10.1007/978-94-017-8672-0_6,

being human") the term "human nature" does not appear, they implicitly cover what we normally assume to be human nature. Even if one grounds rights in universal rationality, morality or dignity, one still does not go far beyond human nature.

While the history of Western ideas concerning human rights is relatively well-known, much lesser interest has been shown in the concept of human nature developed by Confucian tradition and its relation with HR (see Angle 2002; Haiming and Akina 2012). In this paper I want to address a problem of consequences of the classical Confucian view of human nature and possible arguments, conclusions and assumptions concerning rights within Confucianism which exist a priori, are pre-established and grounded in "nature". As we can see below, the lack of any concept of rights—also the naturalistic one—in Confucianism is at first glance quite surprising since Confucian teachings are based on a very optimistic vision of natural goodness embodied in human nature as well as on an idea of continuity between the human realm and the "natural order"—cosmos (天 *Tian*). A reader educated in the West, one who knows theories of natural law, natural rights and, finally, inborn dignity might expect discussions within Confucian tradition similar to Western HR discourse. This "maladjustment" pushed me to reject a dominant approach focused on examining whether "rights" do exist or not in Confucianism and searching for their equivalents in Western tradition. My aim is rather to analyze the impact of a particular vision of human nature developed by Mengzi on the "discursive space of Confucianism". The latter concept covers basic, ready-to-use arguments, the most important issues pre-selected to study, commonly underlying assumptions within this philosophical tradition and discursive "resources which they have each deployed and promoted" (Jullien 2011, 28).

I begin by reconstructing the orthodox Confucian (Mencian) view of human nature. Then, I will shortly juxtapose it with the latest findings in evolutionary studies on human nature (especially regarding the status of morality). Surprisingly, these recent findings in evolutionary sciences confirm basic Confucian assumptions on human nature. Finally, I will analyze these elements of Mencian theory which determine the lack (or weakness) of recognition of rights and even the existence of the discourse on rights within Confucian philosophical tradition. I contend that there is not necessarily a connection between recognizing inborn goodness or inherent morality which are imprinted onto human nature and making the claiming that humans possess rights by virtue of their nature. I want to look at the Confucian vision from the external point of view in order not only to understand this particular perspective better, but also to make a recursive analysis of Western thought. The Confucian example provides me with a starting point to make some preliminary statements concerning status and justification of HR.

6.2 The Mencian View of Human Nature

It was mainly Mengzi (371–289? BC) who developed the orthodox Confucian view of human nature. He devoted himself to expanding and elaborating Kongzi's (551–479 BC) previous ideas. Mengzi worked out a coherent and comprehensive concept which "has exerted profound influence on the shaping of the Chinese mind" (Liu 1998, 33, 55). The significance of the Mencian concept of human nature also results from lack of a systematic analysis of that topic in Master Kong's teaching except implicit faith in man's perfectibility and belief that people are similar by nature but different by habits. Zhuxi (1130–1200) grouped *Mengzi* with *Lunyu, Daxue* and *Zhongyong* as *Si Shu (The Four Books)*, which was the official textbook for a civil services examination in the period between 1313 and 1905. In consequence, every educated man had to study Mengzi's view of human nature. Moreover, Si Shu "were read and memorized by every schoolboy in his first years at school" (Lau 1984, x). Sung-Ming Neo-Confucian philosophers accepted the Mencian theory, which in the spite of variations and modifications, has reminded the firm belief in Chinese.

Confucian philosophers characteristically regard human nature in a holistic way. It means that the problems of human nature and how it can be realized fully are situated within a larger perspective (Ames 2002, 79; see Nisbett et al. 2001). The human realm is an integral part of the immanent, self-creative cosmic organism. Social order is structured and patterned by common, universal regularities ("the rhythms") emanating from Heaven (*Tian*). There is some kind of "regulatory" continuity between heaven, earth and society. All major ancient Chinese philosophical schools tried to propose a set of directives on how to order society according to *Tian's* "regulatory" patterns. Therefore, one should not be surprised that, for Confucians, *ren xing* 人性—commonly translated as "human nature" or "natural dispositions"—is connected with *Tian*. Mengzi only adapted an old conception that links *Tian* with *xing* (*Mengzi* 6A:15; *Daxue*).[1] As we will see, human nature is treated as potentially good, and the connection between *Tian* and the goodness of our *xing* suggests (or even presupposes) that *Tian* is good itself. Ivanhoe takes up the question of *Tian's* goodness by saying that Mengzi "implies that the universe is not indifferent to the course of human events, that we live in a morally charged universe which favours those who are good" (Ivanhoe 2002, 13).

Human nature is rooted in heart/mind (*xin* 心) which is the appearance of *Tian* (Zhang and Guo 2008, 32). If one fully realizes what is in one's *xin*, one can then be said to know *xing*, and, in this way, one knows and serves *Tian*—"He who exerts his *xin* to the utmost, knows his *xing*. He who knows his nature, knows *Tian*. To preserve one's heart-mind and to nourish one's nature is the way to serve *Tian*" (*Mengzi* 7A:1). Interestingly, moral emotions are understood as basic elements of heart/mind and a part of human nature. Mengzi stressed that everybody possesses inborn "four sprouts" (*si duan* 四端). Perhaps the best translation of *duan* is "sprout" because of the fact that the character of *duan* contains an image of a plant emerging from the ground. Generally, agrarian metaphors are strongly represented in *Mengzi*. He wrote about "the sprouts of commiseration", "the sprouts of shame and dislike", "the sprouts of deference and compliance" and "the sprouts of right and wrong"

[1] Translations of Mengzi, Lunyu, Daxue, Zhongyong are taken from Wing-Tsit Chan's *Source Book in Chinese Philosophy* (Chan 1963), sometimes with my slight modification.

(*Mengzi* 2A:6). These four "sprouts" are basic emotions natural to humans, just like "four limbs". They are originally within man's *xin* and they are not imposed by external forces during the lifetime. They generate moral impulses which stem from within an individual (*nei* 内), and are not acquired through conformity with external standards or sources (*wai* 外). They "do not give me a lustre from the outside, they are in me originally. It is only that I have not reflected upon this" (*Mengzi* 6A:6). They function almost automatically and instinctively as the child-at-the-well thought experiment shows:

> When I say that all man have the mind which cannot bear to see the suffering of others, my meaning may be illustrated thus: Now, when man suddenly see a child about to fall into a well, they all have a feeling of alarm and distress, not to gain friendship with the child's parents, nor to seek the praise of their neighbours and friends, nor because they dislike the reputation [of the lack of humanity if they did not rescue the child]. From such a case, we see that a man without the feeling of commiseration (...) shame and dislike (...) deference and compliance and (...) right and wrong is not a man. The feeling of commiseration is the beginning of benevolence; the feeling of shame and dislike is the beginning of righteousness; the feeling of deference and compliance is the beginning of propriety; the feeling of right and wrong is the beginning of wisdom. Man have these Four Sprouts just as they have they four limbs. (*Mengzi* 2A:6)

In the above example, Mengzi did not say that anyone would help the child, but only that anyone would experience a pre-reflective, moral emotion of sympathy (Bloom 2002, 100–101). He believed only that a sudden, pre-reflective and emotional character of the reaction suggests that it is an authentic expression of our nature. Moreover, there is a difference between universal emotional "moral infrastructure" hard-wired in heart-mind and particular cultural conditions which co-determinate moral behaviours. But for Mengzi, culture and *xing* do not oppose each other. Nature represents fundamental human tendencies linked with the holistic cosmic process (*Tian*); culture, on the other hand, represents the condition necessary for their cultivation. One requires another. Trying to understand human nature by imagining what it would be like without culture is clearly at odds with Confucian discursive space.

The Mencian conception strongly stresses the necessity of being under the influence of stimulative culture so as to gradually develop "sprouts" into mature forms—virtues of benevolence (*ren* 仁), righteousness (*yi* 義), observance of the rites (*li* 禮) and wisdom (*zhi* 智). Mengzi, therefore, proposed "a developmental thesis"—four emotional "sprouts" evolve gradually into moral virtues due to a stimulative cultural environment and self-reflective processes. Goodness is a part of human nature which means that we possess inborn moral emotions evolving into virtues through a natural process of development in a normal, nutritive environment. Mengzi frequently used metaphors of agricultural growth suggesting that humans are born with a certain potential, which can be either nurtured or stunted:

> In years of plenty, most young men are gentle; in years of poverty, most young men are violent. It is not that the potential (才*cai*) that Tian confers on them varies like this. They are like this because of that by which their hearts are sunk and drowned. Consider barley.

> Sow the seeds and cover them. The soil is the same and the time of planting is also the same. They grow rapidly, and by the time of the summer solstice they have all ripened. Although there are some differences, these are due to the richness of the soil and to unevenness in the rain and in human effort. (*Mengzi* 6A:7)

It does not mean that everything depends on the quality of the cultural habitat. Mengzi did not want to underestimate the impact of culture, education and commitment to cultural code. Self-reflective, conscious processes are important too. Moreover, moral attitudes can be described as "internal" because the human is autonomous and self-cultivation is taken under one's control. For example, Mengzi says that Sage King Shun was raised in a bad familial environment (*Mengzi* 5A:2) and was originally completely uncultivated. However, he was able to respond to insignificant virtuous aspects in his environment and develop himself morally (*Mengzi* 7A:16). One, as a self-reflective agent, must "expand and fill out" one's own "sprouts" in a particular way. Personal flourishing could be achieved only by one's own effort, discipline, self-control and self-focus. Although culture is a necessary factor, we possess real, deterministic (intellectual) forces by nature. Culture is an indispensable condition, but the developing process is self-guided. In other words, humans should take care to cultivate these pre-reflexive, affective moral impulses in a reflective, self-conscious way. Mengzi put it succinctly: "seeking is relevant to obtaining because what one seeks is in oneself". Moreover, environmental and personal dimensions are inseparable because the cultural habitat influences the process of flourishing and self-directed processes are needed for choosing a right, that is, stimulative environment. As J. Perkins mentioned: "moral cultivation depends on choosing to put ourselves in the circumstances that will evoke the proper feelings in us, and to do this in a way that transfers these feelings to analogous groups" (2002, 215). Thus, when asked to clarify his teaching that "human nature is good", Mengzi responds: "As for what [humans] genuinely are, they can become good. This is what I mean by calling [their natures] good. As for their becoming not good, this is not the fault of their potential" (*Mengzi* 6A:6).

Although "the goodness of human *xing* is like the flowing downward of water", people often forget about these natural predispositions by not reflecting on them and being exposed to inappropriate environmental conditions (*Mengzi* 6A:7–8). The fact that they tend to become bad is not the fault of their potential (*Mengzi* 6A:6). Emotional "sprouts" are common to all men, but only a fully developed man "simply never loses it" (*Mengzi* 6A:10). We all have a potential to develop them: "just as all water has a down-going tendency, all people have a tendency toward goodness" (*Mengzi* 6A:2). To have it and not to develop it is a self-destructive process. However, it may happen that such internal moral "impulses" might be muted, and, to some degree, also forgotten. People differ in the manifestation of this goodness because they have not developed these original endowments and resources equally.

6.3 Mengzi Meets Evolutionary Sciences

Recent findings in evolutionary sciences give a fresh and relatively coherent picture of human nature. Of course, despite the emerging agreement among researchers who represent different fields, but work under the unifying Darwinian umbrella, there still remain many controversies. Surprisingly, looking from the evolutionary perspective, the Mencian view on human nature seems to be fundamentally correct. D. J. Munro has already showed that his position derives remarkable support from recent developments in natural sciences. Especially, there is an emerging consensus in evolutionary sciences that mirrors Mencian views of morality (Munro 2005; Slindegard 2011). F. de Waal—perhaps the world's best known primatologist—said that he "could not agree more" with the Mengzi's position (de Waal 2006, 51). A more conservative thesis tells that his thought seems to be easily reconciled with evolutionary sciences. I will outline five theses acknowledged by Mengzi as well as by most of evolutionary researchers.

Firstly, as I. Bloom argued, the orthodox Confucian theory of human nature is "fundamentally biological" (Bloom 2002, 91). Newborns enter the world equipped with certain "sprouts". It is coherent with a critique on the blank slate (*tabula rasa*) and the ghost in the machine theories (Pinker 2002, 5–13). Both Confucianism and evolutionary sciences oppose radical cultural determinism (cultural constructionist theories of human nature) and monopoly of social learning and socialization theories. They also reject an essentialist approach to human nature (see Ingold 2006). Human nature is seen rather as a mosaic, a plastic attribute of human and not as static, categorically fixed good or evil, rational or affective, individualistic or social. Most leading scientists criticize the "static" (and reductionist) view of human nature which holds that "we are direct, unalterable products of our evolution, our biology, our genes; there can be no intervention to prevent, ameliorate, or enhance this <natural order>". In contrast, "a presupposition of plasticity involves the belief that changes may take place over time in a person's physical, psychological, and social structures and functions. People always exist in the physical and social world which they both influence and are influenced by. Because of the reciprocal relations between people and their worlds one may be optimistic about the possibility of positive changes" (Lerner 1984, 14–15).

Secondly, a thesis that moral tendencies form a part of human nature, so strongly influenced by Confucians, has been accepted by many evolutionary scientists. They emphasize overestimation of the *homo homini lupus* view of human nature (np. de Waal 2006, 3) which suggests that morality is, in some sense, external to human nature. While the evolutionary approach does not negate the existence of the "dark" side of human nature, or even the "grey" one—connected with the "Machiavellian intelligence" theory, it also highlights the "glorious" one. Morality is not a cultural overlay, a thin "veneer hiding an otherwise selfish and brutish nature" (de Waal 2006), or "the temple on the hill of human nature" (Haidt 2003), or something imposed by external forces (Mengzi). Confucians, just as most of evolutionary scientists, strongly criticize the thesis that morality is a system that restrains our most basic "natural" impulses and runs contrary to them. Moreover, as they treat moral tendencies as a part of human nature, any dualistic account must be rejected. For example, such authors as Huxley, Durkheim, Freud, Sahlins thought in a dualistic way: we are partly nature, partly culture rather than a well-integrated whole. In that counter-Confucian perspective, morality has only cultural roots and minimizes brutal, selfish, competitive, "natural" tendencies.

Thirdly, an opinion becoming more and more common is that we cannot explain the emergence and development of morality as a part of human natural endowment in evolutionary terms without examining the function of emotions. Four emotional "sprouts" stand at the center of Mengzi's view. Morality as a whole is built on affective grounds. Mengzi emphasized that "moral concepts owe much more to innate social emotions than western psychologists or ethicist have traditionally recognized" (Munro 2005, 52). Charles Darwin wrote something similar, but was even more optimistic than Mengzi on the role played by moral emotions: "many a civilized man, or even boy, (...) has disregarded the instinct of self-preservation, and plunged at once into a torrent to save a drowning man, though a stranger. (...) Such actions (...) are performed too instantaneously for reflection or for pleasure or pain to be felt at the time" (1971, 110). Today, "basic" emotions are widely acknowledged to be adaptive mechanisms, each designed by biological natural selection to perform a task which involves physiological, psychological and behavioral elements of the organism in such a way so as to encourage it to respond adaptively to recurrent types of fitness-relevant threats and opportunities in the environment. For example, psychology showed that human morality has emotional roots and intuitive foundation (e.g. Haidt 2001, 2003, 2007). Neuroscience proved that moral dilemmas activate emotionally involved brain areas (e.g. Moll et al. 2005; Moll and de Oliveira-Souza 2007; Mobbs et al. 2007). The study of the brain systems "put a final stop to the Platonic legacy that moral reasoning, or any reasoning, can occur without the emotions playing a role" (Munro 2005, xiv; Joyce 2006, 124). However the evidence that moral judgments are driven largely by intuitive emotional responses are strong, but it does not follow from this that emotional intuition is the whole story.

Fourthly, from the evolutionary perspective, the question of innateness of morality should be more precisely discussed. If one wants to be a Darwinian, one necessarily should accept that human morality is an evolutionary, and perhaps adaptive, effect. However, it has nothing to do with possible biological reduction of (moral) norms, judgments, systems and codes. The "thin" thesis—accepted by Mengzi—states that there exist some innate, hard-wired elements of moral faculty. The "thick" thesis accepts the existence of some innate content of (moral) norms. For social scientists, it might be surprising that evolutionary researchers reject moral nativism, the opinion that humans are born with innate knowledge about substance of (moral) norms (Dwyer 2007, 904). For example, de Waal wrote: "we are not born with any specific moral norms in mind, but with a learning agenda that tells us which information to imbibe. This allows us to figure out, understand and, eventually, internalize the moral fabric of our native society" (2006, 166). The content of particular moral norms is "determined not by biological processes but by cultural traditions" (Ayala 2006, 157). There is no such thing as a gene of morality or moral behavior; "as with any behavioral trait, the underlying genetics is bound to be complex, and environmental influences may be large and difficult to pin down" (Berkoff 2004, 25). Simply, "there is no 'morality gene' and it is unlikely that any serious thinker has ever thought otherwise" (Hales 2009, 434). Nowak and Sigmund write that "indirect reciprocity is (...) likely to be connected with the origins of moral norms. Such norms are evidently to a large extent culture-specific, but the capacity for moral norms seems to be a human universal" (2005, 1291; Alexander 1987, 77; Joyce 2006, 44, 181). Also, the theory of innate moral domains departs from the view that we have innate moral rules (Haidt and Joseph 2004). Mengzi's position is quite similar. His theory is strongly biological—four emotional, inborn "sprouts" form a part of human nature—but this does not mean that we posses knowledge about any substantive or procedural (moral) norm only by virtue of our biology (gens). Mengzi would say that, to fulfill our nature, there should exist a system of (moral) norms treated as tools for flourishing, or that humans are well-designed to proceed on norms whose precise content is culturally dependent; or that native moral substance (four "sprouts") should be extended into virtues by external cultural forms (norms).

Fifthly, when analyzing human nature and morality we cannot oppose biology and culture, nature and nurture. The "genius of Mencius (...) was that, as early as the fourth centuries B.C.E. he saw with remarkable clarity the importance of both biology and culture" (Bloom 2002, 91). He not only "saw (...) the importance of both biology and culture", but also recognized their functional co-working and interdependence. Evolutionary sciences stress the importance of a cultural environment which influences even basic biological structures of brain. As a neuroscientist said: "The line between genetics and environment is blurred. The presence of a gene or a biological tendency (...) may not be sufficient to direct brain development or behavior. This development often depends on input or stimulation from human culture" (Tancredi 2005, 24). Another author said: "Evolution gave us the preconditions of morality, but it is only as a result of the cultural elaboration of this raw material that we come to be moral beings". Our biological heritage "enables all the flexible, rational and caring behavior that we could want, and allows us to seek to become ever more moral beings" (Levy 2004, 205). For some scientists, reciprocal altruism "acquires concrete content only in the context of specific cultural values concerning the legitimate rights and obligations of individuals" (Gintis et al. 2008, 234). Evolutionary scientists maintain a role of cultural transmission in determining the content of moral convictions. This is "consistent with there being an innate 'moral sense' designed precisely to make this particular kind of cultural transmission possible" (Joyce 2006, 140). An innate trait requires environmental, cultural input in order to become manifest or developed. According to the developmental thesis, the inborn moral emotions need a nutritive cultural environment and self- reflective intrinsic activity to be developed into virtues. There are specialized innate mechanisms whose function is to enable cultural learning concerning culturally based moral content. We have evolved human nature with a unique capacity for culture creation and the ability to develop inherent emotional "sprouts" into virtues. According to the Mencian position, the role of inherent traits is more than just a general "capacity for culture". Nurture is also a part of human nature, a part of our evolved biological heritage, although its precise content is obviously highly variable. Therefore, in this perspective, morality is not only an emotional and cognitive, but also environmental (culture) and personal (self-reflexive processes) phenomenon.

Simultaneously, there are a few profound differences between the visions of the Second Sage and the general features of evolutionary approaches. I shall point out the most obvious ones. (1) Mengzi stressed that human nature is what links us with the universe *(Tian),* the cosmic order. Most of the evolutionary scientists have not even tried to propose such a grand, unifying conception. Nevertheless, D. J. Munro promoted a shift in traditional Confucian ethics from *Tian* and *xin* connection to an explicit biological basis (2005, 14). He wrote: "The Confucian Heaven is not likely to be understood clearly or be engaged by the international audience" (2005, 17). (2) Obviously, Mengzi did not take evolutionary perspective, gene or biological heredity theory. But I. Bloom argued that he had a concept of "biological continuity of life" (2002, 96). (3) In *Mengzi* there are no considerations concerning building blocks of morality in non-human animals. Researchers of non-human animal behavior have shown that moral emotions, such as empathy and compassion similar to Mencian prerequisites for *ren* and *yi* appear to be shared by other species (see de Wall 1996, 2006; Stępień 2010). (4) Despite profound sociological insight, The Second Sage over-emphasized human potentiality of self-mastering necessary to develop the emotional "sprouts" into virtues. His political and societal considerations were based on a correct but incomplete description and explication of human nature which lead him to idealism and a simplistic, one-sided model of human nature. He seemed to put too much hope on a "noble" part of human nature minimizing influences of the "dark" and the "grey" ones. Mengzi did not propose a balanced, whole-picture view. (5) One should remember that Mengzi developed a descriptive and explicative conception, but his central practical aim was to present a guide on how to self-cultivate one's own nature. It is not completely clear what is logically former to Mengzi: a descriptive and explanatory theory or prescriptive guidelines on self-developing practice.

6.4 The Confucian (Mencian) View of Human Nature and Human Rights

My working thesis is that a well-established Confucian (Mencian) concept of inborn goodness (four, hard-wired emotional "sprouts" and a capacity to develop them), which in many ways preceded scientific discoveries, was rather an obstacle in developing sensitivity for human rights in the Middle Kingdom and not its stimulus. Paradoxically, some arguments and conclusions have not been easy to derive from Confucian "discourse space", precisely because of the particular features of the

Mencian view of human nature. It is not a deterministic argument that, within Confucianism, there were no plausible resources or chances to develop a discourse on human rights or even sensitivity to them. Rather, I will point out that various intellectual and practical issues attracted Confucian philosophers' attention.

Confucians have developed a process-oriented account of human nature instead of an essence-based one. Their propositions have been grounded on a particular vision of anthropology. Being human is seen as a "self-creative process rather than accidental qualities happening to a core unchanging essence, where all value lies either by nature or by the creation of God" (Gier 2004, 174). The self is perceived rather as a process than as a static substance—"the human being is not something we are; it is something that we do, and become. Perhaps 'human becoming' might thus be a more appropriate term to capture the processional and emergent nature of what it means to become human. It is not an essential endowed potential, but what one is able to make of oneself given the interface between one's initial conditions and one's natural, social, and cultural environments" (Ames and Rosemont 1998, 49). This process-oriented vision stresses the inherent ability to transgress from a less developed to a more mature form. The character of this "silent transformation" (Jullien 2011) is best captured by the concept of the Way (*Dao* 道), fundamental not only to Confucianism, but to Chinese culture in general. The questions of (1) how one can fully develop (but not actualize) one's nature and (2) what the unchanging, substantive elements of nature are remain a core issue for the whole Confucian enterprise. The path to human rights in Confucianism could not be traced to some putative essence of man. As a consequence, only a small "discursive space" was left for discussions about rights inscribed in human nature or derived from inborn dignity (see Lee 2008). The focus on various methods of self-development based on inborn moral emotion has attracted much of the Confucian philosophers' attention.

Moreover, according to Confucianism, dignity may be seen rather as a starting point (common endowment of man), or a final stage of human development, than as the pre-existing foundation for any rights. However, more accurately, within "transformational" thinking which emphasizes the developmental character of becoming human, dignity is best described as an holistic, highly valued personal trait which is constantly and incrementally established by the individual itself. Dignity is a feature of what we are becoming. It is not understood in the same way as in Western naturalistic conceptions which stress axiomatic and essentialist character of rights. The framers of international instruments did not even define human dignity, but there is an implicit relationship between human rights and human dignity. Furthermore, although in Confucianism every human being has innate potential to become virtuous—truly and fully human—it is an achievement of very few. Human dignity is understood as an achievement of a small elite group. Consequently, only these few might deserve a full package of rights and privileges.

The non-essentialist and developmental language of inborn human capabilities instead of rights possessed by all humans alike characterizes classical Confucianism. One would say that the fact that the Chinese have never had metaphysics of substance (applied to human nature) might be treated both as an advantage and a disadvantage in terms of practical consequences. The disadvantage is that they have never had a tendency to seek fixed, unchanging, atemporal and asocial (or pre-social) rights. *Xing*, in the Confucian "discursive space", does not indicate transcendental, eternal, unchanging essence shared equally by all humans, but rather naturally arising dispositions. Mengzi held that neither our nature is fixed as good nor that, if left alone, it is to develop goodness. His theory of natural goodness stresses rather that "man is capable of becoming good". Perhaps Mengzi would agree with B. William's following remark:

> The notion of a basic human right seems to me obscure enough and I would rather come at it from the perspective of basic human capabilities. I would prefer capabilities to do the work, and if we are going to have a language or rhetoric of rights, to have it delivered from them, rather than the other way round. (Williams 1987, 100)

This is much in line with Confucian thinking. The capabilities ("sprouts") are universal features of humans and they presuppose a developmental account. A self-reflective, and, at the same time, culturally-driven process of their manifestation is a prerequisite for yielding dignity and respect of an individual. But still, according to the Confucian position, humans may need certain rights so that they can fulfill their nature, but rights cannot be derived from nature/human nature. Mengzi would say that the environment sensitive to human rights might be a stimulative factor serving as a helpful device in the process of self-development. This conclusion is compatible with Confucian "discursive space" and from the beginning was "at hand" for Confucians. On the contrary, the idea that rights are untimely and apriorical features imprinted in human nature was much less possible to derive while considering the Confucian developmental vision of human nature. Of course, as for reaching a stage of a perfectly developed human (*junzi, sheng*) a cultural environment in which human rights are respected and protected might be necessary (Weiming 1998, 305), but environmental factors (HR too) are instrumental in obtaining some fundamental traits and virtues in the process of creative self-transformation. This created a situation in which utilitarian and deontological mechanisms for maintaining social order have been marginalized in favor of the aretological one. Regardless of that, Confucianism, from the beginning, made efforts to address issues similar to those which occupy HR advocates: benefiting the common people as a priority, extending care and love, established humane governance (e.g. *Lunyu* 6:30; *Lunyu* 20:2; *Mengzi* 7B:14).

"Transformational" and developmental visions of human nature determine the main Confucian strategy for maintaining social order, namely nursing the "sprouts" of goodness by educational means. Moreover, the expense of moral education which is necessary when we assume that one should expand and flourish the noble "sprouts" is perhaps a better predictor of totalitarian (authoritarian) politics. In Confucianism, the institutions of political, social and economic life function so as to develop the virtue of being a perfected human being. Because of the focus on the noble part of human nature Confucianism has been too optimistic and one-sided. In consequence—there were no pressures to search for mechanisms that could neutralize the fruits of the "evil" part of human nature. If one acknowledges and takes into account egoistic tendencies, one will be inclined to search for various institutional shields against it. Mencian optimism has created the situation where the question of how to reduce—by institutional means (maybe also recognized and guaranteed rights)—the negative effects resulting from the "evil" part of human nature has been marginalized and even muted. The Mencian view highlighted the problem of developing the emotional "sprouts" into virtues. Mengzi did, in fact, grant that human beings can follow the path of gratifying their senses rather than their moral "sprouts"; nevertheless, nowhere did he suggest that this involves nurturing not-so-good "sprouts". He who clearly employs the sprout analogy, has never discussed the notion of 'weeding out' bad sprouts, but only has studied how to grow/develop/nurture the good ones.

Mencius believed that inborn moral emotions are not experienced with the same intensity by all people: for example, feelings towards one's own child and brother are not the same as feelings to a neighbor's child (*Mengzi* 3A:5; see *Zhongyong* 20, 13–14). The route of development rests, in the first place, on the expansion of the "four seeds" within the family, and then beyond it. According to Confucianism, to demonstrate the same feelings both to one's relatives and strangers is not appropriate—a direction and intensity of moral emotions are conditioned by the degree of kinship. For Confucians, this is a descriptive and prescriptive thesis. This approach led both to a difficulty in adopting the principle of formal equality as well as to mistrust of normative tools which are insensitive to the subtleties of the kinship structure. Preferred means of social ordering must therefore be flexible and contextual—to allow a modification of the content of duties, depending on actors. These several principles lying at the foundation of Western accounts of human rights have important counterparts in Confucian humanism.

6.5 Conclusions

Generally speaking, from the perspective of evolutionary science, the Mencian theory of natural goodness is fundamentally correct. Mencian doctrine accords with the main thesis acknowledged by evolutionary sciences: recognition of the role of biology (genetics), the fundamental role played by (moral) emotions, a critique of normative nativism, culture-nature interdependence and a developmental account. The Confucian picture of human nature is quite optimistic, inborn goodness is a core of human nature. This optimism was not translated into the discourse of rights (see Angle 2002, 74–100; Cao 2006, 40) because of the above mentioned features of the Mencian view of human nature (among other things). The Confucian example has shown that there is not necessarily a connection between the recognition of the natural goodness of human nature and deriving human rights from it. In contrast, Enlightenment theories have accepted some assumptions about essential features of human biology and on those grounds have built a picture of universal human psychology. The next step was to derive certain rights from that, shared by all humans, psychophysical ground. This move has enabled (together with other factors) the rise of the discourse on HR but, on the other hand, it seems to be logically unjustified and it is not consistent with today's discoveries made in evolutionary sciences. Paradoxically, the "wrong", erroneous Western approach to human nature has led to results which are axiologically approved and difficult to challenge nowadays. But the scientifically correct Confucian (Mencian) approach has determined the lack (or weakness) of recognition of rights and even the existence of the discourse on rights within Confucian philosophical tradition. This requires consideration.

Firstly, we should bear in mind the difference between grounds and justifications of HR. It is not so obvious that the genealogy of HR must be identical with its justifications. Moreover, HR may be grounded in various ontological spheres at the same time (human biology: empathy, compassion, care, altruism; social conditions: social consent, well-being, justice, cooperation). Secondly, without doubt, there are proximal (distal) biological causes of the emergence of HR but, at the same time, there are also basically social (cultural) ultimate causes. Human rights understood as social constructions (Gregg 2012, 39) are underpinned by empathy, natural emotions and cognitive abilities which are mostly biologically based. As Gregg puts it: "From a constructivist perspective, the idea that human rights are 'natural 'is itself a construct: a product of history, internally contested, with ambiguous boundaries of definition and application" (2012, 160). They are a cultural product which, to some degree, supervene on natural tendencies. We can construct a normative system in part by using our biological, inborn capacities. However, there is still a long way leading from biological tendencies to an institutionalized set of norms and not to a causal, linear or logical connection between biological and cultural dimensions. Such a distinction helps to connect universal and particular dimensions of HR. From that perspective, far from being exclusively Western, the foundations (grounds) of human rights appear to be shared by both Western and Confucian theories. Thirdly, human rights may still be justified by an appeal to human biology (e.g. common "sprouts").

References

Alexander, R.D. 1987. *The biology of moral systems*. Cambridge, MA: Harvard University Press.

Ames, R.T. 2002. Mencius and the process notion of human nature. In *Mencius: Contexts and interpretations*, ed. A. Chan. Honolulu: University of Hawaii Press.

Ames, R.T., and H. Rosemont. 1998. *The analects of Confucius: A philosophical translation*. New York: Ballantine Books.

Angle, S.C. 2002. *Human rights and Chinese thought. A cross-cultural inquiry*. Cambridge: Cambridge University Press.

Ayala, F. 2006. Biology to ethics. An evolutionist's view of human nature. In *Evolutionary ethics and contemporary biology*, ed. G. Boniolo and G. Anna. Cambridge: Cambridge University Press.

Beitz, C.R. 2009. *The idea of human rights*. Oxford: Oxford University Press.

Bekoff, M. 2004. Wild justice and fair play: Cooperation, forgiveness, and morality in animals. *Biology and Philosophy* 19(4): 1–32.

Bloom, I. 2002. Biology and culture in the Mencian view on human nature. In *Mencius: Contexts and interpretations*, ed. A. Chan. Honolulu: University of Hawaii Press.

Cao, D. 2006. Kay worlds in Chinese law. In *Images in law*, ed. A. Wagner and W. Pencak. Aldershot: Ashgate.

Chan, W. 1963. *A sources book of Chinese philosophy*. Princeton: Princeton University Press.

Darwin, Ch. 1971. *The origin of species*. London: J. M. Dent & Sons Ltd.

Daston, L., and M. Stolleis (eds.). 2008. *Natural law and laws of nature in early modern Europe*. Aldershot: Ashgate.

De Waal, F.B.M. 1996. *Good natured: The origins of right and wrong in humans and other*. Cambridge: Harvard University Press.

De Waal, F.B.M. 2006. *Primates and philosophers. How morality evolved*. Princeton: Princeton University Press.

Dwyer, S. 2007. How not to argue that morality isn't innate: Comment on Prinz. In *The evolution of morality: Adaptations and innateness*, Moral psychology, vol. 1, ed. W. Sinnott-Armstrong. Cambridge: MIT Press.

Gier, N.F. 2004. Whitehead, Confucius, and the aesthetics of virtue. *Asian Philosophy* 14(2): 171–190.

Gintis, H., F. Henrich, S. Bowles, R. Boyd, and E. Fehr. 2008. Strong reciprocity and the roots of human morality. *Social Justice Research* 21(2): 241–253.

Gregg, B. 2012. *Human rights as social construction*. Cambridge: Cambridge University Press.

Haidt, J. 2001. The emotional dog and its rational tail: A social intuitionist approach to moral judgment. *Psychological Review* 108(4): 814–834.

Haidt, J. 2003. The moral emotions. In *Handbook of affective sciences*, ed. R.J. Davidson, K.R. Scherer, and H.H. Goldsmith. Oxford: Oxford University Press.

Haidt, J. 2007. The new synthesis in moral psychology. *Science* 316: 998–1002.

Haidt, J., and C. Joseph. 2004. Intuitive ethics: How innately prepared intuitions generate culturally variable virtues. *Daedalus* 133(4): 55–66.

Haiming, W., and W.K. Akina. 2012. A naturalist version of Confucian morality for human rights. *Asian Philosophy* 22(1): 1–14.

Hales, S. 2009. Moral relativism and evolutionary psychology. *Synthese* 166: 431–447.

Ingold, T. 2006. Against human nature. In *Evolutionary epistemology, language and culture*, ed. N. Gontier. Dordrecht: Springer.

Ivanhoe, P.J. 2002. *Ethics in the Confucian tradition: The thought of Mengzi and Wang Yangming*. Atlanta: Scholars Press.

Joyce, R. 2006. *The evolution of morality*. Cambridge, MA: The MIT Press.

Jullien, F. 2011. *The silent transformations*. Calcutta: Seagull Books.

Lau, D.C. 1984. Introduction. In *Mencius*. Hong Kong: Chinese University Press.

Lee, M.Y.K. 2008. Universal human dignity: Some reflections in the Asian context. *Asian Journal of Comparative Law* 3: 283–313.

Lerner, R.M. 1984. *On the nature of human plasticity*. Cambridge/New York: Cambridge University Press.

Levy, N. 2004. *What makes us moral? Crossing the boundaries of biology*. Oxford: Oneworld Publications.

Liu, S. 1998. *Understanding Confucian philosophy: Classical and Sung-Ming*. Westport: Praeger Publishers.

Mobbs, D., H. Lau, O.D. Jones, and Ch. Frith. 2007. Law, responsibility, and the brain. *PLoS Biology* 5(4): 0693–0700.

Moll, J., and R. de Oliveira-Souza. 2007. Moral judgments, emotions and the utilitarian brain. *Trends in Cognitive Sciences* 11(8): 319–321.

Moll, J., R. de Oliveira-Souza, F. Krueger, and J. Grafman. 2005. The neural basis of human moral cognition. *Nature* 6: 799–809.

Munro, D.J. 2005. *A Chinese ethics for the new century: The Ch'ien Mu lectures in history and culture, and other essays on science and Confucian ethics*. Hong Kong: Chinese University Press.

Nisbett, R.E., E. Peng, K. Choi, and A. Norenzayan. 2001. Culture and systems of thought: Holistic vs. analytic cognition. *Psychological Review* 108: 890–905.
Nowak, M., and K. Sigmund. 2005. Evolution of indirect reciprocity. *Nature* 437: 1291–1298.
Panikkar, R. 1982. Is the notion of human rights a Western concept? *Diogenes* 30: 75–102.
Perkins, F. 2002. Mencius, emotion, and autonomy. *Journal of Chinese Philosophy* 29(2): 207–226.
Pinker, S. 2002. *The blank slate: The modern denial of human nature*. New York: Viking.
Sim, M. 2004. A Confucian approach to human rights. *History of Philosophy Quarterly* 21(4): 337–356.
Slindegard, E. 2011. Of what use are the odes? Cognitive science, virtue ethics, and early Confucian ethics. *Philosophy East and West* 61(1): 80–109.
Stępień, M. 2010. The question of non human primates morality. In *Studies in the philosophy of law. Law and biology*, ed. J. Stelmach, 135–154. Cracow: Jagiellonian University Press.
Tancredi, L. 2005. *Hardwired behavior what neuroscience reveals about morality*. Cambridge: Cambridge University Press.
Wei-ming, T. 1998. Epilogue. In *Confucianism and human rights*, ed. T. de Bary and T. Wei-ming. New York: Columbia University Press.
Williams, B. 1987. The standard of living: Interests and capabilities. In *The standard of living*, ed. G. Hawthorn. Cambridge: Cambridge University Press.
Zhang, P., and Q. Guo. 2008. New insight into Mencius' theory of the original goodness in human nature. *Frontiers of Philosophy in China* 3(1): 27–38.

Chapter 7
The Natural Faculty of Empathy as a Basis for Human Rights

Frederik von Harbou

7.1 Introduction

To most of today's scholars, human rights naturalism appears to be an outdated idea. Instead, scepticism about any transcultural foundation for human rights seems prevalent. Richard Rorty for example, prominently attacked any human rights foundationalism as "outmoded". According to him, the attempt to provide a basis for human rights in rationality or any other supposed feature of "human nature" failed because there were no "morally relevant transcultural facts" about human beings as such at all (Rorty 1993, 116). Michael Ignatieff goes even further when he asserts that human rights are to be defended "in full awareness that they must counteract rather than reflect natural human propensities" (Ignatieff 2001, 80).

However, such positions of radical cultural relativism which suppose culture is the *sole* source for the recognition of human rights, and which oppose any concept of human nature, suffer from several severe shortcomings. First, they hardly ever meet their own demands, namely to abandon any assumptions about human nature. Rorty – although stating we should neglect the question 'What is our nature?' – stresses that human rights correspond to our capacity to sympathize with strangers, thus implying that there are certain morally relevant transcultural facts about human beings, namely the existence of our natural faculties of sympathy and empathy.[1] This is a position we shall try to elaborate in this essay.

Second, the (frequent) reference to the Holocaust as the "historical moment in which [human rights] had been shown to have had no foundation whatever in natural human attributes" (Ignatieff 2001, 80), appears to be rather problematic.

[1] Rorty (1993, 119); for critical remarks on Rorty's approach cf.: Woods (2009, 55).

F. von Harbou (✉)
Trainee lawyer at the Higher Regional Court of Berlin (Kammergericht), Germany
e-mail: frederik.vonharbou@fu-berlin.de

M. Albers et al. (eds.), *Human Rights and Human Nature*, Ius Gentium: Comparative Perspectives on Law and Justice 35, DOI 10.1007/978-94-017-8672-0_7,

To take the Holocaust as proof of the fundamental and ultimate cruelty and overall malice of humans would imply that people are closer to their "real" selves in totalitarian systems or situations than people living under other conditions, e.g., a constitutional democracy. But the opposite assumption seems more likely. Since totalitarian systems are by definition illiberal, there does not seem to be any reason to think that their citizens are freer to live according to their natural inclinations than people living in non-totalitarian states. Thus, if we want to find out about human nature, we should neither neglect historical facts about the atrocities people can commit nor develop our concept of human nature on the sole basis of the most extreme cases of cruelty. We must pledge for a more balanced view of human nature, following Jonathan Glover's analysis of the history of the twentieth century: "We are a species both brutal and sickened by brutality" (Glover 2000, 41).

Third, and most importantly, radical cultural relativism neglects that humans have natural tendencies to care for the well-being of others. These tendencies – as shown in this essay – are based on our faculties to sympathize and empathize with other human beings. These propensities are innate and universally distributed among individuals. Although the modes of expression of sympathy and empathy – for example, the question whether in a situation it is appropriate to show sympathy or not – are to a considerable degree shaped by socialisation, the very capacity to feel empathy constitutes a basic human feature rooted in a mostly automatically activated neurophysiological mechanism of (healthy) individuals.

The purpose of this essay is to expand on the last point. In the first part of our argument (Sects. 7.2 and 7.3), we will explore the analytical connections between the concepts of human rights, morality, altruism, compassion and finally empathy. In the second part (Sects. 7.4 and 7.5) we will evaluate the results of some empirical studies that show that empathy is a natural human faculty. In the last part of this essay (Sect. 7.6) we shall point out that our naturalistic conception of human rights should not be mistaken for any reductionist account.

7.2 From Human Rights to Morality and Altruism

Our argument begins with an analysis of the concept of human rights. The term "human rights", though notoriously controversial and indeterminate, is widely held to refer to certain universal, inalienable and indivisible entitlements held by all human beings, as already implied by the Universal Declaration of Human Rights of 1948. As such, human rights obviously cannot be conceived of as mere legal rights. The criteria, even if only the criterion of universality, formulate an ideal rather than reflect the actual situation of many individuals' legal standing in their respective countries around the world. In fact, the issue of human rights is brought up (e.g. by the media or non-governmental organisations) virtually only in the case of their *violation*, that is in the case of a conflict between (positive) national legal provisions and human rights. Therefore, it seems that the concept of human rights – if it makes sense at all[2] – can only be understood as referring to *moral* rights.[3]

But even if one regards human rights as mere international conventions, one must acknowledge that the *contents* of human rights norms – particularly of those regarded as *ius cogens* – are largely congruent with basic moral principles, such as the prohibition of killing. Of course, human rights traditionally only oblige states while moral principles oblige everybody. Furthermore, moral imperatives are very diverse around the world: norms on sexuality may vary remarkably from one country to another. Human rights regimes, on the other hand, are not meant to be morally comprehensive, that is, they are not intended to substitute for conceptions of social justice or even a "good life." They are rather restricted to a "core" of protections against violations of basic human interests, e.g., protections against cruel or degrading treatment. Acting on this assumption, one can identify four classes of fundamental human rights entitlements:

1. protection of physical and mental integrity (e.g., prohibition of torture, entitlement to a reasonable degree of protection (usually by police forces) against harm from other individuals)
2. protection of subsistence (entitlement to basic social care)
3. liberties (freedom to do what one wants, only limited by other individuals' rights)
4. right to equal treatment (implying a certain degree of political and economic participation)

There are three more rights which – without being original human rights themselves – follow from the objective to maximise the chances of fulfilment of the aforementioned human rights under the present political conditions of statehood: (a) the right to a system of positive legal rights; (b) the right to fair and effective trials; (c) the prohibition of exclusion from citizenship, thereby protecting the application of the aforementioned human rights as civil rights.

Human rights can be regarded as expressing a certain minimal moral standard. States must comply with these minimal demands to be regarded as legitimate by other states. Whether one regards human rights as primarily moral rights or as legal rights merely expressing a certain minimal moral standard, the inherent connection between human rights and morality can hardly be denied entirely.

While the concept of morality may be at least as controversial as the concept of human rights, we need no comprehensive definition of morality in order to proceed in our analysis of the term "human rights". However, the common ground of the four classes of human rights, identified above, may be expressed by a moral imperative not to harm any person, but instead to help everyone if one can (this imperative

[2]There is a long tradition of human rights scepticism: prominent critics of the notion of universal (moral) rights include Bentham (1816/1962, 522 et seq.), Burke (1790, 46), Arendt (1986, 426), and Habermas (1998, 106).

[3]Cf. for example: Feinberg (1973, 85), Sandkühler (2010), Shue (1996, 13), and Tugendhat (1993, 336 et seqq).

may – as, for example, Arthur Schopenhauer (1840/1977, 586) claimed – be the quintessence of ethics, whether from a religious or a secular perspective). Since we assume that human rights express a "minimal morality" (blaming harmfulness and praising helpfulness), we may try to further disclose the premises of such value judgements.

Judgements of vice and virtue depend not only on the conformity with objective criteria. A seemingly helpful action (such as a donation to a poor person) cannot be regarded as virtuous if the underlying intention was evidently selfish (i.e. if the only reason to donate was to enhance the donor's own prestige). This shows that, besides the objective criteria of helpfulness and prevention of harm, there has to be a (decisive) *altruistic* motivation in order to render an otherwise neutral action morally good. Altruism, defined as deliberately striving for the well-being of another person, is a crucial element of moral behaviour and thus an integral part of the motivational basis of human rights protections' moral contents.

7.3 From Altruism to Compassion

The analysis of human rights and morality leads us to the phenomenon of altruistic behaviour. This raises the issue of how a genuinely altruistic motivation, comprising the *intention* to help another person (instead of merely accidental behavior), is possible at all. Or, as David Hume (1740/2008, SB 576; 364) put it in his *Treatise*: Why are we concerned by the misery or the happiness of a stranger?

> After what manner does it [the stranger's happiness] give pleasure? 'Tis certain our own interest is not in the least concern'd; [...] it must delight us merely by communication [...]. We enter into his interest by the force of imagination, and feel the same satisfaction, that the objects naturally occasion in him.

To Hume it was clear that the only explanation for this intriguing phenomenon could be found in our natural inclination to share the feelings of other people – "the happiness of strangers affects us by sympathy alone" (Hume 1740/2008, SB 618).

But why shouldn't it be possible to behave altruistically without feeling sympathy or compassion for another person, for example merely out of (practical) reasoning or applying certain moral principles? There is a long tradition of philosophers who believe that moral judgement does not require any emotional involvement or even hold that following any natural inclination would be contradictory to genuine moral orientation. According to the most prominent of this view's proponents, Immanuel Kant, an action can only have moral worth if it was motivated solely by the duty to act according to the Categorical Imperative – or the 'Moral Law,' as Kant frequently refers to it (Kant 1788/2005, A 127).

David Hume, however, already elaborated on this with a persuasive argument for why sentiments always enter into moral judgement and both, analytical philosophy, and recent findings in psychology and neuroscience support his view. According to Hume, reason is limited to the basic function of discovering the truth or falsehood either of a single matter of fact or of the relation between several matters of fact (Hume 1751/1998, SBN 285 et seq.; 1740/2008, SB 458.). As such, it is incapable of providing any ends on its own: "Reason, being cool and disengaged, is no motive to action, and directs only the impulse received from appetite or inclination, by showing us the means of attaining happiness or avoiding misery" (Hume 1751/1998, SBN 294).

If, for example, we regard utility as the main criterion for morally approving or rejecting any human behaviour, we can make use of reason as an instrument to determine whether a certain observed behaviour is useful or not – but reason alone does not give us an answer to the question whether we should prefer useful actions over useless (or even harmful) ones in the first place. As reason is limited to judgements of truth and falsehood, it is entirely neutral to the *value* of any ends, so that it is even "not contrary to reason to prefer the destruction of the whole world to the scratching of my finger", as Hume declared provocatively (Hume 1740/2008, SB 416). Instead, Hume asserts that

> a sentiment should here display itself, in order to give a preference to the useful above the pernicious tendencies. This sentiment can be no other than a feeling for the happiness of mankind, [...] reason instructs us in the several tendencies of actions, and humanity makes a distinction in favour of those, which are useful and beneficial (Hume 1751/1998, SBN 286).

Given this limited role, Hume believes that "the ultimate end of human actions can never, in any case, be accounted for by *reason*, but recommend themselves entirely to the sentiments and affections of mankind, without any dependence on the intellectual faculties" (Hume 1751/1998, SBN 293). Reason, thus conceived, simply lacks any *motivational* force. According to Hume's famous dictum, reason is only "the slave of the passions, and can never pretend to any other office than to serve and obey them" (Hume 1740/2008, SB 415). Morality therefore can never be entirely derived from pure reasoning or from principles which are not themselves based on sentiment. Hume thus concludes that morality is "more properly felt than judg'd of" (Hume 1740/2008, SB 470).

However, from the perspective of Kantian philosophy, Hume's argument remains contested today. Critics like Thomas Nagel, John Rawls or Christine Korsgaard mainly reproach him for lacking any notion of practical reason and for confining himself to moral *psychology*. On the other hand, we may find support for Hume's view both in analytical philosophy and empirical science. One may, for example, illustrate Hume's comparison of the motivating force of sentiment and the rather neutral role of reason by referring to Elizabeth Anscombe's distinction between the different "directions of fit" of cognitive states (like beliefs) and conative states (like desires) (Anscombe 1957, 56; Döring 2009, 24–29).

While what is expressed by cognitive states is only correct if it "fits the world," the same is not true for conative states, which aim to change the world as it is – or, so to speak, at making the world fit the contents of the conative states, e.g., certain desires. As moral beliefs – for example, the belief in a certain moral principle – are cognitive states, it is hard to see how – as Kantian philosophers (or, more precisely, "cognitive internalists") believe – they are supposed to be at all motivating. As they have opposed "directions of fit," cognitive (thetic direction of fit) and conative (telic direction of fit) states seem to be mutually exclusive. An additional conative state would thus be required to explain the motivational force of moral judgements (the fact that we feel an urge to do what we hold to be morally "good"). As sentiments or emotions (as they may well form a "desire" to change the world, take for example anger) have a telic, not (primarily) a thetic direction of fit, they are very good candidates for an adequate reconstruction of moral behaviour from an analytical perspective.

This view is further confirmed by findings from empirical science. Studies in neuropsychology have shown that areas of the brain associated with processing emotion (e.g., the *posterior cingulate gyrus*) show increased activity when test subjects are asked to consider a moral dilemma (e.g., a different version of the *trolley and footbridge dilemma*)[4] compared to when asked to consider a morally neutral (e.g., logical) dilemma (Greene et al. 2011, 2106). Other studies came to similar conclusions, for example when participants were instructed to classify several short statements as "right" or "wrong," some of which had moral content (e.g. "the elderly are useless"), whereas others were purely factual (e.g. "stones are made of water") (Moll et al. 2003, 301).

We may thus assume that emotions constitute a necessary part of moral judgement and moral behaviour. However, this does not necessarily lead to the conclusion that moral behaviour implies sympathy or compassion[5]; other emotions like anger or disgust may be displayed during moral judgements with no reference to other people. Yet, as we have already found out, moral judgements concerning human rights – perhaps unlike other moral judgements – imply an altruistic motivation, the intention to strive for the well-being of another person. Such an intent cannot be constituted by, for example, mere personal disgust or personal anger not induced by empathy, but requires an emotional state directed towards another. Sympathy or

---

[4]These thought experiments go back to the work of Philippa Foot (1967) and Judith Jarvis Thomson (1976). The trolley dilemma roughly goes as follows: Suppose a runaway trolley is going to kill five men working on a track. The only way for you (as you are in control of the switch) to prevent this from happening is to steer the trolley from one track to another, on which one person is standing who will then surely die. What should you do?

In the footbridge dilemma, you have the parallel case that a train is going to kill five workers. But now you don't have two tracks but only one. Instead of operating the switch, you may now save the five persons by throwing a fat man from a bridge and in front of the train. The fat man will be killed, but the trolley will be stopped by the man's weight.

Interestingly, although the outcome is the same, the majority of people judge the action in the trolley dilemma as permissible or even obligatory and the action in the footbridge dilemma as morally wrong. Greene explains the difference in subject's evaluations by pointing out that the position in the first scenario was rather detached and abstract, while the second situation was "up close and personal," cf. Greene et al. (2011, 2106).

[5]Jesse Prinz, for example, advances a sentimentalist account of ethics while remaining sceptical about the importance of compassion or empathy for moral judgment, cf. Prinz (2006, 30 et seqq.) and Prinz (2011, 211).

compassion is precisely such a state, as it renders another person's interest relevant to ones own and – in addition – has a similar motivational force as other, more personal (or selfish) emotions.

'Compassion,' or 'sympathy,' is frequently defined as an emotional reaction regarding another person, giving rise to a motivation to assist this person; Wispé defines it as an "increased sensibility of another person's suffering as something to be alleviated".[6] While the concern for another person's well-being makes sympathy an evaluative and prosocial emotion, empathy on the other hand is believed to be a more neutral faculty (or just a mechanism), which doesn't necessarily involve the motivation to help (Eisenberg 2008, 574). Thus, empathy does not always lead to compassion or sympathy. However, as compassion or sympathy may only arise on the basis of an initial identification with the other person's situation, empathy is the precondition for any feeling of sympathy. But what exactly is empathy and how does it 'work'? We shall try to answer these questions in the subsequent sections.

7.4 The Concept of Empathy

While there is a strong increase in the recent usage of the word "empathy" in both academic and other (e.g. political), contexts, the precise meaning of the term is often unclear. According to cognitive scientist Daniel Batson, there are at least eight "different things called empathy": (1) knowing another person's internal state; (2) adopting the posture or matching the neural responses of an observed other; (3) coming to feel as another person feels; (4) projecting oneself into another's situation; (5) imagining how another is thinking and feeling; (6) imagining how one would think and feel in the other's place; (7) feeling distress at witnessing another person's suffering; and finally (8) feeling for another person's suffering (Batson 2009, 4–7).

The list shows that empathy operates on at least three levels: cognitive (such as in usages (1), (4), (5) and (6)), affective ((3), (7) and (8)) and behavioural (2).[7] However, although the cognitive and behavioural dimensions of empathy may be of great importance, depending on the observational perspective, none of these alone justifies the attribution of (full) "empathy" to a person's state of mind. As is the case with a psychopath who is gifted in perspective-taking or mind-reading and in knowing another person's internal state (as in usage (1)) but uses this ability to maximise the torment of his victims, calling him empathetic would seem to be contradictory. This indicates that the attribution of "empathy" refers first and foremost to a certain *emotional* reaction.

---

[6] Wispé (1991, 68); for a similar definition cf. Deonna (2006, 344) and Eisenberg (2008, 574).

[7] In usage (2), Batson moreover refers to a fourth—neural—dimension. However, this usage seems fallacious. Although the matching of neural states may well be what's happening on the neural level when somebody feels empathy for someone else, only persons, not brains, can feel empathy. Cf. Bennett/Hacker (2003, 73) on the "mereological fallacy".

Empathy can therefore be defined as an "affective response more appropriate to another's situation than one's own" (Hoffman 1987, 48). Imagining another person's state (perspective taking) can be described as (mere) "cognitive empathy", constituting only a *way* of arriving at an (full, affective) empathetic response. Another way of coming to feel empathy is by mere perception, as when observing a needle being injected into another person's hand usually elicits a slight feeling of pain in the observer. Considering this, we can summarize that "empathy occurs when an observer perceives or imagines someone else's (i.e. the target's) affect and this triggers a response such that the observer partially feels what the target is feeling" (Singer and Lamm 2009, 82).

As a complex phenomenon, empathy can be reconstructed from various perspectives. One can study how cultural parameters (e.g. the belief in religious doctrine, the reading of novels about socially excluded persons) shape the forms of how empathy is felt by the individuals of a given society, and whether feeling empathy is regarded as a virtue.[8] However, as we are concerned with the question whether there is any foundation for human rights in human nature apart from cultural influences, the crucial point for our argument is to identify the *natural* determinants of empathy. This can be done by assembling empirical data from different branches of science, including developmental psychology and brain science.

7.5 Empathy as an Inborn Ability

Empathy has been the subject of numerous studies in developmental psychology. These show that the developmental process is very similar across different cultures.[9] Martin Hoffman distinguishes five stages in this process: (a) reactive newborn cry; (b) egocentric empathic distress; (c) quasi-egocentric empathic distress; (d) veridical empathic distress; (e) empathy for another's experience beyond the immediate situation (Hoffman 2000, 6).

The first stage is marked by the phenomenon that newborns tend to start crying when they hear other infants cry. According to Hoffman, "the newborn is responding to a cue of distress in others by feeling distressed himself. The cry must therefore be considered an early, rudimentary precursor of empathic distress (feeling distress at witnessing another person's suffering) – precursor because the 'other' to which the newborn is responding is probably sensed by the newborn as connected to the 'self', that is, as part of the same global psychological entity as the self" (Hoffman 2000, 65).

[8]Cf., Hunt (2007) for an intriguing historical account of how the rise of the modern novel in the eighteenth century may have nourished a culture of empathy which in turn may have opened the path for the establishment of human rights.

[9]For a study on Chinese and American children cf. Borke (1973).

Towards the end of their first year, infants are not confined to mere passive reactions to other's distress anymore, but start to actively lower their own feelings of distress. However, they behave in the same way in empathic distress and actual distress situations, indicating that they still have difficulty with distinguishing between something happening to the other and something happening to the self. Hoffman calls these reactions – only seemingly contradictorily – "egocentric empathic distress" (Hoffman 2000, 68).

Only a few months later, early into their second year, one can observe the first instances of helpful behaviour in children. They now understand that the true source of distress lies within another person and – for example, in the case of injury – try to alleviate their pain. However, toddlers at this age do not realize that others have their own independent inner states and believe others see things as they do, so the children only use helping strategies they would find comforting in a similar situation (e.g. taking a crying child to one's own mother, not to the other's mother). Hoffman aptly calls this the stage of quasi-egocentric empathic distress (Hoffman 2000, 70). This behaviour disappears around age two, giving way to the fourth stage; Hoffman refers to this as "veridical empathic distress," comprising all the basic elements of mature empathy. Children now understand that others have inner states of their own, and they develop a rudimentary form of perspective taking and adapt their helpful behaviour. This development is closely preceded by the children's formation of self-awareness (as is indicated by their ability to recognize themselves in the mirror) (Bischof-Köhler 1989, 148).

The last step, feeling empathy for another's experience beyond the immediate situation, occurs much later, mostly around the ages of 5–8, and requires imagination and the understanding that people have something like a life story (marked, for example, by poverty or chronic illness) (Hoffman 2000, 80).

Developmental psychology thus shows that the ability to empathize is rooted in very early childhood. Empathy is an integral and universal faculty of healthy[10] human beings. But what are the underlying mental and neural processes of empathy? While the precise neuropsychological mechanisms are yet to be fully disclosed, there is already an impressive array of both empirical research and theoretical work on the topic.

One of the most thorough accounts of the mechanism of empathy was proposed by Stephanie Preston and Frans de Waal. Their "Perception-Action Model" (PAM) of empathy tries to integrate both the proximate (behavioural and neuronal) and ultimate (evolutionary) bases of empathy. Preston and de Waal believe that the perception of emotional expressions automatically activates those neural mechanisms responsible for generating the observer's emotions and suggest that there is a phylogenetic continuum between 'emotional contagion' (the often unconscious

[10]Lack of empathy is most prominently studied in individuals diagnosed with autism spectrum disorder or psychopathy. However, only the latter may be affected by a true lack of (affective) empathy, while the former may only have deficits in cognitive empathy (perspective taking or mind reading), cf. Blair (2005, 711).

process of adopting another's affective state – arguably the most basic form of empathy) and higher forms of empathy, such as perspective taking (Preston and de Waal 2002).[11]

The PAM is supported by findings in neuroscience, which show that "sharing the emotions of others is associated with activation in neural structures that are also active during the first-hand experience of that emotion" (Singer and Lamm 2009, 81). For example, observing another person in a painful situation activates the neural network associated with the coding of the motivational-affective dimension of pain in oneself (Decety and Lamm 2009, 201). These findings are in line with the discovery that experiences and observations are frequently based on partly the same neural structure, that is, that certain brain cells – called "mirror neurons" – are activated both during a subject's performance of a certain action (or experience of an emotion) and during the mere observation of such an action performed (or such an emotion expressed) by somebody else (Rizzolatti and Sinigaglia 2008). Thus, from the perspective of neuroscience, empathy appears to be based on such a mirroring mechanism.

The aforementioned scientific studies suggest that empathy is an integral part of human nature; while developmental psychology shows that, across different cultures, developing empathy is rooted in early childhood, neuropsychological research reveals that empathy is largely the result of a rather automatic activation of certain representations based on one's own emotional repertoire. Taken together, these findings provide a rough idea of how empathy works and how it may be conceived of as a universal and inborn ability of human beings.

7.6 Naturalism Without Reductionism

Acknowledging that human rights have their *foundation* in a natural human quality, namely empathy, is not to say that they can – in any sense – be *reduced* to empathy, so that they may well be replaced by a simple appeal to compassion.

The concrete shape of human rights in their institutionalized forms (i.e. as positive law) have naturally to a large extent to be conceived of as products of historical developments and the evolution of the legal system, with its intrinsic rationality.[12] Traumatic historical experiences like the Thirty Years' War, slavery

[11]On the evolutionary perspective, see also: de Waal (2006). Here, de Waal assumes a functional role of empathy in the process of natural selection (mainly in the relations between mothers and their offspring and between members of groups) and describes how different stages in the development of empathy (e.g. emotional contagion and mental state attribution) may correspond to different stages in the evolution of morality (e.g. moral sentiments and reasoned judgment). Evolutionary psychology, however, remains a controversial field and the value of such arguments may be questioned. For a more detailed discussion see von Harbou (2012) and Mahlmann (2008).

[12]Luhmann (2001, 76 et seqq.), for example, depicts the legal system as largely autonomous and autopoetic, referring to its self-created elements.

and, certainly, the Holocaust have shaped both the contents and forms of our present human rights regimes. Legal traditions and practical requirements have led to a highly complex system of positive (subjective) rights, with its limitations derived from both other people's subjective rights and public interest.

Empathy and compassion on the other hand are rather "raw" feelings; different feelings of empathy and sympathy may be contradictory (e.g. the compassion felt for both the victim of a crime and the culprit) providing no clear criteria of how to solve cases of conflict. As private feelings, they have to be translated into political and legal terms and brought into a certain rational order to render them effective in the social sphere. Any direct reference to them in the political field tends to be either hypocritical or dangerous (or both), as Hannah Arendt pointed out in her famous analysis of the French Revolution (Arendt 2011, 89 et seqq.): Ironically, it was the "pity with the people" that Robespierre and others used to refer to in order to legitimize the revolutionary *terreur*. The inherent danger of such a reference becomes obvious when one considers that as an emotional state, pity (like, for example, rage) is absolute and does not know of any proportionality (which on the other hand a rights discourse may provide). At the same time, Arendt observes, it seems impossible to really feel pity for an abstract entity such as a whole people. Compassion, as it really exists, is always an emotion felt by a singular person towards another singular person, so any use of the term in a political context raises the suspicion of hypocrisy.

Another reason why the social significance of human rights may under no circumstances be entirely reduced to empathy lies in the fundamental value of (subjective) rights for the status and self-esteem of persons. To show this, Joel Feinberg proposed an interesting thought experiment in which he conceived of a hypothetical society which he called "Nowhereville": While people in Nowhereville behave perfectly altruistically, thus fulfilling all relevant moral duties towards each other, nobody has any rights. Feinberg now asks what would be missing in such a society and arrives at the conclusion that (subjective) rights have an inherent value that may not be fully substituted by the mere observation of certain rules. According to Feinberg, rights are empowering persons to "'stand up like men', to look others in the eye and to feel in some fundamental way the equal of anyone" (Feinberg 1980, 151). There is in fact a fundamental difference between having to rely on the compassion of another person and having the power to make claims for the legitimate respect which each person owes the other; human rights express the latter kind of position, and they should not be conceived of as mere gifts for which one must be grateful.

A last argument against any kind of reductionism follows from the different logical structure of human rights norms on the one hand, and observations about empathy (or any other natural human attribute) on the other. The impossibility to derive normative statements from mere descriptive ones was already pointed out by David Hume, who famously argued that one cannot deduce what one 'ought' to do merely from what 'is' the case (Hume 1740/2008, SB 469). Since George Edward Moore, we call attempts to reduce normative concepts (e.g. "good") to certain descriptive qualities (e.g. "pleasure," "utility") as "naturalistic fallacies"

(Moore 1903/2002, 62). Apart from all other constraints, deriving human rights norms directly from our natural faculty to empathize would constitute a logical error – such a fallacy would have the same structure as a derivation of the nullity of all human rights norms from our natural tendencies to be selfish and cruel.

Human rights naturalism, as we understand it here, is thus not to be confused with any kind of reductionism. By pointing out that sympathy and empathy form the basis of human rights norms, we merely reconstruct one part of the intricate phenomenon of human rights. A comprehensive account of human rights would have to take its many dimensions and premises into account. However, even if conceived of as a predominantly social institution, one may ask what the reasons for the formation and endorsement of human rights are – Hume's question of why we care for strangers in the first place – and that's the question we tried to answer by referring to our natural inclinations to feel empathy and sympathy.

7.7 Conclusions

In this short essay, we first tried to show that human rights, whether or not they are regarded as moral rights, express a certain minimal standard of morality, which in turn implies an altruistic motivation. We have seen that both analytical arguments and neuropsychological findings suggest that an original altruistic behaviour may only be explained by compassion, which in turn requires empathy. Empathy was described as a natural human faculty, developed in early childhood and rooted in neurophysiological mirroring mechanisms. Finally, we have seen that a naturalistic account of human rights does not necessarily imply any reductionism, if one considers that the described phenomenon is too complex to be described from a single theoretical perspective.

Exploring the foundations of our approval of human rights does not lead to any logical argument in favour of their validity, as we already remarked in our references to Hume, Moore and the 'naturalistic fallacy'. However, showing that human rights do not only have to counteract, but also to reflect some of our natural human tendencies, may provide some additional discursive support for human rights claims. This may be even more the case since – roughly following the sceptical Humean tradition – we started from a rather down-to-earth point of view, not relying on any contentious concepts of practical reason or metaphysical truth.

We provided an argument against radical cultural relativism, which holds that human rights are nothing but contingent products of "Western" history. The specific human rights regimes we know today *are* in fact mainly products of European and North American history, but they are also something else. They express an underlying quality of human beings, the impulse to recognize oneself in the person of the victim of violence and the urge to prevent atrocities from happening.

References

Anscombe, G.E.M. 1957. *Intention*. Oxford: Blackwell.

Arendt, H. 1986. *Elemente und Ursprünge totaler Herrschaft*. München: Piper.

Arendt, H. 2011. *Über die Revolution*. München/Zürich: Piper.

Batson, D. 2009. These things called empathy: Eight related but distinct phenomena. In *The social neuroscience of empathy*, ed. J. Decety and W. Ickes, 3–15. Cambridge, MA/London: MIT Press.

Bennett, M., and P. Hacker. 2003. *Philosophical foundations of neuroscience*. Malden/Oxford/Victoria: Blackwell.

Bentham, J. 1816/1962. Anarchical fallacies. Being an examination of the declaration of rights issued during the French revolution (1792). In *The works of Jeremy Bentham* (reproduction of Bowring edition, Edinburgh 1843), vol. 2. New York.

Bischof-Köhler, D. 1989. *Spiegelbild und Empathie: Die Anfänge der sozialen Kognition*. Bern/Stuttgart/Toronto: Huber.

Blair, R. 2005. Responding to the emotions of others: Dissociating forms of empathy through the study of typical psychiatric populations. *Consciousness and Cognition* 14: 698–718.

Borke, H. 1973. The development of empathy in Chinese and American children between three and six years of age: A cross-culture study. *Developmental Psychology* 9(1): 102–108.

Burke, E. 1790. *Reflections on the revolution in France and on the proceedings in certain societies in London relative to that event*, 2nd ed. In a Letter Intended to Have Been Sent to a Gentleman in Paris. London.

de Waal, F. 2006. *Primates and philosophers. How morality evolved*. Princeton: Princeton University Press.

Decety, J., and C. Lamm. 2009. Empathy versus personal distress: Recent evidence from social neuroscience. In *The social neuroscience of empathy*, ed. J. Decety and W. Ickes, 199–213. Cambridge, MA/London: MIT Press.

Deonna, J. 2006. Sympathy and empathy. In *Encyclopedia of philosophy*, vol. 9, 2nd ed, ed. D. Borchert, 344. Detroit: Macmillan.

Döring, S. 2009. *Philosophie der Gefühle*. Frankfurt am Main: Suhrkamp.

Eisenberg, N. 2008. Empathy. In *International encyclopedia of the social sciences*, vol. 2, 2nd ed, ed. W. Darity, 574. Detroit: Macmillan.

Feinberg, J. 1973. *Social philosophy*. Englewood Cliffs: Prentice-Hall.

Feinberg, J. 1980. *Rights, justice, and the bounds of liberty: Essays in social philosophy*. Princeton: Princeton University Press.

Foot, P. 1967. The problem of abortion and the doctrine of the double effect. *Oxford Review* 5: 5–15.

Glover, J. 2000. *Humanity: A moral history of the twentieth century*. New Haven/London: Yale University Press.

Greene, J., R. Sommerville, L. Nystrom, J. Darley, and J. Cohen. 2011. An fMRI investigation of emotional engagement in moral judgment. *Science* 293: 2105–2108.

Habermas, J. 1998. *Faktizität und Geltung*. Frankfurt am Main: Suhrkamp.

Hoffman, M. 1987. The contribution of empathy to justice and moral judgment. In *Empathy and its development*, ed. N. Eisenberg and J. Strayer, 47–80. Cambridge: Cambridge University Press.

Hoffman, M. 2000. *Empathy and moral development: Implications for caring and justice*. Cambridge/New York/Melbourne/Madrid: Cambridge University Press.

Hume, D. 1740/2008. *A treatise of human nature*. Oxford: Clarendon.

Hume, D. 1751/1998. *An enquiry concerning the principles of morals*. Oxford: Oxford University Press.

Hunt, L. 2007. *Inventing human rights*. New York/London: W.W. Norton & Co.

Ignatieff, M. 2001. *Human rights as politics and idolatry*. Princeton: Princeton University Press.

Kant, I. 1788/2005. *Kritik der praktischen Vernunft*, 17. Aufl. Frankfurt am Main: Suhrkamp.

Luhmann, N. 2001. *Das Recht der Gesellschaft*, 3rd ed. Frankfurt am Main: Suhrkamp.

Mahlmann, M. 2008. Naturgeschichte, Ethik und die Theorie des Rechts. In *Nach Feierabend. Zürcher Jahrbuch für Wissensgeschichte 4*, ed. D. Gugerli, M. Hagner, M. Hampe, P. Sarasin, J. Tanner, 107–128. Darwin/Zürich/Berlin: Diaphanes.
Moll, J., R. de Oliviera-Souza, and P. Eslinger. 2003. Morals and the human brain: A working model. *NeuroReport* 14(3): 299–305.
Moore, G. E. 1903/2002. *Principia Ethica*. Reprint of Revised Edition, Cambridge: Cambridge University Press.
Preston, S., and F. de Waal. 2002. Empathy: Its ultimate and proximate bases. *Behavioral and Brain Sciences* 25: 1–71.
Prinz, J. 2006. The emotional basis of moral judgments. *Philosophical Explorations* 9(1): 29–43.
Prinz, J. 2011. Is empathy necessary for morality? In *Empathy: Philosophical and psychological perspectives*, ed. P. Goldie and A. Coplan, 211–229. Oxford: Oxford University Press.
Rizzolatti, G., and C. Sinigaglia. 2008. *Mirrors in the brain*. Oxford/New York: Oxford University Press.
Rorty, R. 1993. Human rights, rationality, and sentimentality. In *On human rights. The Oxford Amnesty lectures 1993*, ed. S. Shute and S. Hurley, 111–134. New York: Basic Books.
Sandkühler, H. 2010. Menschenrechte. In *Enzyklopädie Philosophie*, ed. H. Sandkühler, vol. 2, 1530. Hamburg: Meiner.
Schopenhauer, A. 1840/1977. Preisschrift Über die Grundlage der Moral. In *Werke in zwei Bänden*, ed. W. Brede, vol. 1, 523–723. München: Hanser.
Shue, H. 1996. *Basic rights*. 2nd. ed., Princeton: Princeton University Press.
Singer, T., and C. Lamm. 2009. The social neuroscience of empathy. *Annals of the New York Academy of Sciences* 1156: 81–96.
Thomson, J. 1976. Killing, letting die, and the trolley problem. *The Monist* 59: 204–217.
Tugendhat, E. 1993. *Vorlesungen über Ethik*. Frankfurt am Main: Suhrkamp.
von Harbou, F. 2012. Anspruch und Anthropologie. Unparteilichkeit und Universalismus als rechtsethische Herausforderungen. In *Gleichheit und Universalität*, ed. S. Ast, J. Hänni, K. Mathis, and B. Zabel, 151–164. Stuttgart: Franz Steiner Verlag.
Wispé, L. 1991. *The psychology of sympathy*. New York/London: Plenum Press.
Woods, K. 2009. Suffering, sympathy, and (environmental) security: Reassessing Rorty's contribution to human rights theory. *Res Publica* 15: 53–66.

Chapter 8
Human Rights and Basic Needs

Peter Schaber

There is a widespread view that the basic needs of human beings – what one might call our basic interests – are of great moral importance (Brock 2005, 65). Some think that basic needs are what ground moral duties, while others hold that they are what human rights are fundamentally concerned with. According to the latter view, the basic needs are the objects of the rights we have simply in virtue of being human – thus, human rights are nothing other than our rights to have our basic needs met. If, then, we are to determine what rights we have *qua* being human, it is important to know what the basic human needs actually are. As Gillian Brock puts it:

> Knowledge of what our human needs are must be had before we can sensibly have a go at defining the entitlements that will be protected as human rights. In order to draw up a sensible list of our human rights we must have a sense of our basic needs ... A needs-based account is thus more fundamental than a human rights account. It is necessary to drawing up a coherent and comprehensive list of our human rights (Brock 2009, 72).

Brock maintains that this is true not just for some human rights, but indeed for all of them, proposing that all the rights set out in the Universal Declaration of Human Rights are explicable in terms of the account of basic needs which she develops.

At first sight, the claim that human rights should protect the basic needs of human beings appears to a claim about the *objects* of human rights, not about their justification. Yet it is the latter that Brock seems to have in mind when she says that a concept of basic needs is required in order to draw up a list of human rights. The claim here at issue is that basic needs are interests of human beings that are important enough to generate moral obligations regarding their protection. If this is indeed the case, then a normative relation obtains between basic needs and human rights: a basic need constitutes a reason for having the correlative right, and a right becomes a human right in so far as it takes a basic need as its object.

P. Schaber (✉)
Philosophisches Seminar, University of Zurich, Zollikerstr. 117, 8008 Zürich, Switzerland
e-mail: schaber@philos.uzh.ch

M. Albers et al. (eds.), *Human Rights and Human Nature*, Ius Gentium: Comparative Perspectives on Law and Justice 35, DOI 10.1007/978-94-017-8672-0_8,

It is this conception of human rights that will be discussed in this paper. Do human rights protect the basic needs of human beings? Are human rights based on basic needs? Is this true for all human rights or merely for some – or even, indeed, for none at all? I will argue that basic needs do play a role in the justification of certain human rights, but not in the sense that they actually provide a justification of human rights. The specific role that basic needs play is limited to spelling out certain normative conditions which have to obtain in order for human rights to come into play: it will be argued, however, that in so far as basic needs fulfil this role, they are not doing the normative work of providing us with reasons why we have such rights at all. With respect to the actual justification of human rights, another idea is much more relevant – namely, the idea that one should possess normative authority over oneself.

8.1 Basic Needs

In order to clarify the relation between basic needs and human rights, one first needs to say more about the concept of basic needs. What basic needs do human beings possess? The aim here is not to determine a list of the basic human needs, but rather to give a very general definition of what the concept of a basic need refers to. As with other concepts, the concept of a basic need can be construed in different ways, and not all such construals are applicable in the context of providing a justification of human rights. It is therefore appropriate to focus on those particular uses of the concept that may prove fruitful for grounding the notion of human rights.

There are two broad conditions which must be met by a concept of basic human needs if it is to stand a reasonable chance of providing insight into the nature of human rights. First, to be relevant for the explanation of human rights, a concept of basic needs must be informative, and must be explicable in terms of other concepts that do not themselves require extensive additional explication. Thus it is not informative, for example, to define basic needs by means of the concept of 'extensive damage', such that a need or interest is a basic need if its non-satisfaction entails extensive damage for the person concerned (see Thomson 2005, 175). For what is extensive damage? This again stands in need of a definition, and clearly the definition cannot itself make appeal to the concept of basic needs (thus it would not do, for example, to say that extensive damage occurs when a basic need of a human being is being violated).

Second, in addition to the requirement of informativeness, the concept of basic needs we are seeking should also make it plain why basic needs are morally significant, for only then can it play a role in the justification of human rights. This condition is far from trivial, since there are different concepts of needs and many do not refer to anything that might be thought morally significant. The basic needs we are looking for are needs that provide a foundation for human rights, and that are therefore capable of creating a duty for other human beings. But which are these needs? Without doubt, not all needs of human beings are basic needs in this sense. A mountaineer may need a rope to accomplish the climb he has been planning for a long time; yet no matter how urgently he needs the rope, this is not the sort of need that creates obligations for others. A need can only compel others if it is sufficiently important. In this respect, one may say that needs are basic needs if their satisfaction serves important aims; and it may well be that the climb is an important aim for the mountaineer without, however, being a basic need that creates a correlative obligation. Thus, basic needs are not simply those needs that must be met in order to attain important goals.

To deserve the title 'basic', a need has to be *inescapable* for a human being, as Garrett Thomson rightly claims (Thomson 2005, 176; see also Thomson 1987, 29). They are needs we have *as* human beings, independently of who we are, what kind of life we pursue, or what we have decided to do in the past. The inescapability condition is not met by the needs of the mountaineer. This conception of basic needs also corresponds to the fact that the rights that protect these needs are ours in so far as we are human beings. Basic needs, then, are the object of human rights inasmuch as they are needs that are inescapable for human beings: But what does this general idea of basic needs mean in concrete terms? Which needs are inescapable?

8.2 Survival

It may be proposed that the inescapable needs concern what we need for our physical survival, and thus that such needs are inescapable given only that we do in fact want to survive. Someone who is terminally ill and does not want to continue living thereby ceases to have such inescapable needs. In so far as we want to survive, however, we need certain things: oxygen, water, nutrition, clothes that protect us from the cold, etc. Without oxygen we would be dead in a few minutes; without water we would only survive a few days; without nutrition, a few weeks. One might thus say that the inescapable needs of human beings are just those that must be satisfied in order to ensure survival.

Such a conception of basic needs faces problems, however. To begin with, it is not clear which time frame is to be considered with respect to these needs and the entitlements deduced from them (see also Griffin 1986, 43). Human beings, it is proposed, must be enabled to survive – But for how long? And exactly what are human beings entitled to in this regard?

Furthermore, this conception of basic needs is problematic because it cannot achieve what is required of it as concerns the foundation of human rights. Consider, for example, the right not to be subjected to torture, or to cruel, inhuman or degrading treatment or punishment (Art. 5 of the Universal Declaration of Human Rights). This right can be violated without jeopardizing the survival of the victim, and the same is true for other human rights as well, e.g. the right not to be enslaved (Art. 4), or the right not to be discriminated against (Art. 7). Even the social human rights – such as the right to an adequate standard of living (Art. 25) – concern more than ensuring mere physical survival; for example, the right to an adequate standard of living includes the right to support in the event of unemployment, sickness, disability, widowhood, old age or other threats to one's livelihood. A conception of basic needs that is relevant to the foundation of human rights must thus encompass more than simply what is required for physical survival.[1] Although it may well be the case that not all human rights enshrined in official documents prove to be human rights in this sense, it must nevertheless be possible to ground some central human rights through appeal to a preferred account of basic needs; but someone who conceives of basic needs in the way just outlined will be unable to do this convincingly. None of the social human rights can be shown to be human rights on this account, and this undoubtedly diminishes the appeal of a conception of basic needs couched in terms of minimal survival.

8.3 A Minimally Good Life

Basic needs can only be relevant for a theory of human rights if they are defined more broadly, as something that goes beyond what is important for our survival. According to Gillian Brock, the importance of meeting our basic needs lies in the fact that they enable us to live a good life in a minimal sense: '[H]uman needs are what we require to function minimally well as the kinds of creatures we are' (Brock 2005, 65; see also Miller 2007, 184). Thus, we have to satisfy our basic needs not in order to survive, but in order to survive *well*. The things Brock has in mind are those that pertain to all human beings, not just some, in respect of living well. The significance of those things is thus independent from the contingent properties of particular human beings; they are important for human beings due to the qualities they have *qua* human beings – and independently of what else happens to be good for human beings, a human life cannot be good if basic needs are not being met. This, at least, is what Brock proposes.

This concept of basic needs is morally relevant. If we are to take the wellbeing of others into account in any way, then their basic needs are to be understood as the necessary conditions for a good life. Such a concept of basic needs also seems far more promising with respect to providing a justification of human rights.

[1]Jeremy Waldron writes in this vein: 'We want a philosophy of human need, not just biological need. Certainly, no such minimalist theory could do anything like the work rights have done in political debate – signaling concerns about threats to self-respect, religious convictions, free expression, control of body and sexuality, choice of career, cultural and political participation (. . .)' (Waldron 2000, 121).

8.4 Basic Needs and Rights

Brock makes the following comment with respect to the things that human beings need in order to live well:

> One will need at least a certain amount of (1) physical and mental health, (2) sufficient security to be able to act, (3) a sufficient level of understanding of what one is choosing between and (4) a certain amount of autonomy ... and (5) decent social relations with at least some others (Brock 2005, 63).

Independently of whether we accept this list or prefer another in its place, the question arises as to whether *all* or merely *some* of one's basic needs have to be met if one is to live a minimally good life. Is the satisfaction of an individual basic need a necessary but not sufficient condition for such a life? An affirmative answer to this question is suggested by the fact that the satisfaction of a single basic need is not, after all, enough to be living a good life. Someone who is merely healthy, for example, is not necessarily and by that fact alone living a good life.

On the other hand, it is also uncertain whether the satisfaction of each and every basic need is a *necessary* condition for a minimally good life. It may well be that certain basic needs are not being met and yet nobody would say that the person concerned is not living a good life. Let us consider an ascetic, who devotes her life to religious introspection and maintains no social relations with other human beings. We would not call this life bad simply due to the lack of social contact; we would do so only if the lack of social contact was uncompensated by some other goods. And we certainly would not consider the ascetic's life to be bad in the sense that we would consider intruding into her sphere of freedom so as to protect her from herself. This is not simply because the missing social contact may be compensated for by other things: it also has something to do with the fact that the ascetic has freely chosen this type of life. Although there is a basic need of this human being that is not being met, her rights are nevertheless not being violated – or, as one might prefer to put it, none of her rights remains unfulfilled.

This provides a provisional insight into what the true relation may be between basic needs and the rights that protect them. Human rights are realized not simply if the basic needs of human beings *are* being met; rather, they are realized when the *possibility* is guaranteed that the basic needs *can* be met. The human rights of our ascetic would thus only be violated if she were being hindered in establishing social relations – if she had in some way been forced to desist from meeting others. In this vein, one might therefore try saying that human rights are what ensure *access* to the satisfaction of basic needs. According to this conception, they are what protect the possibility of living a good life.

Is this really what human rights protect? The first objection that comes to mind with respect to this proposal is that only those basic needs whose protection can be made obligatory shall thereby be considered to be protected by human rights. If there were, for example, a basic need for love, there would still be no human right to be loved, since nobody can be obliged to love another – at least if love is thought to mean a particular emotional disposition towards others (see also Miller 2009, 257). Thus, only those basic needs that can also be obligations to others are thereby protected by human rights.

Is it then these basic needs that are the object of human rights? Let us consider Art. 5 of the Universal Declaration of Human Rights: 'No one shall be subject to torture or to cruel, inhuman or degrading treatment or punishment.' Does this protect a basic need in the sense that was just sketched?

Human beings do not want to be subjected to cruel, inhuman or degrading treatment. But is this a basic need? If it were, such treatment would have to diminish or completely destroy the possibility of them living a good life. Is this the case when we degrade others? A degrading act – such as those prohibited by article 4 – may conceivably diminish the wellbeing of the victims without reducing the possibility of their living a good life. It is true that where people are systematically degraded, their possibility of living a good life is diminished; but this is not the case for many degrading acts if they occur just once.

Suppose Paul is being degraded by his colleagues at work,[2] but that he takes up another post and quickly forgets about this unpleasant episode. It is clear that wrong was done to him, and that a basic moral right, the right not to be degraded, was violated by his colleagues. That this is morally bad is, however, not due to the fact that one of Paul's basic needs in the above sense has been violated. Despite this degrading treatment, which Paul quite soon puts out of his mind, we may reasonably imagine that he goes on to live not only a minimally good life, but one that is positively good.

Someone may object that although degradation may not affect some individuals, it nevertheless does reduce the wellbeing of other – and indeed most – individuals to a greater extent. It is precisely this fact that gives rise to the right not to be degraded. This is a right which protects human beings from a general or group threat, not one which should be evaluated at the level of individuals.[3]

Admittedly, degradation does affect the wellbeing of human beings. However – and this is the key point – the wrong that is done to human beings when they are degraded does not consist simply in the fact that their wellbeing is diminished. Degradation aims to show human beings that they do not count for as much as someone else, or perhaps that they do not count at all, and that others can treat them as they please. That wellbeing is also thereby diminished is only a side effect of such degradation. Degradation does not aim at diminishing someone's wellbeing, but rather at showing them that they have lower status.

[2]Some may be of the opinion that degrading behavior only violates human rights if it is intended by the state or, at least, if the state intentionally lets this behavior go unanswered. It is not the purpose of this article to discuss such a conception of human rights. What is relevant in this context is only what degrading acts do to human beings, not who carries out those acts nor who must be performing them in order for them to be violations of human rights.

[3]I thank Andreas Cassee for this objection.

Consider a state in which some of the citizens are assigned the status of serfs. Although they have no rights, the government makes an effort to ensure their physical and mental health, and the serfs have a certain degree of autonomy inasmuch as they are allowed to choose within a rather constrained array of options and are allowed to establish relations between one another. They are not, however, allowed to demand anything of the state. The serfs may politely ask for things, but they cannot demand that they be treated well, nor that certain things should not be done to them. In other words, they have no rights vis-à-vis the state to any particular treatment.

According to Brock's list, the basic needs of these serfs are being met. Furthermore, the possibility cannot be excluded that the serfs are in fact happy, since they have grown up in this situation and are accustomed to it. Yet independently of the serfs' subjective perceptions, this state is doing a moral wrong to these individuals. In so far as their moral personality is not recognized, a basic human right is being violated. The breach consists in their being treated as beings who cannot demand anything from others – thus, a human right is being denied them. This is morally wrong, independently of whether the satisfaction of basic needs is affected or not.

8.5 Normative Authority

By treating individuals as beings without rights one denies the existence of any normative authority on their part. It seems to me that this idea of normative authority is more central to the understanding of human rights than basic needs.

'Normative authority', in the sense relevant to the present discussion, is the authority to do what one wants with regard to whatever falls within the purview of one's authority.[4] It is not simply the capacity to do what one wants, but the right to do so, and so encompasses the right not to be obstructed in one's actions as carried out within the sphere of one's authority.

What is it that falls under our normative authority? The obvious answer is: anything that is important in my life – for example, who I live with, who I marry, what profession I choose, what projects I take up, etc. That human beings have normative authority about these things means that they have the right to determine who they live with, who they marry, etc., and that they may not be hindered in their enjoyment of these rights.

Various rights included in the Universal Declaration of Human Rights protect exactly this normative authority: the right to freedom and security of the person

[4]Normative authority should not be mistaken for 'normative agency', a term coined by James Griffin (Griffin 2008, 45). According to Griffin, normative agency consists in the capacity to choose and pursue what one considers right. It is a capacity and – unlike normative authority – not a right. This capacity is not necessarily affected when human beings are subject to degrading behavior. After such degradation one may well be able to live one's life as one sees fit.

(Art. 3), the right not to be held in slavery or serfdom (Art. 4), the right not to be subjected to cruel, inhuman or degrading treatment (Art. 5), the right to be recognized as a person before the law (Art. 6), the right to marry and to form a family (Art. 16), the right to religious freedom and the freedom of opinion (Art. 18), to name just a few.

These rights protect the normative authority of human beings, not their basic needs. Of course, it is undoubtedly correct that it is not good for human beings to be subjected to inhuman behavior or to be denied their choice of spouse. But even if it were good for a person to be denied this choice – if, for example, the marriage she envisages would inevitably make her unhappy – one would do her wrong by denying her agency. This is so because the right to marry the person you choose forms part of the normative authority everybody has over his or her life.

One might want to argue that while this may be true for civil rights, it is not true as regards social and cultural rights. These latter rights, it might be proposed, draw their legitimacy from the idea that each protects a basic need that is manifestly at risk. In this vein, article 25 (everyone has the right to a standard of living adequate for the health and well-being of himself and of his family), article 26 (everyone has the right to education), and article 27 (everyone has the right to freely participate in the cultural life of the community) of the Universal Declaration serve to protect basic needs of human beings. However, upon closer examination it becomes clear that these social and cultural human rights also protect the normative authority of human beings. Let us consider the right to adequate living conditions, including food, clothing, housing, medical care and necessary social services, and the right to security in the event of unemployment, sickness, invalidity and widowhood (Art. 25.1). This right protects basic needs in the absence of which one cannot lead a minimally good life, but the attribution of this right is not justified simply on the grounds that these are needs that must be accommodated. Rather, human beings have the right to adequate living conditions in order to be able to exert their normative authority: human beings should be able to live a life as they please, in so far as this is possible, but this requires that they possess certain goods in the absence of which their scope for expressing their authority would be severely curtailed. For example, someone with no access to nutrition or water would be unable to exert her normative authority.

The right to an adequate standard of living, one element of which is the right to nutrition, does not concern the mere survival of the person. Normative authority can only be exerted when there are acceptable options from which to choose. Someone who is made to choose between options that are unacceptable to her is not living life as she sees fit. This is why the list of goods that are necessary for an adequate standard of living contains more than those goods which are necessary for mere physical survival. Article 25.1 rather aims at safeguarding a standard of living which can sustain an acceptable range of substantive options.

One might argue that it is not evident why the idea of normative authority, as opposed to that of basic needs, should be what is thought to ground the right to an adequate standard of living, particularly since it is clearly impossible to live a good life in the absence of the assets mentioned in article 25. Further consideration will show, however, that this right cannot be derived simply from a consideration of basic needs. Human beings have a right to the goods mentioned, but they may also relinquish that right. As we have seen, the religious ascetic does this out of her own free choice, thus liberating others from the obligation of ensuring that she can enjoy the complete range of basic goods. This is permitted. People do not need to exercise their right to an adequate standard of living. Article 25 requires that people should have a standard of living with acceptable options in the specific sense that it requires that they possess the opportunity to enter into a situation in which they have those options. Thus, on closer inspection, it is clear that it is the right, not the goods, which is protected, as well as the right to choose these goods if the person so desires. To reiterate: the right in question is the right to choose between acceptable options, and this is exactly the right that we have already referred to as the normative authority of human beings. A concomitant of this is that people have the right to those basic goods which enable them to have this choice, and it is in this sense that the right to basic goods laid down in article 25 should be understood.

Thus the possibility of meeting basic needs has a specific role with respect to enabling a person to access a meaningful range of options, and the two should not be confused. The right to basic goods protects only those basic goods that we need in order to live a life that allows for our choosing between acceptable options. In other words, the right to basic goods is simply a necessary condition for the protection of normative authority.

This is not to say that all human rights are grounded in the idea of normative authority over oneself; rather, my aim is merely to show that, contrary to what some scholars have proposed, basic needs cannot ground key human rights. If basic needs were indeed the ground for the ascription of human rights, it would have to be possible to justify the civil liberties cited above, as well as social rights, by appeal to the notion of basic needs. But, as we have seen, this is not the case.

8.6 Normative Authority as a Basic Need?

One could argue that these two ways of grounding human rights (via basic needs and via normative authority) are not so different after all, since – so it might be claimed – normative authority over oneself can also be understood as a basic need. According to this position, to live a minimally good life human beings must be able to shape their lives as they see fit; thus they have a basic need for normative authority.

In my view, this objection is unconvincing on a number of grounds. First, it betrays an unsatisfactory grasp of the relationship between normative authority, basic needs, and the good life. It may well be correct that normative authority is necessary for living a good life, and it is certainly plausible that basic needs have to be met in order to live a minimally good life (Wiggins 2005; see also Wiggins 1987, 14). But these premises do not entail that authority over oneself is a basic need. The fact that normative authority pertains to having a good life is not the reason why such authority commands respect. Disrespecting the normative authority of others is not wrong because it limits or destroys their possibility of living a good life – since, as we saw earlier, although degradation may indeed have this effect, it would also be wrong even if it did not. It is wrong to degrade others because in so doing one impugns their normative authority. This is the first point that can be made to block the counterargument introduced above.

The second point turns on seeing that the normative authority of human beings is fundamentally different from their basic needs, and thus that respecting the normative authority of human beings is also different from meeting their basic needs. Normative authority is logically related to our ability to demand things of others – we have it when we can *request* certain things from others. Normative authority refers to the precisely *normative* property of being entitled to demand and request things from others; basic needs, however, are *natural* properties which we have as human beings – since as a matter of fact we need certain things to be able to live a minimally good life. Statements about the basic needs of human beings are thus purely descriptive. To say that human beings need nutrition and water to live well is not to make an evaluative/normative proposition; it is simply to observe something that is the case.

This highlights a third distinction between normative authority and basic needs. Since basic needs are *natural* properties, and propositions about basic needs are descriptive, it is uncertain what normative significance the latter actually have. One might try saying that it is their relation to good life which makes basic needs normatively significant, but this reasoning begs the crucial question of why this creates an *obligation* on others to ensure their satisfaction. Why, one may ask, do we have the right to have our basic needs met, such that there is a corresponding obligation on the part of others? But as regards the normative authority of human beings, this question simply does not arise. Normative authority is the rightful entitlement to demand things from others. It is clear from the fact that it is a rightful entitlement that it entails an obligation on others to respect it. Everyone who recognizes this rightful entitlement therewith also recognizes the obligation it engenders.

Last but not least, presenting oneself as a being with basic needs is different from presenting oneself as a being with normative authority. As a being that has normative authority, I approach others not as a supplicant hoping for a favorable reaction, but as someone who can make legitimate demands. As a being with basic needs, I may ask others to behave in a certain way, and I might point out that it would be good if they would consider my basic needs. But I am not thereby entitled to demand that others do – or desist from doing – specific things. As Waldron points out, a person can shamelessly make such demands.[5]

[5]See Waldron (2000, 130): 'rights are the claims a person can put forward for her own sake and on her own behalf without the moral embarrassment usually associated with assertions of self-interest.'

There are thus at least four distinctions between normative authority and basic needs. The objection that normative authority is a basic need – and thus that grounding human rights in the concept of normative authority is ultimately no different from grounding them in basic needs – can therefore be rejected.

8.7 Conclusion

Contrary to what certain authors have claimed, human rights are not grounded in basic needs. Rather, I have claimed that core human rights are based on the person's basic entitlement to exercise normative authority. In so far as basic needs are protected through certain human rights instruments, they are protected because they are instrumental in human beings exercising their normative authority. Both the civil liberties and the social rights analyzed in this paper are intended to protect the exercise of this normative authority. I have not tried to answer the question of whether the idea of normative authority can ground all or even most human rights; it is clear, however, that it grounds more human rights than the idea of basic needs. As we have seen, the notion of a basic need, lacking the requisite underlying normative character, is not the sort of thing that can ground any rights at all.

Translated by Charlotte Baumann and Ben Young. I am grateful to Andreas Cassee and Anna Goppel for their helpful comments.

References

Brock, G. 2005. Needs and global justice. In *The philosophy of need*, ed. S. Reader, 51–72. Cambridge: Cambridge University Press.

Brock, G. 2009. *Global justice. A cosmopolitan account*. Oxford: Oxford University Press.

Griffin, J. 1986. *Wellbeing. Its meaning, measurement, and moral importance*. Oxford: Oxford University Press.

Griffin, J. 2008. *On human rights*. Oxford: Oxford University Press.

Miller, D. 2007. *National responsibility and global justice*. Oxford: Oxford University Press.

Miller, D. 2009. Spoonful of sugar helps the medicine go down: Gillian Brock on global justice. *Journal of Global Ethics* 3: 253–260.

Thomson, G. 1987. *Needs*. London/New York: Routledge & Kegan Paul.

Thomson, G. 2005. Fundamental needs. In *The philosophy of need*, ed. S. Reader, 175–186. Cambridge: Cambridge University Press.

Waldron, J. 2000. The role of rights in practical reasoning: Rights versus needs. *Journal of Ethics* 4: 115–135.

Wiggins, D. 1987. Claims of need. In *Needs, values, truth*, 1–58. Oxford: Oxford University Press.

Wiggins, D. 2005. An idea we cannot do without. In *The philosophy of need*, ed. S. Reader, 25–50. Cambridge: Cambridge University Press.

Chapter 9
Human Rights Beyond Naturalism

Arnd Pollmann

The notion of human rights seems to deduce those rights from the mere fact of belonging to our biological species. It is because we are *human* – nothing more and nothing less – that we are supposed to simply 'have' those rights. That is why philosophers tend to justify these claims in reference to 'human nature' and to name them 'natural rights'. And in fact, traditionally typical for the philosophical rhetoric of human rights are iterated references to an 'inborn' human worthiness and equality, which are obviously intended to already guarantee a certain kind of timeless and transcultural validity as well as a philosophical justification of these rights on almost ontological grounds: It is because we *are* human by nature that we already have those normative claims. Take, for example, this popular philosophical definition which is representative for many others: "Human rights are a species of moral rights: they are moral rights which all persons equally have simply because they are human" (Gewirth 1982, 1).

So there is a significant normative evocation resulting from a terminological conjunction of 'human' and 'rights' which seems to be both self-evident and highly contentious at the same time. Indeed, this terminological conjunction raises an expectation which might have serious political and juridical consequences: On the one hand, there is the assumption of a certain sort of rights reserved not only to some special persons or groups of persons but to *all* human beings. On the other hand, these are rights of human beings *as such*, which means that these are claims and entitlements completely independent from any individual characteristics or personal properties which might make a difference between individual human beings in many other walks of life. These two assumptions are highly contested, for they both directly contradict a well-accepted philosophical common place, namely David Hume's warning that philosophers have to avoid 'is-ought-fallacies': There

A. Pollmann (✉)
Institute for Philosophy, University of Magdeburg, Post Box 4120,
39016, Magdeburg, Germany
e-mail: arnd.pollmann@ovgu.de

M. Albers et al. (eds.), *Human Rights and Human Nature*, Ius Gentium: Comparative Perspectives on Law and Justice 35, DOI 10.1007/978-94-017-8672-0_9,

is no path that leads from descriptively stating an empirical fact to an evaluative or moral statement about a desirable state of affairs or action-guiding norms (Hume 2004, ch. 3.1.1).[1] Nothing follows from the simple fact that something is what it is; it follows neither that it is good or should be this way nor that it is bad or should not. The ontological sphere of 'being', therefore, must not be mixed up with the sphere of 'ought'. And if we now assign this dichotomy to human rights discourse: The idea of rights which human beings *should* have just because they *are* human beings by nature seems to be odd, as this idea is notoriously aiming to bridge a conceptual gap that is indeed unbridgeable in principle.

But what exactly is wrong with references to human nature when it comes to human rights? And if philosophical allegations of facts regarding the nature of the human 'being' do not already guarantee the human 'ought' of possessing human rights, what else could be the reason why human beings simply have those rights by virtue of being human? Let me start with a critical survey on contemporary trends of justifying human rights with respect to human nature. We can distinguish here between philosophical attempts to base human rights on assumptions with respect to the biological or vital 'first nature' of human beings, which will be discussed in the first section of the chapter, and alternative approaches justifying human rights from the viewpoint of our 'second nature', presented in the next section of this chapter; a second nature, which is said to historically and culturally reshape and overwrite our basic biological material. Even though all of these approaches will turn out to be philosophically deficient, since a plausible justification of human rights may, at its conclusion, abstain from *substantive* references to human nature, all of the approaches are based on two important premises that still seem to be indispensable within human rights discourse. Hence, in Sect. 9.3 of this chapter I will take a closer look at what will be called the 'presupposition of species-membership'. Then, in Sect. 9.4, I will reconstruct the central 'axiom of equality'. In the last section of the chapter I will present my own account of human rights. They will be interpreted there as 'constitutional' rights from the conceptual outset, and this approach will be a thoroughly political one *beyond naturalism* which refrains completely from appeals to human nature.

9.1 Justifications on Grounds of the 'First' Human Nature

It does not matter which language we speak – whether we talk of 'human rights', 'droit de l'homme', 'Menschenrechte', 'derechos humanos', 'insan hakları', 'права человека' (prava cheloveka) etc. – since as long as one uses the term 'human' or its linguistic equivalents, the speaker appears to rely on basic common features of all human beings or on universal characteristics and properties of human life. These seem to justify the talk of human rights by way of referring to some natural or basic

[1]For a critical discussion, see Hudson (1969).

needs, typal capabilities, species-specific vulnerabilities, inborn qualities or human essences. But what exactly are those dispositions or attributes of the human being that should be taken into account here? To answer this question one can choose from at least three different argumentative patterns referring to the natural 'equipment' or even the biological 'substrate' of human beings.[2]

In some way traditional albeit somewhat 'unseasonable' are theoretical approaches that rest on *metaphysical or theological* grounds (Maritain 1943; Spaemann 1987; Pontificial Commission 2011). These accounts often build on the history and tradition of natural law deducing the possession of 'inborn'[3] human rights from a certain cosmology, from divine ordinance or a given ontological order of being – and thus from a 'exceptional position' which is given to the human race as such within that superior order. For this reason and on this ontological level an 'absolute value' of all human beings can be stated which does not allow for any weighting or tradeoffs and directly leads to natural rights as well. Every human being has human rights, since the given metaphysical or divine order of being privileges all human beings in the identical manner. This approach obviously has the conceptual advantage of deriving the validity of human rights from a 'higher' ground of authority, so that this validity is irrevocably given and therefore protected from all kinds of arbitrary access. But this metaphysical or theological approach also has the disadvantage that the assumption of an absolute value is then also protected from all kinds of philosophical criticism: Where do we know that all human beings are of the same exceptional worth? These justifications can only be accepted by those who already *believe* in that kind of natural order, metaphysical 'logic', or a specific religious system.

To avoid this sort of contentious postulate which is impregnated against criticism, *anthropological and materialistic* accounts seek to rely only on empirical 'facts' about the human being; e.g. natural interests, basic needs, or universal threats that are supposedly empirically evident because of their biological or vital urgency (Finnis 1980; Shue 1996; Galtung 1994).[4] According to these accounts, all human beings show by nature certain common traits or deficits which afford a claim to corresponding natural rights which ought to protect and support all members of our species. Hence, the decisive reason for having those rights by nature results from a typal violability of the human animal given by the *conditio humana* which rests on our natural constitution. Apart from the fact that these accounts obviously do not mind the is-ought-gap, if one only looks into the main human rights treaties on the United Nations level, one might be doubtful as to whether all these rights can, in

[2]This survey will aim at identifying systematic differences between arguments rather than philologically sorting out various authors. These authors mostly mix up at least some of the following patterns.

[3]However, the term 'inborn' is quite misleading here. These approaches often extend human rights protection to prenatal life as well.

[4]I will pass over the methodological question as to whether these approaches are really easier to validate.

fact, seriously be called 'natural'. There are, e.g., rights to a fair trial, to privacy of correspondence, to form and join trade unions, to protection against unemployment. It is uncontroversial that these rights rest on general human interests. But can we think of any use of the term 'natural' which would allow for calling them 'natural rights'? Have there been law suits, postal traffic, trade unions or wage labor in the so-called state of nature?[5]

Transcendental and rationalist accounts in the Kantian tradition put forward a significantly different argument by reflecting on the 'conditions of possibility' for leading a well-lived, humane life (Gewirth 1982; Bielefeldt 1998; Griffin 2009). According to these accounts, human beings may not naturally have human rights as such, but rather possess a highly appreciated and species-specific attribute originally constituting the legal status of equal respect; e.g. 'reason', 'rational agency', or 'human dignity'. At the same time, this typal attribute is philosophically formulated as a *moral* capacity from the beginning. This helps to explain why *rights* result from the reciprocal binding power of mutual worthiness. The idea is that every individual that is endowed with reason and rationality will necessarily also respect the right of any other individual to be treated in a way that acknowledges species-specific worthiness. However, all of these accounts show a major shortcoming: They are far too demanding with respect to their normative claims by drawing not just on humanity as such but also on a higher-value *personhood* (Heun 2004). If – from the outset – reason, rational agency, or human dignity is constructed as an already ambitious faculty for mutual respect, one also has to concede that this faculty does not apply to *all* human beings. Therefore, one might wonder whether, e.g., human embryos, severely mentally disabled persons, or people suffering from dementia should be excluded from the circle of rights bearers.

9.2 Justifications on Grounds of the 'Second' Human Nature

A kind of second-order naturalism[6] is represented by theories which do not search for the grounds of equal legal treatment within our natural or biological constitution but in the way human beings transform their first nature historically, culturally or habitually and also by the way in which they morally and legally frame it. According to this, *virtue ethics and perfectionist* accounts draw on teleological and substantial conceptions of the human 'essence' and therefore on a purposive idea of a 'second' nature which has to be realized first (Vlastos 1962; Food 2001[7]; Nussbaum 2006).

[5]One could draw the opposite conclusion, namely that theses rights cannot be 'proper' human rights and should therefore be dropped from the UN-lists. But the global acceptance of these rights in international law empirically contradicts this conclusion.

[6]For differences between these two sorts of naturalism see McDowell (1996).

[7]Although Food did not touch upon the subject of human rights, her book had a great impact on the discussion.

Thus, human rights call for necessary conditions of *ethical flourishing* and thereby realize the purposive idea of a virtuous or successful, well-lived life. It is just this common potential to follow that demanding ideal which constitutes the status of equal moral and legal respect. Nevertheless, these theoretical approaches cannot avoid the critical argument that they exclude all those kinds of human beings that simply do not possess this ethical capacity. And what is even more doubtful is whether substantial ethical and even teleological concepts of 'the good life' are compatible with the well accepted assumption that human rights should leave room not only for individual but also cultural pluralism (An-Na'im 1992).

That is why justifications based on *contractualism and discourse ethics* lift the argument to an intersubjective meta-level (Höffe 2007; Habermas 1996; Forst 2010). Here, the burden of proof no longer rests upon a substantial and somehow paternalistic account of a human being's essence but with a constructivist concept of imaginary bartering or a discursive agreement with which the parties *themselves* mutually bind each other by guaranteeing the fulfillment of their basic human interests. Accordingly, human rights lie within an 'overlapping consensus' of all possible bartering or communicating partners who mutually – and by representing those who cannot themselves take part – deliberate upon the question on which quasi-natural claims they can reach agreement, these having been given by the simple fact of being human. Nonetheless, these justifications reveal a serious conceptual deficit. The imaginary parties of that bartering or discursive covenant will only then think of themselves as already recognizing the fundamental rights of all other parties if these latter parties, for their part, simultaneously recognize the rights of the former. But that would mean that the validity and possession of human rights would be dependent on mutual approval and adjudication, which is clearly against the quite traditional and well-established idea of human rights as being 'inalienable' and 'unconditional' (Brown 1955; Schiller 1969).

The final alternative justification draws on arguments from *philosophy of history and evolution theory*. According to these arguments human rights can be interpreted as a historically necessary result of the conflicting, but also cooperative 'social' nature of human beings (Bobbio 1995; O'Manique 2002; Fukuyama 2011). Whereas the confrontational side of human nature repeatedly causes cooperation problems and legal regulatory gaps, the cooperative side of human nature provides us with the motivation to follow new regulatory arrangements. Therefore, human rights can be seen as the institutional outcome of a somehow linear and directed learning process which permanently leads to ever more demanding regulations on the national and meanwhile also international level. This implementation process as such might not be considered 'natural', but human rights are still justified in naturalist terms. These new legal arrangements seek to prevent humanity from a relapse into the 'state of nature'. If human rights are interpreted in such a manner, they form a kind of historical "minimal teleology" (Apel 1990, 370–74) which allows for the philosophical expectation that – despite all the terrible setbacks – the history of mankind still makes some meaningful 'sense' as it leads to legal progress. But this hopeful idealistic expectation ignores the important historical insight that our contemporary thinking about human rights should rather be seen as the result of a totalitarian and monstrous catastrophe, which can hardly be interpreted as the 'necessary' product of a teleological course of history.[8]

9.3 The Presupposition of Species-Membership

However problematic all six of these naturalistic justifications may seem, they each rest on two presuppositions which are both central yet still in each respective case subject to a different justification. In the next two sections of this chapter, it will be shown that they, in fact, are indispensable to the contemporary discourse on human rights. At the same time, it shall be demonstrated that this discourse can indeed give up the concept of using these two premises as a *foundationalist grounding* by deducing them from assumptions referring to human nature. Instead, it will be shown that a plausible concept of human rights does not need such a justification since these two premises are a conceptual *positing* incapable of being justified in principle.

The first of these two presuppositions might already be self-evident, namely the central precondition that one has to belong to the human species to be a bearer of human rights. Despite all individual differences concerning age, biological sex, ethnic, social, religious, cultural, or geographical backgrounds all human beings can make a claim to fundamental human rights simply by being a member of the species *Homo sapiens* whether it be on account of

(a) the 'exceptional position' and naturally 'inborn' worthiness of human mankind (metaphysics and theology),
(b) all human beings sharing the same basic natural interests or standard threats (anthropology and materialism),
(c) all humans have a highly appreciated typal attribute in common (transcendentalism and rationalism),
(d) every human being ethically dependent on these rights to fully unfold the potential of his or her essence (virtue ethics and perfectionism),
(e) us thinking of ourselves as partners in an imaginary agreement to the mutual guarantee of fundamental human possibilities (contractualism and discourse ethics),
(f) an ambivalent human nature which forces mankind into a legal progress in the long run (philosophy of history and evolution theory).

But by taking a closer look at these six justifications, one might realize that all of these naturalist accounts *presuppose* what they are willing to justify, namely

[8]It shall not to be denied that human rights can be understood as the result of an historical learning process, but certainly it cannot be a directed or linear one. See introduction to Menke and Pollmann (2007).

the idea that all human beings have human rights *a priori* or from the outset. So they all get caught up in circular reasoning (Menke and Pollmann 2007, 63–5). The formal structure of the argument is in each case – whether intentionally or just unconsciously – the following: (1) Human rights rest on factor X (exceptional position, natural interests etc.); (2) All human beings have factor X; (3) Therefore, all human beings have human rights. But this argument is begging the question. Because, after all, this conclusion is already implicit in the original term 'human rights' which initially shall be based on factor X in major premise 1. That means that this factor X is basically superfluous since conclusion 3 has already been drawn implicitly just as soon as one uses the term 'human rights' as in premise 1. This term simply does not *mean* anything else but the adjudication of basic rights to all human beings. That is why we should not wonder that at the end all of the six theoretical approaches come up with the result that every human being has human rights.

The philosophy of human rights could now draw the conceptual conclusion to go on searching for an explicitly non-circular justification. But what would such an account look like? It would need to be an account using the term 'human rights' without at the same time presupposing that these, in fact, are the rights of human beings. Such a justification seems to be neither possible nor useful, since the conclusion, at which the theory would aim, is already taken for granted on terminological grounds; and, thus, the deduction of a legal 'ought' from the simple 'is' of species-membership is also taken for granted. One might find this deduction philosophically implausible – for reasons already presented at the beginning – but still one will not be able to use the term 'human rights' without relying on that conclusion from the conceptual outset. In other words, one cannot affirm the idea of human rights while at the same time denying that we have those rights just by being human. One *can* indeed question the idea of simply having these rights by being a member of the human species. One can, furthermore, be convinced that there are rights that not *all* human beings possess. And one could envision some basic rights that can be adjudged not *only* to human beings but also to – at least some – animals. But by doing so one would no longer be speaking about 'human rights' in the strict sense.[9] Just by affirmatively using this term one necessarily sets the far-reaching 'generic'[10] implication that the bearers of those rights are *all* members of the species *Homo sapiens*. This 'all' has two different but interconnected meanings: on the one hand it says that no one can be a rights bearer without, at the same time, being human. And on the other hand, it means that no one can be human without at the same time being a rights bearer. In other words, the set of human beings and the set of human rights bearers are coextensive.

[9]There are important differences between the political idea of human rights and the interpersonal scope of 'morality', which can include non-human beings as well. See Menke and Pollmann (2007, ch. 1).

[10]'Generic' statements are generalizing statements about the typical qualities inherent in examples of a certain type (lat. *genus*) which 'exemplify' this type by manifesting those qualities *normally*, though not necessarily constantly or completely. I am grateful to Thomas Hoffmann for this clarification.

Therefore, the 'solution' to the justificatory problem posed above will not be to *insist* that there has to be a non-circular justification for the possession of human rights. Instead, we simply should accept the argumentative circle by initially *stating* that the attribute of species-membership is both the necessary and also sufficient condition for taking part in the human rights game. It is a conceptual 'setting' in the strict sense: if philosophy wants to explain, systematically, what human rights are, then philosophers must presuppose – even if it only be this one point[11] – that these are rights of all human beings. This basic presupposition only then becomes philosophically problematic if one wants to add a presumably constitutive and also 'natural' factor X whose universal validity is highly contested. The premise of mere species-membership is far less demanding in that respect since it goes without such a substantial and supplementary factor X. One may accuse those who hold this presupposition concerning species-membership of resting on a biological and therefore still naturalist assumption which may be philosophically no less an issue than the more substantial ones.[12] But is that true?

It is not obligatory to interpret the simple fact of species-membership *merely* biologically, i.e. in the spirit of genetic descent, for there is certainly far more to natural human life than just the reproduction of its biological substrate (Roughley 2011). But, nevertheless, what exactly might be contained within this 'more' can be put into conceptual brackets. Since, in conceptual contrast to the highly controversial problem as to whether *one* single aspect of human life guarantees the possession of human rights, the question as to whether a given being belongs to the human species is quite easily answered and only in very rare cases to be left open.[13]

Accordingly, by referring to mere species-membership, a naturalistic remainder appears to exist, but we do not require a precise description of our first nature to accept and understand that remainder nor do we require a substantial interpretation of our second nature. That is why it seems rational to completely refrain from any problematic references to 'nature' as such. We have two more reasons for doing that. Firstly, the term 'human nature' suggests a timeless invariableness that is empirically contradicted by both human evolution and the cultural fact that human beings can artificially modify *themselves*.[14] The idea of 'natural' rights is, therefore, misleading, because it excludes the possibility of any 'new' rights coming up

[11]Many other important justificatory problems remain if we do accept this conceptual first premise: What, for example, speaks in favor of 'human dignity', for the concrete rights in detail, their 'universalism' or their 'indivisibility'? What about their legal implementation, the legitimacy of the international human rights regime, or the question of humanitarian intervention?

[12]For a thoughtful discussion see Hey (2001).

[13]What is unclear in cases of, e.g., embryos, mentally disabled persons, or people in comas is *not* whether they are members of the human species but how strictly this membership should be treated with respect to human rights; in other words, how consequently one is willing to accept the normative implication that indeed all human beings should count as (full) members from the start.

[14]See the contributions to Schleidgen et al. (2011).

through the progression of social and legal history.[15] And secondly, as we will see in the last section of this chapter, the question as to the 'purpose' of human rights need not only be answered independently from naturalist references but rather only by reference to social and political life conditions which are almost *against* nature. But before that, the presupposition of species-membership has to be complemented by a very important second conceptual premise.

9.4 The Axiom of Equality

Contemporary accounts of human rights commonly rest on a second central premise which is closely connected to the first by way of *equalization*. Almost every human rights document adopted by the United Nations starts with a confession similar to the following: "All human beings are born free and *equal* in dignity and rights".[16] So, beginning on this rhetorical level, it seems to be self-evident not just that all human beings are included in the circle of 'rights-holders' but that all of them are in exactly the *same way*. The important difference is: Even if one is willing to subscribe to the first presupposition of species-membership, this does not necessarily mean to already award an *equal* worth to all human beings. For you could hold the conviction – which is still quite popular in some traditional religious groups – that men, women, and children might likewise be members of the human rights community. But still, divine providence has provided the male sex with some *additional* rights.[17] From inclusion in the human rights community and its corresponding entitlements, *equal entitlements* do not inevitably follow. Therefore, a proper theory of human rights has to adopt a second generic and strict egalitarian premise: Not just *all* members of the species Homo sapiens are bearers of those rights, but all these members in *equal respect*.

As is well known, the notion of equality is notoriously ambiguous and unclear (Gosepath 2011). We have to decide what it is supposed to mean in the philosophical and also political context of human rights. The first important distinction refers to the fact that equality is often misconceived as a logical, ontological or qualitative 'identity'. If two entities, e.g. two holders of rights, are not just equal but identical, then they are simply *so* similar that they are strictly the *same;* hence, we could no longer compare them and speak of equality. For if we want to compare two equal entities there has to be at least some difference between them. And their equality can only be the result of a comparison with reference to a certain criterium, also

[15]Think, for example, of legally quite new but widely accepted human rights like the right to intellectual property or the right to informal self-determination with respect to laws governing data protection, private data security, and the private sphere.

[16]Article 1 Universal Declaration of Human Rights (1948) [italics inserted by author].

[17]One might remember George Orwell's *Animal Farm* (1945): "All animals are equal. But some animals are more equal than others." The pigs were meant.

called the *tertium comparationis*.[18] In the case of human rights, this egalitarian reference point is obviously the mere fact of belonging to the human species, as discussed in the former section of this chapter. In other words, one *can* understand this membership as an equal membership right away and therefore speak of equal human rights. But we do not have to, if we do not want to believe in human rights.

That is why the concepts of equality and identity or sameness have to be distinguished. However, concepts of identity and sameness also play a crucial role in human rights discourse. And this causes some other terminological confusion. On the one hand, the equal entitlements associated with human rights result from a common membership within precisely the *same species*. On the other hand, the normative assumption of the equal worth of all human beings does not only lead to the political and legal claim for equal treatment but also to the adjudication of *identical rights*.[19] To put it succinctly, all human beings have the *same* human rights because they all are of *equal* worth, given that they all are *equal* as human beings, which follows from them all belong to the *same* species. This legal provision already leads to a second differentiation within the notion of equality, namely the empirical or descriptive question as to whether human beings already 'being' equal must be distinguished from the normative or prescriptive question regarding to what extent they 'should' be equal (Williams 1973). In the first case, equality or inequality should be a given fact caused by either natural or social circumstances.[20] In the second case, we deal with a political postulate which argues for the elimination of – at least some – empirical inequalities. Although we must differentiate between these two aspects analytically, when it comes to human rights, we again are confronted with a peculiar combination of these two aspects. Since the normative claim that all human beings *should* be legally treated equally results from the empirical fact that these human beings already all *are* equal as members of the same species. In other words, the human rights concept of normative equality calls attention to empirical inequalities of a fundamental legal kind[21] that must be overcome given that all human beings are supposed to be of equal worth just by being human.[22]

But the question of course is: *Why* are human beings equal factually as humans and *also* normatively in terms of worthiness? Is there a plausible philosophical explanation or even justification for this? Since equality in terms of human rights is – at least also – a postulate, we need to accept that other postulates should

[18]See Radbruch (1993, 259): "Equality is not a given fact. Things and human beings are unequal like one egg varies from the other. Equality is always an abstraction from given inequality seen from a specific reference point" [translated by author.].

[19]Compare this with the well-known differentiation by Dworkin (1978, 272), that "the right to equal treatment" has to be distinguished from the "right to the treatment as an equal".

[20]It was Rousseau (1984, 77) who clarified this additional difference between two forms of *empirical* inequalities.

[21]This stands in contrast e.g. to egalitarian theories of 'social justice' which claim for a far more demanding idea of equality by including not just basic rights but also resources, chances etc. that are not (or should not be) protected by human rights.

[22]Vlastos (1962).

be possible as well. One could adopt the first premise of species-membership while at the same time grading different levels of entitlement for different kinds of human beings. Incidentally, that was quite typical for the historical human rights declarations of the revolutionary eighteenth century. They may have declared 'human' rights rhetorically, but at the time they were only meant for the white, male, free and taxpaying citizen. In this respect, the idea of equality still had to be enforced *within* human rights discourse and against traditional and persistent forms of unequal legal treatment. This enforcement was the result of political and sometimes revolutionary struggles for "non-discrimination" (Gutmann 2012), which therefore must be interpreted not merely as an inclusion of more and more people into the circle of human rights bearers but also as their strict legal equation. Human rights equality seems to be something that is historically left over when "all fixed, fast-frozen relations, with their train of ancient and venerable prejudices and opinions, are swept away" (Marx and Engels 2002, 552) and social practices or traditional forms of unequal treatment increasingly lose their legitimacy. But what exactly is meant by saying that legal non-discrimination is an historical 'left over'?

There is much evidence that the increasing acceptance of the legal idea of equality is not so much the result of philosophical writings or justifications but of historical, cultural and also individual experiences of injustice as well as corresponding learning processes. These learning processes implement neither an historical or linear 'logic' of development nor have they been experienced, simply put, by all societies or even individuals yet (Rorty 1993). If one is already persuaded to believe in equal respect, then one is obviously dependent on being *accustomed* to this "attitude" or "practice" of equal respect (Menke and Pollmann 2007, 62–8). However, there are no strictly compelling or irrefutable reasons for adopting this attitude. Some people might already have it because of personal or historical experiences related to discriminatory forms of injustice whilst other people may not. So this standpoint of equality is somehow '*unverfügbar*' (beyond one's command). And if one is an opponent to this idea of equality she will not be convinced to change over to the opposite camp just by listening to philosophical arguments. It is, indeed, impossible to believe in contemporary concepts of human rights and to simultaneously deny the equal worth of all human beings. But one could deny the equal worthiness of human beings, even though one could then no longer hold the modern human rights attitude.

Following this, the egalitarian notion of human rights seems to be dependent on a second conceptual postulate, namely that if we want to understand what equality means when it comes to human rights – namely, not just referring to the worth of human beings as such but to their equal worthiness –, then we already must be *willing* to see ourselves as equals and therefore to subscribe to that standpoint. But since this premise is no terminological implication but an intentional normative postulate, we ought to call this premise an "axiom" of human rights discourse. An axiom is a basic principle of scientific theory which is taken for granted without it having been proved or deduced. So within every contemporary account of human rights, this premise – whether explicitly or only implicitly – plays the somehow paradoxical role of a "groundless ground" (Menke and Pollmann 2007, 87).[23] There is no "*Letztbegründung*" (ultimate justification) for this idea. The only way to lay the philosophical foundations of that idea may be by justifying it *ex negativo*; that means by demonstrating that all ways of justifying the opposite idea of a generic inequality will fail (Tugendhat 2001, 180–2). Therefore, the basic human rights claim to be even-handedly respected and treated as an equal would be the result of the absence of compelling reasons for discriminating legal practices since discrimination would mean contradicting the postulate of equal entitlements.

Consequently, we ought also to accept this second argumentative circle as a conceptual necessity, namely that all human beings have the same human rights if and only if we are willing to respect them as equally entitled members of the legal community.[24] There is no 'deeper' or more fundamental justification possible – so philosophers can stop looking for that. Since it is and was a *decision* to base national and also international human rights law on nothing more than this idea of human equality – a decision from which the recognition of equal human rights *follows* – it still must be said that this decision could have been made differently. This renunciation of any further attempt to justify equality may indeed lead to the hasty criticism that it is normatively 'arbitrary'; this abstinence, however, has the undeniable advantage that it is no longer dependent on any 'higher' authority in which we would then have to 'believe' (e.g. god, cosmos, nature, reason as such, etc.) (Tugendhat 2001, 167). It is the reverse: the idea of human rights should rest on the – in a positive sense – disillusioning insight that there is *no* such higher authority than us human beings when it comes to human rights.

9.5 Human Rights as Constitutional Rights on Non-naturalist Grounds

But beside these two presuppositions of species-membership and equality, most philosophers are still convinced that human rights are 'natural' rights – as they result from some kind of pre-political 'state of nature' or at least from the realm of morality which stands 'above' the positive law of the state. So human rights commonly are interpreted as inborn rights prior to the political and legal system. But how exactly should this priority be interpreted? And do we necessarily have to think of this priority as somehow 'natural'? What comes now in the last section of this chapter will advocate for a fundamentally different version of human rights. If we take a closer look at the historical and also political purpose and function of human rights, we will see that the necessity of *claiming* human rights is given

[23]Very similar: Feinberg (1973, 93).

[24]Additionally, the six naturalistic accounts presented above will only then come to the result of equal rights if they presuppose an equal worthiness as already given.

only in cases in which a political and legal system is already in the making or even established. In other words, human rights are neither natural nor moral rights but political claims from the start (Menke and Pollmann 2007; Beitz 2009).

Think, for example, of the 1776 Virginia Bill of Rights. Article 1 proclaims: "That all men are by nature equally free and independent, and have certain inherent rights, of which, when they enter into a state of society, they cannot, by any compact, deprive or divest their posterity". And Article 2 declares: "That all power is vested in, and consequently derived from, the people; that magistrates are their trustees and servants, and at all times amenable to them." Although this declaration rhetorically speaks of freedom and independence "by nature", the point of interest here is the imaginary and almost 'unnatural' moment in which we "enter into a state of society". It is this moment in which we become aware of inalienable and unconditional rights derived from the common power of the people against the "magistrates" as "their trustees and servants". And if seen from this historical and also contractualist perspective, the adjudication and validity of human rights is paradoxically both a prerequisite and also the result of the political act of founding a new constitution (Wellmer 1999, 270). That is the explanation of the, in some way, 'pre-state' but by no means 'pre-political' validity of human rights and, therefore, of their unnatural rather than natural origin. The need to insist on human rights emerges only if political power is interpreted, organized and legitimized in a specifically *modern* way. Human rights would be fundamental legal provisos with respect to the political officials; provisos that would be implemented as integral parts of the constitution, if this constitution would, as it is said in the Virginia Declaration, "consequently" be derived from the power of the people; that means that this act of constitutional drafting would in fact be left to all equal citizens.[25]

In this modern political interpretation, human rights do function as 'constitutional rights' from the conceptual outset.[26] Human rights are part of the modern progress of civilization which turns illegitimate forms of arbitrary rule into legitimate state power by means of juridification.[27] As soon as they achieve juridical reality, human rights are constitutional protections against the arbitrary misuse of political power. But on the conceptual level of philosophical theory, they need to be interpreted as 'virtual' or 'intended' rights within a legal thought experiment of collectively and democratically founding a new political order. So if we want to identify the very purpose of those rights, we have to imagine a process of sovereign constitution-making in which all those people who afterwards will be *subjected* to political power are originally the *subjects* of democratically constituting

[25]By persistently using "would" here, it is indicated that this is a thought experiment and not a historical narrative. Nevertheless, such acts of constitution-making have taken place; e.g. in the revolutions of the late eighteenth century, around the year of 1989 or in so-called Arab Spring.

[26]From French *constitutionnel* derived from Latin *constituere*, 'to arrange, settle, found, set up'.

[27]Not just human rights are part of this modern progress but also e.g. the new idea of a division of powers, the implementation of political participation or the humanization of criminal law.

and legitimizing it. Within this thought experiment, human rights play the crucial role of constitutive reservations through which the original act of founding is performed.

The philosophical explanation for this is that the twofold conceptual postulate of equal species-membership, as it was presented in the last two sections of this chapter, must be interpreted politically. In the imaginary process of democratic self-constitution, this twofold postulate is reiterated as the constitutional claim that all those individuals that will be subjected to political power must, at the same time, be seen as equal subjects – and not as 'objects' – of this power merely on account of the fact that they are all human. That means that all those human beings that will be brought under the rule of law will thereby 'keep' a general right to equal treatment as well as specific human rights that tell the powers what they should do and should not with respect to individuals. Accordingly, human rights should not be misunderstood as the result of benevolent acts of grace by progressive leaders or governments who suddenly remember and realize their commitments to a dubious 'natural law' (Beitz 2009, ch. 3). People themselves declare those rights democratically and therefore *prescribe* them to any further execution of political power – as concrete respects in which this power must not be misused. In other words: Human rights are no concessions *to* the people but provisos *by* the people. Only these provisos provide the political and legal system with democratic legitimacy, since modern citizens would never – if they were asked – voluntarily and unanimously subscribe to a constitutional order leaving them (or at least certain minorities) without those fundamental rights (compare Habermas 1996).

Seen from this modern democratic perspective, it is conceptually impossible for a political and legal system not to provide its citizens with codified constitutional rights and yet still consider its power relations legitimate. Of course, state officials *can* misuse their power empirically. But they cannot do so without simultaneously 'biting the hand that feeds' and losing their legitimacy. This also helps to explain the rather irritating fact that we delegate the responsibility to protect human rights only to those – the state officials – who are the most serious violators of these rights. Since their double role of guaranteeing and violating rights is conceptually constitutive for human rights theory, there would simply be no human rights at all if their *primary* purpose did not consist in forcing the political system to legal self-restraint. And only because history teaches us that nation states fail in protecting the human rights of their citizens do these rights have the *secondary* purpose to also force the international community to assume some kind of 'deficit guarantee' (compare Beitz 2009, ch. 5).

But to conclude, the still popular idea that human rights are natural rights is more than misleading. Although every philosophical account of human rights must be based on the presupposition of species-membership on the one hand and on the axiom of human equality on the other, the very purpose of human rights is not to transform pre-political claims already given or possessed in a state of nature but to implement power restrictions that only then become necessary if modern citizens understand themselves as the legitimizing subjects of political power. It, therefore, would be highly deceptive to call these constitutional rights 'natural' because a political system regulated by basic constitutional rights is quite artificial or even 'against' human nature. The subject of human rights is not the pre-political or natural human being but the anticipated, democratically transformed, and also revolutionary subject of his or her own future. Therefore, a plausible human rights approach can and should abstain from any substantial references to human nature not just while clarifying the twofold presupposition of human equality but also by way of identifying the purpose of these claims which should be interpreted as constitutional rights from the very beginning.

References

An-Na'im, A.A. (ed.). 1992. *Human rights in cross-cultural perspectives*. Philadelphia: University of Pennsylvania Press.

Apel, K.-O. 1990. *Diskurs und Verantwortung*, 370–374. Frankfurt am Main: Suhrkamp.

Beitz, Ch. 2009. *The idea of human rights*. Oxford: Oxford University Press.

Bielefeldt, H. 1998. *Philosophie der Menschenrechte*. Darmstadt: Primus.

Bobbio, N. 1995. *The age of rights*. Oxford: Polity Press.

Brown, St. 1955. Inalienable rights. *Philosophical Review* 64: 192–211.

Dworkin, R. 1978. *Taking rights seriously*. Cambridge: Harvard University Press.

Feinberg, J. 1973. *Social philosophy*, 88–94. Englewood Cliffs: Prentice Hall.

Finnis, J. 1980. *Natural law and natural rights*. Oxford: Clarendon Press.

Food, Ph. 2001. *Natural goodness*. Oxford: Clarendon.

Forst, R. 2010. The justification of human rights and the basic right to justification. *Ethics* 120: 711–740.

Fukuyama, F. 2011. *Origins of political order*. New York: Farrar, Straus and Giroux.

Galtung, J. 1994. *Human rights in another key*. Oxford: Polity Press.

Gewirth, A. 1982. *Human rights*. Chicago: University Press.

Gosepath, St. 2011. Equality. In *The Stanford encyclopedia of philosophy*. http://plato.stanford.edu/archives/spr2011/entries/equality/.

Griffin, J. 2009. *On human rights*. Oxford: University Press.

Gutmann, T. 2012. Normenbegründung als Lernprozess? Zur Tradition der Grund- und Menschenrechte. In *Von der religiösen zur säkularen Begründung staatlicher Normen*, ed. T. Gutmann et al., 295–313. Tübingen: Mohr Siebeck.

Habermas, J. 1996. *Between facts and norms*. Cambridge: MIT.

Heun, W. 2004. Gattungszugehörigkeit oder Personsein als Anknüpfungspunkt der Menschenrechte? In *Menschenrechte und Bioethik*, ed. E. Klein and C. Menke, 24–41. Berlin: BWV.

Hey, J. 2001. *Genes, categories, and species*. Oxford: Oxford University Press.

Höffe, O. 2007. *Democracy in an age of globalisation*. Berlin: Springer.

Hudson, W.D. (ed.). 1969. *The is-ought question*. London: Macmillan.

Hume, D. 2004. *A treatise of human nature*. Mineola: Dover.

Maritain, J. 1943. *The rights of man and natural law* [1941]. New York: Sribner.

Marx, K., and F. Engels. 2002. *Manifesto of the communist party* [1847], MEGA, vol. 31. Berlin: Akademie Verlag.

McDowell, J. 1996. Two sorts of naturalism. In *Virtues and reasons*, ed. R. Hursthouse, G. Lawrence, and W. Quinn, 149–179. Oxford: Clarendon.

Menke, Ch., and A. Pollmann. 2007. *Philosophie der Menschenrechte. Zur Einführung*. Hamburg: Junius.

Nussbaum, M. 2006. *Frontiers of justice*. Cambridge: Harvard University Press.

O'Manique, J. 2002. *The origins of justice*. Pennsylvania: Pennsylvania University Press.

Pontificial Commission "Justitia et Pax". 2011. *The church and human rights* [1975]. Rome: Vatican.
Radbruch, G. 1993. *Rechtsphilosophie, Collected works*, Bd. 2. Heidelberg: Müller.
Rorty, R. 1993. Human rights, rationality, and sentimentality. In *On human rights*, ed. S. Shute and S. Hurley, 111–134. New York: Basic Books.
Roughley, N. 2011. Human natures. In *Human nature and self-design*, ed. S. Schleidgen et al., 13–33. Mentis: Paderborn.
Rousseau, J.-J. 1984. *A discourse on inequality* [1755], 77. Harmondsworth: Penguin Books.
Schiller, M. 1969. Are there any inalienable rights? *Ethics* 79: 309–315.
Schleidgen, S., et al., (eds.). 2011. *Human nature and self design*. Paderborn: Mentis.
Shue, H. 1996. *Basic rights*, 2nd ed. Princeton: University Press.
Spaemann, R. 1987. Über den Begriff der Menschenwürde. In *Menschenrechte und Menschenwürde*, ed. E.W. Böckenförde and R. Spaemann, 295–313. Stuttgart: Klett-Cotta.
Tugendhat, E. 2001. Wie sollen wir Moral verstehen? In *Aufsätze 1992–2000*, 163–184. Frankfurt am Main: Suhrkamp.
Vlastos, G. 1962. Justice and equality. In *Social justice*, ed. R. Brandt, 31–72. Englewood Cliffs: Prentice-Hall.
Wellmer, A. 1999. Menschenrechte und Demokratie. In *Philosophie der Menschenrechte*, ed. St. Gosepath and G. Lohmann, 265–291. Frankfurt am Main: Suhrkamp.
Williams, B. 1973. The idea of equality. In *Problems of the self*, 230–249. Cambridge: Cambridge University Press.

Chapter 10
Human Rights, Human Nature, and the Feasibility Issue

Jörn Reinhardt

10.1 Introduction

In the opening chapter of *The Social Contract*, Jean-Jacques Rousseau describes his own undertaking that is, at the same time, the task of political philosophy as he conceives it: "to consider if [. . .] there can be any legitimate and sure principle of government, *taking men as they are and laws as they might be*" (emphasis added) (Rousseau 1762, 49). Rousseau's famous statement contains an important caveat in its final clause according to which principles of government must not only be legitimate, but also feasible, i.e., realizable politically ("laws as they might be"). In order to meet this requirement, they must start from realistic premises and not from idealized or abstract assumptions based on human nature and ability ("taking men as they are"). It is not without irony (although it shows how difficult the task is) that Rousseau – despite his awareness of the problem – is widely regarded as someone who failed to comply with this standard.

This chapter explores the relevance of arguments from human nature for dealing with the feasibility issue. The focus is on questions of fundamental rights. Feasibility concerns are not only relevant for contemporary debates on international human rights and constitutionalism; they have already shaped natural right thinking in its various traditions. The first two sections deal with the natural right tradition. As the short quote from Rousseau above indicates, here the function of reflections on human nature is twofold. "Human nature" does not refer only to a standard of legitimacy. It is also a reminder that normative projects must not go beyond human ability. Arguments from nature are used to justify and, at the same time, to specify and limit substantial rights positions. Regarding the range of possible arguments from human nature, I will start out with some clarifying remarks (Sect. 10.2).

J. Reinhardt (✉)
Faculty of Law, University of Hamburg, Rothenbaumchaussee 33, 20148 Hamburg, Germany
e-mail: joern.reinhardt@uni-hamburg.de

M. Albers et al. (eds.), *Human Rights and Human Nature*, Ius Gentium: Comparative Perspectives on Law and Justice 35, DOI 10.1007/978-94-017-8672-0_10,

For the sake of clarification, I will make use of the common distinction between "classic" and "modern" natural right concepts and distinguish them further from contemporary naturalistic approaches. It is no accident that questions of feasibility became particularly important for the modern natural rights tradition. To the extent that the dimensions of "is" and "ought" are differentiated, the relation of "ought" and "can" needs clarification. While classic natural law theory could still make use of a perfectionist understanding of human nature and a rather stable anthropocentric teleology, the teleological premises became questionable in the course of critical enlightenment thinking.

The role that human nature plays in Kant's concept of right is exemplary in this respect (Sect. 10.3). Kant offers a reason-based justification for the principle of right. At the same time, he makes use of a teleological perspective on human nature. From the *Idea for a Universal History with a Cosmopolitan Aim* onward, Kant elaborates on a philosophy of history that draws on the idea of a "natural goal" or a "natural plan" (*Naturabsicht*) in order to support his constitutionalist aspirations. It serves as a self-assertion that the concept will eventually be brought about. Even though arguments from nature are widely used to deal with the problem, it became evident that the answers were to be found not so much in human nature, but in society and history. With these reflections Kant points to the space, as it were, that will eventually be taken up by social theory.

The feasibility issue is also central to the contemporary discourse on human rights (Sect. 10.4). The ideas of a "natural right" (in its many variations) and post-1945 human rights are two quite distinct phenomena. However, the problem that modern natural rights theory treated as a question of human nature and abilities reappears in the context of contemporary debates on human rights and international constitutionalism. A central premise in both discourses is that a concept of rights must be realistic (or rather realizable). Here, too, the debates on political feasibility, realism, and utopianism veer away from the "last questions" often linked to meditations on human nature and treat the problem in a more piecemeal and pragmatic way. Apparently the feasibility issue requires a more complex set of argument than the recourse on "human nature" can provide. Nevertheless, characteristic traits of the argument from human nature prevail in the debate over what is sometimes referred to now as the "realistic utopia" of human rights.

10.2 From Natural Right Theory to Naturalism

Notwithstanding the scientistic connotations that the term 'natural' has adopted in the context of modern natural science, 'naturalistic' conceptions of human rights inherit their main features from natural rights in the classic and modern natural right tradition (cf. Beitz 2009, 48–72). Despite the common intuition, the tradition itself is heterogeneous to a degree that it is not clear as to whether all the relevant, more or less related concepts can be traced to a core of common assumptions. Since the concept of a natural right is often presented in rough and distorting sketches, which portray it as something completely incompatible with modern legal thought, I will first try to exclude some unproductive interpretations. Exclusion does not mean that these interpretations are unmotivated or that there is no evidence to be found in the vast body of material detailing natural right thinking. But they appear as neither essential for nor particularly characteristic of the concept. First, natural right theories do not necessarily claim that a conflict with the natural law renders the positive law invalid. The natural right sets a standard of legitimacy for the political order. But this standard has no necessary implications for the law's validity. It is also important to note that the idea of a natural law, already in its classic variations, is not identical to the conventional way of life of the community. It does not rely on the idea that all men and women agree upon the right path either. On the contrary, reason is required to perceive it. Therefore, everyone will not necessarily know the natural law and mere authority cannot determine its content (Strauss 1953, 84). The idea of a "natural right" provides the conceptual space necessary to articulate a difference that can hardly be ignored. It is always possible (and meaningful) to ask whether the positive law is right or wrong, whether it is legitimate or not. Judgments of this kind refer to a standard that is not identical to the positive law as it is. In the language of "natural right", demands on the social and political order can be formulated. It is a starting point for critique. It can come in a revolutionary or in a gradual reformatory mode. It has a foundational as well as a subversive component.

Finally, the natural right does not need to be understood as unchangeable and ahistoric, even though it is often presented that way. In fact, the natural right standard can be articulated in various ways. It was the ambition of dogmatic metaphysics to bring these standards into a coherent system that gave rise to this idea. But the interpretations that are closer to the Aristotelian origins of the idea of natural right allow for changeability and variation.[1] Natural rights are standards that every political system must respect (or promote). Therefore, the idea of a natural right is related to (but obviously not coextensive with) the modern idea of human rights.

10.2.1 Classic and Modern Natural Right Theory

Important differences in the natural right tradition are often brought under the rubric of both classic and modern natural rights approaches. The distinction is

[1] Cf. Gadamer's reading of the Aristotelian understanding of natural right as an essentially "critical" concept (Gadamer 1960, 316–317). Cf. further Joachim Ritter's interpretation of Eth. Nic. V 1134 (Ritter 1963, 157–158). See also Strauss (1953, 156–159). In contrast, Leo Strauss has apparent difficulties with the Aristotelian idea that the natural right is changeable, preferring instead the Thomistic model: "The Thomistic doctrine of natural right [...] is free from the hesitations and ambiguities which are characteristic of the teachings, not only of Plato and Cicero, but of Aristotle as well. No doubt is left [...] regarding the immutable character of the fundamental propositions of natural law" (Strauss 1953, 163). For the various threads of the natural right tradition compare now Contreras (2013).

the result of a generalization in order to highlight certain important conceptual differences. It must be understood more in an analytic than a historiographic sense.[2] These differences concern the different models of political freedom, the role of the individual, and the basic methodological question of how to identify the human species in its "natural" state.

What is characteristic for natural right theories in their classic variations is a comprehensive conception of human life. The idea of a natural right is derived from the natural law that signifies – in its most abstract designation – a human life-form that is "in order".[3] Since the human life-form is – in contrast to other animals – characterized by its capacity for reason, the cultivation of reason is considered to be an essential aspect for the development of human nature. As human nature can only be realized within the social and political order, the polity or *polis* is of particular importance. "Reason" and "polis" are the main coordinates of this thinking. While the latter refers to the actual constitution, the mode of life and the actual practice and conduct as the starting point of natural right thinking, the former indicates that it cannot remain there. The human life-form provides the standards that are essential for individual self-fulfillment and intact social relations, but practical reasoning goes beyond their contingencies and possible deficits. Political theory as *practical philosophy* is a reconstruction of actual practices and practical realities. The relevant sources of normativity lie in actual human practices. Therefore, it was important for Aristotle, in thinking about the right structure of the polis, not to start from scratch, but instead from the actual constitutions and written laws of the Greek world (see Ritter 1963, 148–52 with reference to Eth. Nic. X 1181 b 6 and 17). The standards are not deduced from an abstract notion of human nature. As a practical philosophy (in the Aristotelian sense), political theory has to develop its principles from the given political situations.[4] However, the approach must not be confounded by a mere comparative approach that chooses the best among the existing concepts. It includes a critical evaluation of the established practices and does not result in the acceptance of the given situation.

---

[2]For Leo Strauss, it was the emergence of modern natural science that lead to a highly visible break with tradition. The replacement of a teleological by a materialistic worldview had a profound impact on social and political life. According to Strauss, the protagonist "who was the first to draw the consequences for natural rights from this momentous change was Thomas Hobbes" (Strauss 1953, 166). One of these consequences was a turn toward an "individualistic" or subjective conception. Michel Villey sees already in William of Ockham the advent of the modern notion of a "subjective right" in contrast to the Aristotelian-Thomistic variation of classic natural law (Villey 1964; cf. also B. Tierney's objections to Villey's reading of Ockham in Tierney 1997, 18).

[3]Therefore, it is disputed whether the classical tradition had the concept of a "right" at all. The idea that natural rights are derived from the natural law is deeply rooted in natural rights thinking (cf. Finnis 2002). However, the term refers to a state of affairs that is "right" rather than to individual entitlements in the modern sense of the term right (for a critique of the idea that subjective rights can be extracted from the natural law, see e.g., Villey 1983).

[4]This methodological starting point is not self-evident for the classic concept of natural right. It differs already from the Platonic approach, and even within the tradition that explicitly refers to Aristotle it has been lost by the time in scholarly orthodoxy.

Classic natural law concepts – to a wider or lesser degree – presuppose a "perfectionist" understanding of human nature. While the standards (or basic goods) might vary with the respective understanding of the natural right, the assumption is common that non-compliance with these standards leads to distortions in social life.[5] The anthropocentric teleology has two important implications: On the one hand, human nature appears as a source of normativity from which authoritative standards can be derived. On the other hand, it ensures that these normative standards will eventually be realized. Of course, classic natural right conceptions did not assume that the world would develop according to philosophical thinking. Discrepancies were obvious and had to be dealt with. But the presupposition of a teleological structure of human nature was a powerful scheme through which these discrepancies could be understood and explained.

The formation of critical modern thinking has called the teleological premises into question. In the absence of a natural teleology not only does the status of arguments from human nature change, but the feasibility issue becomes relevant. On the assumption of an anthropocentric teleology the question of feasibility could hardly arise. But if it is essentially up to human efforts to bring about a social or political order, then the question of whether it is realizable or not becomes an essential component for the justification of a political conception. Therefore, the problem of feasibility is high on the agenda of modern natural right thinking.

Modern natural right theories still rest on the knowledge of human nature, and the term nature refers to what is not a mere product of practical reasoning, but its condition. Yet this condition is now interpreted in a different way. "Nature" does not refer to the perfect *form* of human life, but the imperfect *matter* that has to be dealt with in imposing the form upon the world (cf. also Korsgaard 1996, 3–5). The human condition, natural impulses, and the natural dynamic of the human species are understood as the relevant factors in the process of realizing the form. The notion of human nature is not only detached from the metaphysical presuppositions, but also from the social and institutional embedding. Human nature refers to the "state of nature" as a state of life anteceding civil society, or, more or less, independent of social relations and political institutions.

The outcome of these reflections could not be more divergent. The two extremes are probably Hobbes and Rousseau. In the outline of the *Social Contract*, Rousseau holds that a community of equals can dispense with a dimension of fundamental rights altogether ("aliénation totale"); Hobbes, on the contrary, insists on fundamental guarantees against the sovereign but reduces these to a core concept that could hardly be more minimalist. Hobbes' narrow notion is already an expression of a reflection on the problem of feasibility (cf. Strauss 1953, 182–187). In contrast to the perfectionist interpretations of human nature, he explicates the idea of a natural

[5] Whether this idea presupposes that an "ought" is build into the constituents of human nature, is controversial, cf. Finnis (2002).

right from the natural impulse of self-preservation.[6] This narrow notion of human nature leads to a narrow understanding of the natural right. The result is a core concept of a defensive right. Even though contractualist on the outset, participation in political power is not even within the horizon of the concept. The motivation for this restriction is methodological: The standard is deliberately lowered to make it more attainable. The less ambitious the anthropological premise, the more probable that the concept will eventually be brought about. But if human nature is understood along these lines, one seems to gradually cut off the resources needed to formulate a coherent concept of rights altogether. Not that the animal conditions of the human being, bodily needs, and the capacity for suffering could not play a role in justifying subjective rights positions, but there is no direct way from a "disenchanted" nature to the concept of a right.

James Griffin observes anxiously:

> Natural law began as part of a teleological metaphysics capable of supporting strong interpretations of how morality is rooted in nature, and it ended up at the close of the eighteenth century in something approaching vacuity. It is not that the strong, non-vacuous conceptions of natural law do not have their own considerable problems. Still, many scholastic conceptions of natural law gave us at least *something* to go on in deciding what natural rights there are. Once the metaphysical and epistemological background that they provided is abandoned, as it was in the course of the Enlightenment, what is left? Is enough left? [emphasis in the original] (Griffin 2008, 14).

Torn between these two alternatives, the issue of human rights and human nature seems to present itself, as Leo Strauss put it, "as a matter of party allegiance". On one side stand the "Catholic and non-Catholic disciples of Thomas Aquinas" (Strauss 1953, 7). On the other side stand those who have accepted the perspective of modern natural science all the way down, and are thus always in danger of cultivating nihilism as a practical consequence.

10.2.2 Arguments from "Second Nature" and Anthropology

Already the set up of this alternative is challenged by contemporary versions of naturalism and Neo-Aristotelianism. These accounts do not leave the concept of human nature to the natural sciences, but insist instead that human nature cannot be reduced to its physical determinants. Therefore, nature is largely re-incorporated into the realm of the *logos*. Naturalistic theories in this sense start from a broad

[6]This kind of methodological reservation finds an echo, for example, in Benthams's criticism of the guiding political principles of the French Revolution ("nonsense upon stilts") and in Burke's critique of the "spirit of speculation" that animated French revolutionary thinking (on the latter, see Strauss 1953, 294–304). For Kant, it is important that "even a race of devils, granted that they have intelligence" can bring about the constitutional idea. A concept of rights that is built on Machiavellian territory is more likely to be realized, even though it might – given the circumstances – remain an ambitious enterprise.

notion of human nature. On the basis of our natural capacity for moral experience and judgment, they develop an understanding of the capabilities and potentials of the human being that can provide arguments for general norms and standards including human rights.

The ontological background of these theories is a "second nature"-naturalism as introduced by John McDowell; McDowell's concept tries to avoid the fallacies of traditional empiricism ("myth of the given") without giving way to the suggestions of idealism or mere social constructivism. His central motivation is to develop a realistic account of intentionality and perceptual experience (McDowell 1998). The concept explicitly includes the perception of fundamental values and ethical properties of a situation or action. "Nature" on this account refers to what is not merely a product of our practical reasoning, but is instead its object, namely the fabric of the world, including the social world that makes our judgments true or false. But it begs the question of whether – as McDowell himself states – it is not merely the "rhetoric" of ethical realism that is being used.[7] Despite a different façade, Neo-Aristotelians seem to be in the grip of the very same difficulty that plagues all other positions. This difficulty begins when the immediate perceptions of something as good or valuable are called into question. Not that natural law theories could not account for diverse, plural, and even incommensurable interpretations of the basic good and the value and telos of the human life-form. As we have seen, natural law positions are not committed to the idea that each and everyone perceive the common good in the same way. But the repertoire of a "second nature"-account does not seem adequate to deal with the problem: If the self-evidence of propositions is called into question, the lost evidence cannot be restored by pointing to common experience, because this is what is missing. In the face of lacking accordance in the evaluation of even the most basic human rights issues (see e.g. Cassese 2012b, 138 on criminal law provisions in Sudan), a common understanding of the human life-form and its telos (on which an argument for human rights can be based) is hard to identify. This affects the presupposition of a shared and stable basis for our moral judgments. In the same way that perceptions and anthropological arguments

[7]Cf. McDowell (1998, 192): "To use the rhetoric of ethical realism, second nature acts in a world in which it finds more than what is open to view from the dehumanized stance that the natural sciences, rightly for their purposes, adopt. And there is nothing against bringing this richer reality under the rubric of nature too. The natural sciences do not have exclusive rights in that notion; and the added richness comes into view, not through the operations of some mysteriously extra-natural power, but because human beings come to possess a second nature". McDowell also makes use of the German word Bildung to characterize the process by which each human being achieves his or her "second nature" (McDowell 2003, 87). This reference shows the affinity McDowell's understanding has towards the hermeneutical tradition in general and Gadamer's work in particular. But the notion of nature is not among the guiding concepts used by Gadamer to "recover" the fundamental hermeneutic problem. The term refers merely to the "nature of things" or the "nature of the subject matter" (*Natur der Sache*), which is something completely different. In order to demarcate a naturalistic approach from competing, e.g., constructivist, explanations it must be made clear as to why this development is an expression of human nature and not a cultural achievement or even a repression of nature.

lose their status as evidence once they are seriously contested.[8] As an essentially contested category human nature is not a stable source of normativity anymore. As Charles Beitz correctly observed, "inferences drawn from naturalistic conceptions about the contents and basis of international human rights are normative positions requiring a defense; it is a mistake to regard them as analytic" (Beitz 2009, 72).

This does not mean that arguments from nature are irrelevant for the idea of human rights and the defense of particular human rights positions. Claims for the recognition of human rights rely on ideas of the value of the human person and draw on ideas of the integrity of the human life-form as it is spelled out by anthropology in general and its philosophical variations in particular. But anthropological findings – even if they are widely shared – cannot be used without mediation for the justification and interpretation of fundamental rights. Rather they must be understood as political claims in need of a political justification. For example, if anthropological findings suggest that working is an essential aspect of individual self-fulfillment, then it does not follow that there is or should be a right to work. The extent to which anthropological arguments can be transferred into a subjective right (evoking a corresponding obligation) is a juridical-political question and no longer a question of anthropology; part of the juridical-political justification is to provide answers to the question of feasibility.

10.3 *The Role of Nature in Kant's Concept of Right: Social Theory* avant la lettre

In order to further clarify the interdependency of the justification of rights and feasibility considerations, I will now turn to the role of nature in Kant's concept of right.[9] Despite his reason-based approach, feasibility considerations shape the form and content of Kant's concept of right. Because Kant cannot entirely separate the justification of a concept of right from the question of realizability, he makes use of human nature as an argument to support the claim to feasibility.

[8]The different anthropological accounts do not offer a "neutral" access to human nature, but are expressions of a certain ethos. On the contribution of (empirical) anthropology to the human rights discourse and the role and activities of the American Anthropological Association, see Goodale (2009).

[9]I refer to the volume (Roman) and page numbers of the Prussian Academy Edition (Kant 1902) and of the respective translation (Arabic numeral). For Kant's Idea for a Universal History, I used the Allison translation in Rorty and Schmidt (2009). The other translations are taken from Reiss (1970). The following abbreviations are used to refer to specific works: PP – Toward Perpetual Peace; Idea – Idea for a Universal History with a Cosmopolitan Aim; ThP – On the Common Saying: That May be True in Theory, but is of No Use in Practice; RL – Metaphysics of Morals, Part I ("*Rechtslehre*"); CJ – Critique of the Power of Judgment; CoF – The Contest of Faculties; Religion – Religion within the Limits of Reason Alone.

Like most of his enlightenment contemporaries, Kant's aim was to detach natural law theory from dogmatic metaphysics and to re-establish it as a science. Since the very foundation of dogmatic metaphysics was challenged, simply renewing the untrustworthy sources of metaphysical speculation was not plausible. Neither was it possible to simply depart from tradition and example, since this approach was, to a large extent, brought against the established traditions and, moreover, positive examples were missing. In this situation, the recourse to reason was a useful invention. Sufficiently abstract, it could have appeared to be a promising medium through which political philosophy could be reoriented in order to arrive at valid legal and political principles. The *Rechtslehre* (the first part of *The Metaphysics of Morals*) introduces the universal principle of right as an expression of practical reason. According to this principle, "every action which by itself or by its maxim enables the freedom of each to coexist with the freedom of everyone else in accordance with a universal law is *right* [Recht]" (RL, VI, 230/133).[10] At the centre of this conception lies "the one sole original, inborn right belonging to every man in virtue of his humanity" (RL, VI, 237), the right of freedom insofar as it is compatible with the freedom of all others (autonomy).

Kant's concept of natural right is not based upon an idea of human nature, but on reason and rationality. That Kant still refers to it as a "natural" right cannot conceal that it is more a right of reason than a right of nature.[11] This qualification is necessary. It does not mean, however, that reflections on human nature and ability are of no importance to Kant.

The Kantian concept of rights contains important specifications that are already an expression of feasibility considerations. This concerns the content of the right and the institutional framework in which the right is to be realized. The understanding of freedom that is characteristic for the modern natural rights tradition differs from the ancient notion of freedom. The autonomy it promises is different from the notion of the classics. It contains an irreducible dimension of negative freedom and, therefore, implicitly accepts a fundamental change in the formation of society.[12] With the emphasis on subjective rights, Kant completes a central change from classic to modern natural right, "the shift from the metaphysics of natural *law* to that of natural *rights*" (Haakonsen 2006, 280). Kant insists on the difference between legality and morality, leaving individual motives and motivations, to a large extent, aside. Even though the structural similarities of the categorical imperative and the principle of

[10]The German noun Recht can mean either "law" or "right", and Kant often uses the word in both senses. In fact, they refer to each other. Subjective rights can only be guaranteed in a state of law, and the positive law must not contradict essential rights in order to be legitimate.

[11]Cf. only the extensive use that Kant makes of the language of natural right theory at the beginning of *Rechtslehre*. The term "natural" signifies only that the main principle can be recognized a priori by reason (see Kant's remarks in RL VI 237). The "state of nature" is something that is to be left behind: Exeundum esse e statu naturali (cf. Religion, VI, 97). On the reception and critique of a natural right by Kant, the transformation of a natural right in a law of reason and its implications, see already Wellmer (1998, 85–106).

[12]For a sociological reconstruction of the natural law tradition, see Thornhill (2013).

law are obvious, because both originate from practical reason Kant does not derive his conception of rights from moral theory. The idea of autonomy is central not only for morality but also for the principle of law, but it takes on a very different form. It is only concerned with the compatibility of the outer spheres of individual freedom. The regulative idea of the law is not the "kingdom of ends" (cf. Reinhardt 2009, 57–78). Consequently, it does not promise happiness, only peace. It might not be as attractive as other enlightenment ideas of emancipation, but it is, according to Kant, realistic.

Feasibility considerations also affect the institutional context for the realization of the concept of right. The basic right is not understood as a human right in an international sense, but as a domestic right. It does not mean that Kant renounces the claim to universality. The formulation of the right is absolute, but as to the possible options that exist to protect the fundamental right, Kant opts for the domestic and not the international possibility. He does not extend the demand for a constitutionalization beyond the nation state. Beyond the sphere of the nation state, individual rights come down to a right to visit.

Of course, this picture can be made much more complex. While Kant's modified account can escape some of the problematic aspects of dogmatic metaphysics, it certainly brings its own difficulties. These difficulties concern the question of whether a "universal principle of right" can be justified through a reason-based approach and – if presented as a mere requirement of reason – whether it has sufficient content. Regardless of these difficulties, however, Kant could have left it at that. According to his own premises, he had deduced a universal normative standard of legitimacy that every legal system must respect and that is neither affected by the non-compliance of empirical governments nor the actual course of history. He could have treated the problem of whether this normative standard will actually be realized or not as an open empirical question.

However, besides this reason-based deduction, Kant is following a very different line of argument. From his early articles in the *Berlinische Monatsschrift* onward, Kant tries to sketch a universal history along the lines of a natural teleology, portraying mankind in progress. It was the contrast, or rather the gap, between theory and practice that prompted him to deal in further detail with human nature and the course of history. The discussions of the relation of 'ought' with 'can' show how much Kant was aware that doubts concerning the viability of political and legal concepts can easily lead to a disintegration of the normative concept itself.[13] The appeal to nature is supposed to "stabilize" the normative expectations in the face of constant backlashes and disappointments.

The idea of a natural teleology is significant for Kant's theory of rights because it provides an idea of progress in terms of a possible development of mankind towards a civil state of law. Kant attempts to understand the conflicted development towards

[13]For the different versions of the principle 'ought implies can' in Kant's writings see Stern (2004, 52–55). Stern favors a weak reading of the principle.

a republican constitution as the unfolding of a natural potential.[14] The notion of nature that Kant relies on does not denote the natural disposition to act reasonably. The driving force behind this movement is the natural instinct of self-preservation. The medium of progress is the "unsociable sociability" (Idea VIII 20/13) of human beings. Rivalry, greed, vanity, and antagonisms of all kind, even simple violence are supposed to further the slow development towards public or cosmopolitan law. In *Idea of a Universal History* nature operates as an "invisible hand" that harmonizes the destructive activities of individuals "behind their back". Kant portrays it as an unconscious, almost mechanical force.[15] Kant uses this argument in many variations and contexts, but the general theme is always the same: The destructive potential of human nature and the actual experience of destruction bring to mind how vitally important is the idea of a constitutional state that secures individual rights. Kant tries to detach the possibility of progress from individual reasoning, motivation, and an optimistic anthropology. This concept relies on the effectiveness of the institutional framework, and, as such, the force of institutions is a decisive factor for a positive development.[16] From this perspective, even the struggle of all against all can have a moralizing force. Out of mere self-interest, the members of a polity will eventually regulate their conflicts by establishing a republican constitution.

Kant's assumption is not that historical progress is a gradual process. He concedes that regression is always possible and that enlightenment and reason have to stay involved if the mechanism is take effect in the long run. But whether it reaches its potential or not remains dependent upon collective efforts and subjective dispositions.

Even with these reservations in mind, Kant's account of a nature-driven progress can easily look like an expression of pre-critical thinking. Already the word "idea" in the title of the article raises the question of how such a concept fits into Kant's critical philosophy (Allison 2008, 24). One has to remember that *Idea of a Universal History* is a relatively early work and Kant reconsidered his argument several times. As many commentators have noted, the work has to be examined through the lens of the Third Critique (see Allison 2008, 25; Yovel 1980). Read in this way, the idea

---

[14]In the Idea for a Universal History (1784), Kant develops this teleology from the overall premise that all natural predispositions of a creature "are determined sometime to develop themselves completely and purposively" (First Proposition; Idea – VIII, 18/11). According to the second proposition, those natural predispositions develop completely "only in the species, but not in the individual" (Idea – VIII, 18/11). The telos and at the same time the greatest problem for the human species "is the achievement of a civil society universally administering right" (Fifth Proposition; Idea – VIII, 22/14).

[15]Cf. PP VIII, 366: "But now nature comes to the aid of the universal, reason-derived will which, much as we honour it, is in practice powerless. And this she does, by means of these very self-seeking propensities." See also Idea VIII, 21/14: "The human being wills concord; but nature knows better what is good for his species: it wills discord."

[16]Especially in Perpetual Peace, Kant ascribes to institutions a constitutive role in realizing the common good, "for it only depends and so much lies within the power of man – on a good organization of the state" (PP VIII, 366).

of purposiveness is not a full-fledged metaphysical statement about the course of history. Instead, Kant takes up a view of history "as if" there was a teleology at work, a continuous development (though not free from reversals) towards a constitutional state. The presupposition of a natural teleology of human nature has come to an end within critical enlightenment thinking, but it reappears simultaneously as an applied idea for practical purposes. In the absence of a substantial anthropocentric teleology, the evolution of the law and the gradual constitutionalisation of the domestic and international sphere can only be consolidated in a continuous practice. The idea of historical progress is a narrative used to support it. This is a visible flaw in the idea of a "natural" teleology.

Still, the (counterfactual) reconstruction of the course of history along the lines of progress remains difficult. The difficulty of such a presupposition is that the actual course of history shows little evidence of such a *telos*. As Kant himself puts it:

> Since human beings in their endeavors do not behave merely instinctively, like animals, and yet also not on the whole like rational citizens of the world in accordance with an agreed upon plan (as e.g. of bees or of beavers) appears to be possible. One cannot resist feeling a certain indignation when one sees their doings and refraining on the great stage of the world (Idea VIII, 18/10).

This observation seems valid even today, more than 200 years after Kant's diagnosis and despite the development of international law and the "evolution" of human rights law.[17] The reconstruction of history along the lines of progress remains, to a considerable extent, auto-suggestive. As Kant suggests, the history of human rights-progress must be brought about, and writing the history is one way of doing just that. Kant explicates this mechanism in the ninth proposition of *Idea of Universal History*: "A philosophical attempt to work out a universal world history according to a plan of nature that aims at the perfect civil union of the human species, must be regarded as possible and even *as furthering* this aim of nature [emphasis added]" (*Idea* VIII 29/231).

The narrative of a natural teleology is supposed to further the development that it is simultaneously describing. But Kant is not advocating a mere exercise in speculative thinking. In order to not remain on the level of the "as if", a history of progress of humanity must "start from some sort of experience" (CoF VII, 84/181).[18] It needs support in the actual course of history. Therefore, Kant is

---

[17]Even though the protection of human rights has become a matter of the world community, one cannot ignore that these developments are fragile, deeply conflicted, and marked by enormous setbacks and inconsistencies (cf. only Pogge 2005). The attempt to write a history of human rights "from the stone age to the globalization era" (Ishay 2004) requires a lot of retroactive interpretation in which nonlinear events and inconsistencies are brought in line. On the function (and the seeming paradox) of the claim of "self-evidence" of human rights for the "invention" of the concept, see Hunt (2007).

[18]Otherwise the narrative will lack the enthusiasm that is "of considerable anthropological significance" (CoF VII 86/183). Identifying "constitutional moments" (B. Ackerman) is also an important element for forming the various narratives of international constitutionalism.

looking for "historical signs" (CoF VII, 84/181), like the French Revolution, that could support and further the narrative of a moral tendency in human history. The normative criteria and the narrative of progress refer to each other: the historical signs are never unequivocal and they can only be identified in light of the normative perspective that they are supposed to support. In a circular process the description of the possible development will eventually help to bring it about. This theme of a speculation that is never fully counterbalanced by reason is not only of importance for Kant's critical thinking; it is of equal importance for contemporary debates on international constitutionalism and possible developments in international human rights law.

10.4 Realizing Human Rights

Human rights, which are largely a post-1945 phenomenon, have subjected international law to the perspective of the individual. After the disintegration of all broader promises, arguably the "last utopia" (Moyn 2010), human rights have gradually transformed the parameters of transnational and international law. Even though it would be an oversimplification to place the modern concept of human rights in an unbroken continuum of what has been the natural rights tradition, the concepts share important structural features. Human rights have become the language with which to articulate experiences of injustice and, at the same time, claims for a political order that benefits the individual human being. They occupy to a large extent the idealist dimension of international law. The project of mastering power through law is in and of itself an ambitious enterprise, but the idea of human rights points beyond the conventional approach to "civilize" the conduct of sovereign nation states by binding them to general rules. Human rights take a prominent place in most long-term projects of global or international constitutionalism (Gardbaum 2008). These projects aim to realize a legal order that has not been realized before. In some form or another, they are engaged in imaginative thinking (Cassese 2012a, XXI; Delmas-Marty 2004).

At the same time, feasibility concerns are relevant. Human rights transgress the sphere of the sovereign nation state. The step beyond the nation state is as necessary as it is difficult. States were not only responsible for the most severe infringements of individual rights, but could also guarantee these rights as domestic rights. For their effective realization, human rights presuppose institutional conditions that to a large extent have to be brought about.

In the following I will try to analyze in further detail how feasibility concerns apply to human rights and to what extent arguments from human nature are relevant to specify them. In order to do so, I will first have to explicate what it means to "realize" human rights. If the feasibility requirement – as I have suggested so far – is recognized as the requirement that the respective conception of rights must be "realizable", then it begs the question of how this requirement fits with normative concepts such as rights, in particular, human rights, because human rights will – in important aspects – always remain unrealized. Therefore, the notion of feasibility itself needs specification.

Subsequently, I shall turn to the methodological implications. An institutional agenda that is to a certain degree idealistic or even utopian will have to take the realist objections into account systematically. The advocacy for ambitious ideas and the call for institutional reforms necessarily go along with reflections on the method of norm-building. The notion of a "realistic utopia" is – often tentatively – used to describe this perspective whereas human nature arguments are used to sharpen it.

10.4.1 Human Rights Unrealized

As Christian Tomuschat notes, "human rights protection is a process which cannot be finalized once and for all at some point in time" (Tomuschat 2003, 319). This structural incompleteness of human rights has two aspects. On the one hand, human rights protection cannot be finalized because a violation of human rights remains a permanent threat. As a consequence, the effective realization of human rights remains a continuous task. On the other hand, the idea of human rights cannot be finalized because the concept, as such, is incomplete. What has been said above about the idea of natural right also applies to human rights. Human rights discourse provides a language that makes it possible to articulate claims for basic standards and a legitimate social order. It is a "re-opener" of the necessary formal closures inherent to national and international law (cf. Reinhardt 2009, 41–47; 250–252). In this process a reformulation and specification of particular human rights and even a revision of central aspects of the concept are always possible.

Beyond these structural reservations the "realization" of human rights is measured in degrees. First, it is important to note that human rights in a juridical sense apply only directly against government actors or actors exercising public authority. Even though the language of human rights is increasingly applied within moral discourse, it does not capture the common juridical sense of the concept. That does not mean that human rights do not regulate relations between private actors, but they do so only indirectly, namely insofar as they impose duties on governments to enact legislature or other measures ("horizontal effect"). Of course, this view is not undisputed, and the attempts to broaden the picture or, rather, to reframe it, are perfectly understandable.[19] Still one has to remember that this concept of human rights "emerged as a minimalist, hardy utopia that could survive in harsh climate" (Moyn 2010, 121). Alternative proposals will have to stand the same test.

---

[19] For the idea of human rights protection through binding transnational law or "societal" constitutionalism, see Teubner (2012, 124–149) with further references.

In order to come to terms with the feasibility problem, it is important to further distinguish between different dimensions of human rights. Despite the affirmation that human rights are "interrelated, interdependent, and indivisible" (reaffirmed at the 1993 Vienna World Conference), there are obvious differences – regarding the political feasibility – between civil liberties that grant negative freedoms and social, welfare, or labour rights. The realization of the latter is more ambitious. That does not disqualify them as human rights. Especially limitations of economic resources alone are not arguments against the human rights character of the respective positions (cf. for the discussion Griffin 2008, 176–189). But a right is an entitlement. It gives rise to a corresponding obligation. Therefore, the question must be answered who can be held liable and who shall protect the right effectively against possible infringements (cf. Alston 2005 for the discussion on labour rights).

Realizing rights requires effective remedies against infringement as well as a functioning legal framework and enforcement mechanisms. Of course, a right can be formally adopted and incorporated into a charter or multilateral treaty. But the formal incorporation of a right in the body of law is only a condition necessary for its realization, but is not itself sufficient. If it remains without noticeable consequences, then it will only nurture the persistent doubt as to whether international human rights law makes a difference at all. It is just the gap between international human rights treaty law and human rights reality that is often the starting point of a fundamental critique of the whole idea of international human rights (cf. Burchill 2008). Even though questions of feasibility and justification have to be distinguished, they cannot be treated independently. There might be reasonable grounds to deny a position the status of a human right for reasons other than infeasibility. But failures of feasibility impact the possible justification of a right. Whether a concept of human rights is politically feasible is not merely an empirical aspect that leaves the justification as it is.[20]

The dilemma of international human rights results from the limits of a state-centered legal order and national governance on the one hand and the insufficiency of existing mechanisms to fill this gap on the other. Apart from the different dimensions of rights, it is important to distinguish between the different levels from which human rights can be realized. On each of the various levels (national, transnational, international), particular enforcement problems arise.

On the international level, the possible enforcement mechanisms differ to a large degree from those available at the nation state level. Further, the degrees to which the rights are enforced diverge greatly (ranging from a committee's competence, or lack thereof, to make only general and non-binding comments after the examination of national reports to the competence of courts to make legally binding and enforceable decisions to address individual petitions or complaints). Despite the monitoring and reporting systems that examine to what extent the treaty parties comply with their obligations, the UN human rights treaty system presupposes that the member

[20]Consequently, Charles Beitz treats the objection of infeasibility as "another kind of skepticism" (c.f. Beitz 2009, 3–4).

states will eventually incorporate the agreed upon standards into domestic law.[21] The reluctance of states to fully implement their treaty law obligations (or to ratify the respective treaty in the first place) lastingly hinders the effective realization of human rights.

The functioning nation state ensures human rights as domestic rights vis-à-vis its own citizens. It can extend its jurisdiction to non-citizens, but this extension often proves problematic.[22] Not only are there obvious enforcement difficulties, but even if sanctions can be applied effectively, the particularity of the enforcement measures is likely to contravene the universality of the human right in question.

The ambivalence of human rights enforcement often provokes two opposite reactions. Human rights advocates take the existing shortcomings and institutional deficits as an incentive to reform in order to bring the given situation closer to the ideal of human rights. They point to the *potential* inherent in the emerging forms of the global legal order.[23] Human rights skeptics, on the contrary, understand the deficits as an expression of the misconception of the respective idea of international human rights. For them, the allusion to potential and possible developments is no more than an uncovered check on the future.[24]

---

[21]More robust reactions towards human rights violations such as the collective use of force or international criminal law measures (understood as a "subset of international human rights", c.f. Stacy 2009, 171) only apply to large scale human rights violations and are often highly selective. Helen Stacy distinguishes three sources of shortcomings: domestic implementation problems, international bureaucracy problems, and credibility problems resulting from a cultural critique (Stacy 2009, 37–75).

[22]The latitude of states to address international human rights concerns in a domestic context is highly contested. A possible way of addressing the problem of world-wide human rights infringements from a domestic starting point are instruments like the United States Alien Tort Statute ("ATS"). The statute gives non-US citizens the possibility to ask for remedy in cases of human rights violations before U.S. federal courts. Two recent decisions in *Kiobel* have limited its scope significantly. The Court of Appeals for the Second Circuit excluded tort liability of companies involved in human rights violations holding, inter alia, that private companies are not directly bound by international law (Kiobel v. Royal Dutch Petroleum Co., 621 F.3d 111 (2d Cir. 2010)). The decision was upheld by the Supreme Court on April 17, 2013 (Kiobel v. Royal Dutch Petroleum Co, 569 U.S. ___, 2013). The majority ruled further that the ATS does not apply to human rights violations committed on foreign territory relying on a general "presumption against extraterritorial application" (*Id.* at 4).

[23]See Cassese (2012a, XX): "brought to fruition *in the next two or three decades*" (emphasis in the original).

[24]A similar divide characterizes the debate about the status quo and possible developments of international constitutionalism. While the constitutionalists are ready to abandon central elements of domestic constitutionalism to be able to make use of the concept on an international level, the opponents insist on the traditional interpretation and, therefore, abstain from describing and evaluating the ongoing developments in terms of constitutionalism. The latter do so for the sake of conceptual clarity and in order to mark the differences between mere legality and constitutionalism; consequently, they accuse the international constitutionalists of "revisionism." The former insists that normative concepts have to be developed and revised in light of changing conditions and respond with the reproach of "nostalgia" (cf. Kumm 2010, 201).

10.4.2 *The "Realistic Utopia" of Human Rights*

In order to further illuminate this controversy and the unsolved problem of feasibility, it is productive to turn to the concept of a "realistic utopia" that is increasingly deployed in political theory and in the context of international law (Rawls 2001, 12; Habermas 2010; Cassese 2012a). The concept is directed against an idealism that does not reflect upon or simply marginalizes the given conditions. Ideals, even attractive ones, might raise false expectations or even have negative or counterproductive effects if their feasibility is not taken into account (cf. Reinhardt 2007, 231–233). Therefore, a "realistic utopia" introduces a twofold perspective designed to express that the normative goals pursued are ambitious, but not utopian in a pejorative sense. Even though the goals are unlikely to be realized at the given moment, they are supposed to indicate a possible development because they are based on realistic assumptions.

In the debates on international constitutionalism and human rights, feasibility concerns are usually implicitly acknowledged, but they are hardly ever explicitly analyzed. It is widely acknowledged, for example, that the parameters of modern constitutions that have been formed in the context of the nation state cannot simply be implemented on a transnational or global level. They are usually reformulated or even replaced by functionalist equivalents. While it is conceded that constitutional aspirations have to be lowered in a global context, the degree to which this is necessary is highly disputed and the very terms of the dispute often remain unclear.

The concept of a "realistic utopia" tries to spell out the relationship of idealistic to realistic elements in more general terms. It was John Rawls who made the term popular, using it in *Political Liberalism* and *The Law of Peoples*. It is doubtful though, whether the choice of words was a fortunate one. From its original literal meaning, utopias point to the impossible ("no-where"). It seems only of limited use to demarcate a state of affairs that is, at the same time, supposed to be realistic and politically feasible (cf. also Cassese 2012a, XXI: "oxymoron"). Not that utopias could not have a critical dimension (cf. Forst 2013), but they are nothing that can "realistically" be expected to be achieved. However, the aim of the concept is clear. It affects both the method and the criteria employed by political theory and legal scholarship. Like in natural right discourse arguments from human nature come into consideration to specify the perspective.

10.4.2.1 A Question of Method: Interpretation vs. Invention

Methodological reservations concern the model of critique and the ways of norm-making. The demarcations that are being used to capture the conceptual differences are, notably, "inventive" versus "interpretative" or rather "constructivist" versus "reconstructivist" approaches.[25] Behind these simplified headings a bundle of complex problems awaits. While a constructivist approach, to a large extent, abstracts from the given conditions to arrive at justified principles for the social and political order,[26] a reconstructive model of critique departs from the historically formed institutional framework. Since the former necessarily retains an "ideal" dimension, it seems promising, in order to ensure the "realism" of the project, to begin in a more interpretative or hermeneutical manner. In this spirit, Martti Koskenniemi writes that a "realistic utopia can only begin with the critique of present institutions" (Koskenniemi 2012, 3). Instead of "inventing" human rights regimes and conceptions of international constitutionalism, a reconstructive critique formulates normative ideals in light of the structural problems and inadequacies of the present social and political order.

Despite the differences, neither of these models can escape the feasibility issue. As we have already seen in the discussion of Aristotle above, a normative reconstruction of actual practices and institutions necessarily goes beyond the status quo.[27] The different approaches might influence the degree of realism of the respective conception, but the problem of feasibility could only be avoided if the normative standards were completely adapted to the given conditions. Even a critique that starts from the present institutions cannot bind the interpretation to the state of affairs as it actually is. The point seems almost obvious in the context of human rights. The various claims to human rights are reactions towards experiences of injustice and humiliation. Hence, the desirable institutional arrangement may be quite different from the state of affairs as it stands. While trying to avoid expanding into the realm of mere wishful thinking, reconstructive models of critique also aim at extending the limits of practical political possibility – and are, eventually, lead back to the problem of whether the normative standards are realizable and whether they provide practical orientation.

[25]Michael Walzer distinguishes three distinct paths for critique: the paths of discovery, invention, interpretation (Walzer 1987, 3). On the differences between the various models of political and social criticism (construction versus reconstruction, interpretation versus invention, etc.), see also Honneth (2009). In his introduction to *Realizing Utopia*, Antonio Cassese characterizes the appropriate method of idealist international law scholarship as "critical positivism" (Cassese 2012a, XVII–XVIII). It is not clear, though, how the commitment to positivism establishes "solid ground" and pursues "realistic and viable avenues" (ibid). As an innovative legal scholar and "judicious reformer", one must necessarily move beyond positive law. On Cassese's approach and its limits see Feichtner (2012).

[26]Cf. John Rawls' attempt to write "ideal theory." Rawls introduces idealized procedural elements ("veil of ignorance," rational agency, etc.) to spell out principles of justice.

[27]For a detailed defense of the reconstructive or "interpretative" approach against the charge that it undercuts the very possibility of social criticism, see Walzer (1987).

10.4.2.2 From Human Nature to Social Science

Therefore, further criteria would be necessary to draw a line between an "abstract" and a "realistic" utopia. For conventional expositions of political realism arguments from human nature are central. According to such a conventional exposition (that is attributed to thinkers in the line of Machiavelli, Hobbes, or Schmitt), human nature is violent, humans act according to power interests, and humans cannot be made perfect. Relations between nation states are modeled along these same lines. While it might be possible to tame the destructive forces within the framework of the state, the relations beyond the state are understood as necessarily destructive. Nation states appear like individuals incapable of regulating their relations in the absence of a higher force. In *The Concept of the Political* Carl Schmitt claims that every political theory can be traced down to an "anthropological confession of faith":

> One could test all theories of state and political ideas according to their anthropology and thereby classify these as to whether they consciously or unconsciously presuppose man to be by nature evil or by nature good (Schmitt 2008 [1932], 58).

The former view is, according to Schmitt, the realistic view. The implications for political theory are obvious, especially for those with a universalistic ambition and the idea of human rights. Already the attempt to ameliorate the situation must appear as abstract and utopian.[28]

Since the categories of good and evil, the doctrine of the fall, and other theological dogmas are too airy to answer concrete questions of political feasibility, the discussion moved away from those "last questions" that C. Schmitt dwells upon. That does not mean, however, that one necessarily leaves considerations of human nature behind. John Rawls, for example, shares the assumption that arguments from human nature are highly relevant to account for the realism of a political conception. But he attempts to minimize the perception that the question of realism comes down to a matter of party allegiance, in other words a case of which side to choose. In order to provide criteria that an "utopia" must fulfill to be considered realistic, Rawls goes back to the quote from Rousseau's *Social Contract* according to which principles of government must take "men as they are and laws as they might be" (cf. Rawls 2001, 13). Just like Rousseau, his aim is to show that his concept of political justice is not only ambitious, but also realizable. The possibility that Rawls is aiming at "is not a mere logical possibility, but one that connects with the deep tendencies and inclinations of the social world" (Rawls 2001, 128). Regarding his aim, Rawls' criteria remain, somewhat colorless.[29] In the end, he offers a pragmatic solution of the feasibility problem. The assumption that "we can to a greater or lesser extent

---

[28]For a more complex and nuanced view of political "realists", especially of mid-century realist positions of thinkers like E.H. Carr and Morgenthau, see Scheuerman (2012).

[29]The first condition is that a political conception must rely on "the actual laws of nature" (Rawls 2001, 12). Second, that it must be "workable and applicable to ongoing political and social arrangements" (Rawls 2001, 13). But given that it is highly contestable as to how far the "laws of nature" extend and what it would mean for a concept to be "workable", for, as Rawls puts it

change political and social institutions and much else" (Rawls 2001, 12), is already a careful option for progress. Whether things will eventually move in the normatively envisaged direction is, according to Rawls, an open empirical question. Like his numerous critics, he stresses the importance of the institutional embedding of human efforts. Those who argue partly against Rawls (who is not only an advocate of a "realistic" utopia, but one of the favorite targets of a realist critique) and in favor of a "real-world approach" to political and legal theory (Brennan and Pettit 2005, 278) start out with more modest assumptions on human conduct, rationality, and individual dispositions for compliance to normative standards. Their approach is more "realistic", not in the sense that it is pessimistic or that it would categorically exclude the possibility of significant social or institutional change, but that it appears to phrase the ideal institutional setting in more moderate terms. As a consequence, it will, for example, include more incentive-compatible elements in the institutional design. Here it is not significant to deal in further detail with this discussion (that would lead us back to questions regarding method and the relation of ideal and nonideal theory). It is important to note only that the kind of argument remains relevant for the feasibility of human rights because it informs about the actual conduct of political actors and the way "how nations behave" (L. Henkin).

At the same time the limits of arguments from human nature become visible. Political theories that emphasize feasibility concerns equally emphasize the importance of taking the results of the social sciences (in a large understanding) into account (see Swift and White 2008; Feichtner 2012, 1155). However, the necessary integration of the social sciences must not lead to the supposition that the realism of a project could be deduced from its findings or justified in a strong sense with recourse to it. The various concepts of international constitutionalism and human rights rely on narratives that are complex enough to integrate possible setbacks. Therefore, they need to be understood as projects and as an essential part of a constitutionalist practice. They contain normative elements, but are not merely prescriptive in character. Rather they have a creative role. They are shaping the terms of the development to come and, by doing so, contemporaneously further the envisaged development.

10.5 Conclusion

Feasibility considerations have shaped natural right thinking from the very beginning and are of equal importance for contemporary human rights discourse. An essential element in the justification of fundamental rights as *legal* rights is to account for their feasibility. Modern natural right theory especially drew largely on arguments from human nature to substantiate their claims: the need for fundamental

himself, we can change the existing institutions and "much else", these criteria can hardly provide an orientation for deciding which standards are utopian in a problematic sense and which are not.

rights including their scope and content. In order to come to terms with the feasibility problems of international human rights, the conceptual scheme of human nature arguments had to be extended significantly. However, characteristic traits of the debate about human nature prevail. The concepts involved are – to a certain extent – auto-suggestive. They indicate the way towards constitutionalizing the world community and are, at the same time, part of the movement towards it.

References

Allison, H.E. 2008. Teleology and history in Kant: the critical foundations of Kant's philosophy of history. In *Kant's Idea for a universal history with a cosmopolitan aim. A critical guide*, ed. A. Rorty and J. Schmidt, 24–45. Cambridge: Cambridge University Press.

Alston, P. 2005. Labour rights as human rights: The not so happy state of the art. In *Labour rights as human rights*, ed. P. Alston, 1–24. Oxford: Oxford University Press.

Beitz, C. 2009. *The idea of human rights*. Oxford: Oxford University Press.

Brennan, G., and P. Pettit. 2005. The feasibility issue. In *The Oxford handbook of contemporary philosophy*, ed. F. Jackson and M. Smith, 258–279. Oxford: Oxford University Press.

Burchill, R. 2008. International human rights law: Struggling between Apology and Utopia. In *Human rights in crisis*, ed. A. Bullard. Aldershot: Ashgate Publishing Ltd.

Cassese, A. 2012a. *Realizing Utopia. The future of international law*. Oxford: Oxford University Press.

Cassese, A. 2012b. A plea for a global community grounded in a core of human rights. In *Realizing Utopia. The future of international law*, ed. A. Cassese, 136–143. Oxford: Oxford University Press.

Contreras, F.J. (ed.). 2013. *The threads of natural law. Unravelling a philosophical tradition*. Dordrecht: Springer.

Delmas-Marty, M. 2004. *Les forces imaginantes du droit (Tome 1): Le relatif et l'universel*. Paris: Éditions du Seuil.

Feichtner, I. 2012. Realizing Utopia through the practice of international law. *European Journal of International Law* 23(4): 1143–1157.

Finnis, J. 2002. Natural law: The classical tradition. In *The Oxford handbook of jurisprudence and philosophy of law*, ed. J. Coleman and S. Shapiro. Oxford: Oxford University Press.

Forst, R. 2013. Utopia and irony: On the normativity of a political philosophy of 'No-where'. In *Justification and critique: Towards a critical theory of politics*. Cambridge: Polity Press.

Gadamer, H.-G. 1960. *Truth and method*. London: Continuum. 2006.

Gardbaum, S. 2008. Human rights as international constitutional rights. *European Journal of International Law* 19(4): 749–768.

Goodale, M. 2009. *Surrendering to Utopia. An anthropology of human rights*. Stanford: Stanford University Press.

Griffin, J. 2008. *On human rights*. Oxford: Oxford University Press.

Haakonsen, K. 2006. German natural law. In *Cambridge history of eighteenth-century political thought*, ed. M. Goldie and R. Wokler, 251–290. Cambridge: Cambridge University Press.

Habermas, J. 2010. The concept of human dignity and the realistic Utopia of human rights. *Metaphilosophy* 41(4): 464–480.

Honneth, A. 2009. Reconstructive social criticism with a genealogical proviso: On the idea of 'critique' in the Frankfurt School. In *Pathologies of reason*, ed. A. Honneth, 43–53. New York: Columbia University Press.

Hunt, L. 2007. *Inventing human rights. A history*. London: Norton.

Ishay, M. 2004. *The history of human rights: From the stone age to the globalization era*. Berkeley: University of California Press.

Kant, I. 1902. *Kants gesammelte Schriften.* The Prussian Academy Edition. Berlin: De Gruyter. The English translations cited are listed separately.
Korsgaard, C. 1996. *The sources of normativity.* Cambridge: Cambridge University Press.
Koskenniemi, M. 2012. Projects of world community. In *Realizing Utopia. The future of international law*, ed. A. Cassese, 3–13. Oxford: Oxford University Press.
Kumm, M. 2010. The best of times and the worst of times: Between constitutional triumphalism and nostalgia. In *The twilight of constitutionalism?* ed. P. Dobner and M. Loughlin, 201–219. Oxford: Oxford University Press.
McDowell, J.H. 1998. Two sorts of naturalism. In *Mind, value, and reality*, 167–197. Cambridge, MA: Harvard University Press.
McDowell, J.H. 2003. *Mind and world.* Cambridge, MA: Harvard University Press.
Moyn, S. 2010. *The last Utopia: Human rights in history.* Cambridge, MA: The Belknap Press of Harvard University Press.
Pogge, T. 2005. Recognized and violated by international law: The human rights of the global poor. *Leiden Journal of International Law* 18(4): 717–745.
Rawls, J. 2001. *The law of peoples.* Cambridge, MA: Harvard University Press.
Reinhardt, J. 2007. Kontrafaktische Annahmen. In *Normativität und Rechtskritik*, ed. J. Bung, B. Valerius, and S. Ziemann, 226–237. Stuttgart: Franz Steiner Verlag.
Reinhardt, J. 2009. *Der Überschuss der Gerechtigkeit. Perspektiven der Kritik unter Bedingungen modernen Rechts.* Weilerswist: Velbrück Wissenschaft.
Reiss, H. (ed.). 1970. *Kant. Political Writings.* Trans. H.B. Nisbet. Cambridge: Cambridge University Press.
Ritter, J. 1963. ›Naturrecht‹ bei Aristoteles. Zum Problem einer Erneuerung des Naturrechts. In *Metaphysik und Politik*, ed. J. Ritter, 133–179. Frankfurt am Main: Suhrkamp, rev. ed. 2003.
Rorty, A., and J. Schmidt (eds.). 2009. *Kant's idea for a universal history with a cosmopolitan aim. A critical guide.* Cambridge: Cambridge University Press.
Rousseau, J.J. 1762. *Du Contrat Social ou Principes Du Droit Politique.* The translation is taken from M. Cranston, *The social contract.* London: Penguin Classics 2012 reprint.
Scheuerman, W.E. 2012. Reconsidering realism on rights. In *Philosophical dimensions of human rights*, ed. C. Coradetti, 45–62. Dordrecht: Springer.
Schmitt, C. 1932. *The concept of the political.* Chicago: University of Chicago Press. 2008.
Stacy, Helen M. 2009. *Human rights for the 21st century: Sovereignty, civil society, culture.* Stanford: Stanford University Press.
Stern, R. 2004. Does 'Ought' imply 'Can'? And did Kant think it does? *Utilitas* 16(1): 42–61.
Strauss, L. 1953. *Natural right and history.* Chicago: University of Chicago Press.
Swift, A., and S. White. 2008. Political theory, social science, and real politics. In *Political theory: methods and approaches*, ed. D. Leopold and M. Stears, 49–69. Oxford: Oxford University Press.
Teubner, G. 2012. *Constitutional fragments. Societal constitutionalism and globalization.* Oxford: Oxford University Press.
Thornhill, C. 2013. Natural law, state formation and the foundations of social theory. *Journal of Classical Sociology* 13(2): 197–221.
Tierney, B. 1997. *The idea of natural rights.* Atlanta: Scholars Press.
Tomuschat, C. 2003. *Human rights. Between idealism and realism.* Oxford: Oxford University Press.
Villey, M. 1964. La genèse du droit subjectif chez Guillaume d'Occam. *Archives de philosophie du droit* 9: 97–127.
Villey, M. 1983. *Le droit et les droits de l'homme.* Paris: PUF.
Walzer, M. 1987. *Interpretation and social criticism.* Cambridge, MA: Harvard University Press.
Wellmer, A. 1998. Natural right and practical reason: On the aporetic development of a problem in Kant, Hegel, and Marx. In *Endgames. The irreconcilable nature of modernity*, 77–133. Cambridge, MA: MIT Press.
Yovel, Y. 1980. *Kant and the philosophy of history.* Princeton: Princeton University Press.

Part II
Species Manipulation and the Transformation of Human Nature – Challenges to Human Rights

Chapter 11
How to Protect "Human Nature" – By Human Dignity, Human Rights or with "Species-Ethics" Argumentations?

Georg Lohmann

The developments of medical genetic engineering (e.g., Preimplantation Genetic Diagnosis (PGD), stem cell research, and therapeutic and, possibly, reproductive cloning) are steps on the way towards *possible* procedures of both negative and positive eugenics. Taken together these developments are characterized by a change in the methodical attitude towards human nature, and at the same time express a change in the evaluation of human nature. The methodical scientific attitudes towards the "inner nature" of humans become more and more aligned with those focusing on the outer nature; the "inner nature" also becomes conceptualized and methodically grasped as an objectified and purposefully manipulable nature. Thus, human nature loses its traditional, intrinsic value according to which it was never just seen as a means, but always also as an indisposable end in and of itself. It becomes a good that can be principally manipulated, utilized, and exploited for other purposes. With regard to this process of transformation I would like to ask the question: Could or should we protect human nature by human dignity and human rights, or do we need other normative standards?

As part of this vivid dispute, some liberal positions grant so-called liberal eugenics, while conservative positions deny them (Buchanan et al. 2000). In Germany, Jürgen Habermas, in particular, took a stand with this argument (Habermas 2003). I will (1) explain the basic concepts and test the hypothesis that human nature can be protected by human rights and the "untouchability" (*Unantastbarkeit*[1]) of human dignity, (2) introduce the argument by Habermas, who first distinguishes the new

[1]In German "unverletztlich" (inviolable) is different than "unantastbar"; therefore, here I use the literal translation "untouchable".

G. Lohmann (✉)
Institute for Philosophy, University of Magdeburg, Universitätsplatz 2,
39106 Magdeburg, Germany
e-mail: Georg.Lohmann@ovgu.de

M. Albers et al. (eds.), *Human Rights and Human Nature*, Ius Gentium: Comparative Perspectives on Law and Justice 35, DOI 10.1007/978-94-017-8672-0_11,

genetic manipulations from the traditional changes of human nature, to then (3) examine against this background three normative arguments for the undisposabilty (*Unverfügbarkeit*) of the human nature; (4) then compare Habermas' species-ethical argument for the undisposabilty of human nature, as I see it, to (5) similar "post-metaphysical" challenges like an intercultural understanding of human dignity as a justified principle of human rights.

11.1 Human Nature, Human Dignity, Human Rights

Although all three thematic notions talk of human beings, they do so in different ways: of the nature of man, of his dignity, and of his rights that he has, because and insofar as he is a human being. One might assume that this order expresses a certain sequence of justification and/or understanding. First, we need to know what the nature of man is, to then, second, be able to determine his dignity. We then, third, understand the dignity of man as a "basis" (justification basis or derivation basis) of human rights. Human rights would then, one might assume, mediated through human dignity, protect human nature. To me, such an order of justification and explication seems both too simple and inconsistent in substance.

People have neither dignity nor rights "naturally", but dignity and rights are meaningful constructs that are cultural and historical creations of man. It is exactly in these constructs that human beings have exceeded what they "naturally" "are." "Culture" is understood here as refinement of nature or semi-naturally predetermined. In the historic development of the human species we can identify the historical periods—and in retrospect make them comprehensible through historical narratives—in which such historical redefinitions of what is human or "human nature" have been conceptualized and configured through practices and institutions.

The term "nature" has a complex history during which its meaning changed many times.[2] Concerning our question, there are two basic meanings of the term (physis, natura) since ancient times that are of interest: (a) "consistency, essence", and (b) "becoming, growth, form."[3] In order to characterize the differences in the meanings, the first basic meaning of "nature" (as a designation of the nature of a thing or a fact) will hereinafter be referred to as "essential nature," while the second basic meaning of "nature" (as the name for everything that is not manmade and is encountered by him) will hereinafter be referred to as "grown nature" or "encountered nature".

Since the writings of Aristotle, "human nature" is understood to be an entanglement of these two basic meanings and I will follow this view; human nature

[2]Compare the still excellent artical "Natur" of Hager (1984).

[3]Cf. Ritter et al. (1984, col. 421); *Oxford dictionaries* also distinguishes two basic meanings: (1): "the phenomena of the physical world collectively, including plants, animals, the landscape, and other features and products of the earth, as opposed to humans or human creations"; (2): "the basic or inherent features, character, or qualities of something".

corresponds to the essential nature of a human being to relate to his encountered nature and the way in which he transforms and creates it according to his purposes. As given and grown nature, human nature is, for human beings in one respect, something non-made and unavailable. In another respect, however, it is something he designs and which he can command according to his interests. As with in this second respect, and as Karl Marx would agree, he is referring as well to both its "external" and its "internal" nature. Nature is something that determines, in a limiting as well as in an enabling sense, him in its cultural achievements and social conditions. In this view of human, existing nature makes itself recognizable in two ways: (a) as enabling human action and experience, and (b) as limitation and restriction of human activity. By that the (inner) human nature causes or determines, on the one hand, the special need of humans, rooted in their respective bodiliness and physicalness, and the related vulnerability and need for protection. On the other hand, it allows (in confrontation with external nature) the development of his special abilities, especially his ability to a deliberate, communicable self-determination and, thus, the creation of human culture.

In the course of human development both the enabling and the limiting aspects of human nature changed. The "essential human nature" gets a socially mediated time index; it appears, although indispensable in a given situation, as designed and modifiable (see the convincing article Habermas 1958). We then talk of the *natural history* of the human being that we can distinguish from the *histories* (plural) of humans and of the *history of all humans*. The later ones have been brought about by people through their intentional, individual actions, which then, however, have cumulated in unplanned interactions and experiences and are comprehensible in the form of *narrative* history.

In this historical process human rights protect and enable the appreciation of each individual person in an equal, inalienable, and categorical manner that counts for all people as it is contained in the concept of "human dignity". These are essentially juridical rights that first oblige, in the form of "legal rights" of the individual, the respective state, then oblige all the states and, finally, oblige the external behavior of human beings indirectly and mediated through a 'horizontal effect'. Thereby for their legal-normative assumptions they claim moral justifiability and for their legal settlement and practical enforcement they need policy decisions by legitimate legislative bodies (cf. Lohmann 2010b for more details). From these complex characteristics of "human dignity" and "human rights" follows that their respective relation to "human nature" must also be dealt with from a moral, legal, and political-historical perspective. I will first consider the question of the title from a legal perspective.

11.2 On the Human Rights Protection of "Human Dignity"

Only since the Second World War has the concept of "human dignity" as used in international human rights documents taken on a new (only gradually clarified) constitutive role for human rights. The concept of human dignity I speak of here stems from the international response to barbaric and horrific atrocities, especially the "crimes against humanity" perpetrated by the Nazis (see Lohmann 2011b). This post war idea of "human dignity"—initially defined as a legal notion—declares humans self-understanding and recognizes that every person is a bearer of essential human rights. Before 1945, the ascribing or possessing "dignity" was associated with meeting obligations (towards others and towards oneself) but not with having rights. When this new aspect to the meaning of "human dignity" is combined with the traditional substantive designations of "dignity" the result is understood as appreciation of freedom (= deliberate self-determination) of the human being, his self-esteem (as awareness of the recognition of equal status with all human beings), and appreciation of his claim to a "dignified life" (as protection and fulfillment of his basic needs) (see Lohmann 2011a). Therefore, "human dignity" is, just as the concept of "human rights", universal, egalitarian, categorical, and individual.

"Human dignity" functions as a pre-constitutional principle justifying human rights. At the same time, the protection of human being by human rights refers to his dignity and to a human life in dignity. The bearer and object of protection by human rights and point of reference of human dignity is, therefore, always a single human being in his given body and physicality, and ultimately his given personal individuality. Constitutionally, this (strange and complicated) legally defined concept of "respect" (understood as both esteem and respect) of the individual is in the German Basic Law (Grundgesetz für die Bundesrepublik Deutschland = GG) the "untouchability" of "human dignity" ("Human dignity is untouchable", Article 1 GG).

Of course no single human being exists without a variety of social relationships with other people, and he only lives in the manifold interaction with nature. But neither are communities constitutive for the individual, nor is nature subject to the legal protection of human rights and human dignity. Communities and nature as conditions of human life are included and mediated only through the protection of the bearer of "individual rights" and through the individuality of the justifying human dignity (see Lohmann 2012b, 2013).

So that would answer our initial question in a certain respect: Through human dignity and human rights a single person has its irreplaceable (personal) individuality protected, not a (general) human nature. Only his *personal* human dignity is "untouchable". Nevertheless, one may wonder whether this does not also protect human nature as a natural condition of a humane life.

Historically "human nature" has always been exposed to processes of historical change; these were, however, not carried out purposefully as such, but were the effects of cultural development. Nowadays, however, genetic engineering and other biomedical research (via Preimplantation Genetic Diagnosis, PGD, stem cell research, cloning, etc.) allow us to change people's natural constitution, yes to "breed" them. And it is the possibility of eugenic interventions that seriously changes our existing assumptions about the role of human nature and our understanding of us as human beings (or would, if eugenics were allowed and performed). We now have reached the point of departure from Jürgen Habermas' considerations.

11.3 Concepts of Valuation for a Justification of Manipulative Interventions into Human Nature[4]

Habermas deals with the practical attitudes and valuations that are used to judge and justify interventions into nature. In doing so, he first studies the traditional attitudes of healing, caring, and breeding that have more and more changed to favor a utilitarian scientific mechanization of human nature. The traditional attitudes, however, have not simply vanished but instead function rather as ordinarily anchored conceptualizations for justifying the different scientific mechanizations of nature.

Habermas first reminds at the quasi-Aristotelian constitution of "our" *lifeworld* (*Lebenswelt; theoria, praxis, poiesis, techne*). Common to the practices of caring, healing, and breeding was the "respect for the inherent dynamics of autoregulated nature" (Habermas 2003, 45). The modern scientific-technical development has shaped and restructured those practices through the logic of applying scientific technologies. Nature itself is demystified and stripped of all self-value; it is regarded as principally objectifiable and manipulable, experimentally explored, and utilized for whatever purposes. Nevertheless, as the punch line of Habermas' explications holds, the "architecture of the ways of acting itself remained intact", i.e., as the commonplace grid of legitimizing technical disposals. Habermas sees a dialectic of development: The more the purely instrumental scientific-technical interventions replace the practical care, healing, and breeding, the stronger the "legitimizing power" of these "in the widest sense 'clinical' ways of acting" become. With regard to the area of the development of genetic modifications in particular, Habermas notes that bio-political aims are being justified with "nourishment, health and extension of life".

But, and this is the critical essence of Habermas' diagnosis, the genetic-technical revolutionizing of the practice of breeding is no longer an "*adjustment* to the self-dynamic of nature", but instead replaces "what has come to be by nature" with "thus of manufactured" (Habermas 2003, 46). This interpretation becomes critical because Habermas intends to point out a loss within this replacement. Together with the "*cognitive* consideration for the self-dynamic of the process of life", the "*practical* consideration, a kind of respect" we have towards living nature gets lost. The closer the self-organizing organisms are to humans, the more important this practical respect becomes and manifests itself in a gentle treatment of the organic nature. Habermas sees the reasons for this "empathy", or "accompanying understanding for the vulnerability of organic life, that constitutes a threshold of scruple within practical dealings" as lying in the "sensitivity of the own body and the distinction of a however rudimentary subjectivity from the world of manipulable objects".

[4]In the following I use parts of Lohmann (2008).

I can interrupt my report of Habermas' explications at this point because the way in which culturally mediated valuations play a role within the assessment of genetic-technical processes has become apparent. Habermas' thesis is not that the empathic respect for living nature is ontologically founded, but that it maintains a difference in valuation that is founded in valuing self-experiences of humans with their own nature. I want to stress the conceptual differences in his position a bit further.

It is decisive for this position that valuations of "we" refer to ourselves as nature and at the same time differentiate ourselves from the outer nature. Habermas hints at the categorical ontological difference between the artificially "thus of manufactured and what has come to be by nature" (Habermas 2003, 46). A purely objectifying, methodically observing, and manipulative attitude towards ones own nature would suspend this categorical difference. This conceptual difference is probably possible in every culture; important for Habermas' argument is the subjective and, in its claim, the anthropological modification that he gives this categorical difference in the relation of the human towards its inner nature. In accordance with Helmut Plessner (Plessner 1981), Habermas claims that we experience our own nature in two different ways: We experience in ourselves a specific difference between the sensitive "being a body" (*Leibsein*) and the disposable "having a body" (*Körperhaben*). Fundamental and "primary" for the development of the "subjectivity of the human person", however, is the "mode of experience of being a body" (Habermas 2003, 89). Having a body is an achievement of the later development during adolescence that comes with the ability to objectively look at one's being as a body, which is primary. This, however, is formulated as conceptually compelling and, with a general anthropological claim, the experience is naturally grown in which a subjectively indisposable nature expresses and manifests itself. Habermas connects with the modal difference between the subjective self-experience of being a body and having a body, and the methodical difference between self-perception from the perspective of the participant and the perspective of the perceiver, respectively. Again, the thesis is that we are only able to objectively, "externally", observe ourselves because at the same time we experience and perceive ourselves from a fundamental "internal" perspective of the participant that is constitutive for subjectivity and cannot be given up.

These three conceptual pairs:

- The ontological difference between the Grown and the Produced,
- The anthropological difference between Being a Body and Having a Body, and
- The methodical difference between the perspective of the participant and the perspective of the observer

define, according to Habermas, the justifying assessment of an adequately manipulative treatment of our own nature, which we are supposed to experience and accordingly respect as *indisposable* (*unverfügbar*) in these different regards.

11.4 The "Indisposability" of Human Nature

Habermas regards PGD, stem cell research, and the possibilities of cloning as pacemakers for positive eugenics. With them the possibility of a step-by-step transformation of our attitude towards pre-personal life becomes real.

> The performative attitude towards a future person that in the embryo-state is already *treated as if* being a person that can say Yes and No is replaced, in the case of positive eugenics, by the attitude of a builder that combines the aim of the classical breeder, i.e. to improve the hereditary characteristics of a species, with the mode of operation of an engineer that is intervening purposefully according to his own master plan – and thus *works with* embryonic cells as material (Habermas 2002, 296).

This is clearly an attitude of having nature at one's disposal, and the normative request to treat the genetic natural constitution of humans as *indisposable* is directed (first and foremost) against exactly this attitude.

Habermas follows the position that, by the legal protection of human rights, the "untouchable" normative value of human dignity, as in the German Constitution (*Grundgesetz für die Bundesrepublik Deutschland*), refers only to the dignity of a person. But it is open to him if, from a moral perspective, the normative argumentation shows the same restriction. Therefore, he distinguishes between the "untouchability" of human dignity and the "indisposability" of human nature. And the question is now, from a normative point of view, how we can argue to treat the genetic natural constitution of humans as *indisposable.*

With Habermas one can distinguish three types of normative arguments:

- moral arguments, in which universal moral standards are applied, as they are needed to justify the normative claims of human rights,
- ethical arguments, in which particular and plural religious, ideological, and/or cultural valuations are applied that hold only for those who share the respective values, and
- "Species-ethical" (gattungsethische) arguments, in which valuations are applied that can claim to be acceptable for all humans without being morally obliging. This includes, first and foremost, motifs that speak in favor of a universal and individual morality of respect for all humans.

Habermas names two essential arguments that refer to morality. Questionable is this change of attitude because it changes (1) the natural prerequisites of universal morality in a way that has a negative impact on the possibilities of unimpaired autonomic self-being, and (2) the conditions of respecting all persons as equal (cf. Siep 2002; Birnbacher 2002). Moreover, (3) there are more general arguments (of the so-called "slippery slope") that refer to the anticipated effects of genetic practices in general. The first, in my opinion, carries most of the argumentative burden, with (2) and (3) building upon that; I will limit myself to a discussion of (1) and will only briefly come back to (2) and (3) at the end.

Habermas first described his intuition that an eugenic manipulation of the genetic hereditary factors, even positively intended, constitutes a breach of individual autonomy because it replaces the grown with the produced, following Plessner, as a disruption of being able to be oneself.

> The realization that our hereditary factors were, in a past before our past, subjected to programming, confronts us on an existential level, so to speak, with the expectation that we subordinate our being a body to our having a body (Habermas 2003, 54).

The argumentative content of this Plessnerian anthropological diagnosis does not seem to satisfy him, however, as the reversal of being a body and having a body does not necessarily imply a self-alienation to an unacceptable degree. "Why should the human not, with a shoulder shrugging 'So what?' get used to this as well?" Such a reversal, which intervenes with the kind/genus-nature of humans, would only be *generally* unacceptable if it could not also be accepted with regard to moral criteria. Therefore, Habermas tries, in a renewed approach, to "clarify the standards that could be breached by such an instrumentalization". In "[t]he future of human nature," this aim is explicated first in an interpretation of Kant's *End-in-itself-formula* and *Law-formula* of the *categorical imperative*, and second, following Hannah Arendt's theory of the "natality" (Habermas 2003, 58) of human life. In the former, the indisposability of the natural preconditions of human life is a condition for the development of a consciousness of freedom, in the latter, the moral-constitutive intertwining of moral universality and authentic individuality requires the indisposability of the natural conditions of ones own body. Habermas admits, however, that it still remains open what exactly is "questionable in a moral sense" here (Habermas 2003, 60).

This is probably why Habermas, in an essay on Adorno, complemented and partially corrected his arguments (Habermas 2005). Habermas places his interpretation of Adorno in the philosophical context of the naturalism debate within bioethics and philosophy of mind. He begins (with Adorno) with an undisguised phenomenology of the individual consciousness of freedom.

(a) The practical consciousness of the freedom of an acting person can only understand itself as authentic when the person can understand the impulses of her inner nature as her own. In order to do so she does not have to presuppose that they are produced or controlled by her. She can, in contrast, acknowledge them as being both indisposable to her (i.e., as something that she now has power over) and as nature-given because they are not preventing her practical (well considered) will, but are rather preventing its *enabling* conditions. Acknowledging identification with one's own nature must, however, presuppose that it is indisposable to others. My own inner nature, being indisposable to me, has to be *indisposable to others as well* for it to be possible to be integrated into my practical consciousness of freedom. Otherwise, I could not accept the impulses pressing themselves on me out of my own nature as my own that I have to arrange myself with in the course of my life, but rather had to, knowing of myself as being genetically modified, regard myself as being determined by others. Even this could be acceptable for humans, but only against the price of a fundamental change of their understanding of freedom.

The request for an indisposability of the inner nature thus means that it has to be indisposable *to others* because, otherwise, the culturally highly appreciated concept of being oneself would no longer be possible. It is thus an *ethical* (in the sense above) argument. This way of understanding oneself as an autonomous and authentic individual person developed within the European enlightenment and is not equally developed in other cultural traditions.

(b) As *mutually* responsibly-acting persons, humans presuppose themselves *as equal*, free, and considerate persons, but, at the same time, refer to their human dignity also *as individuals*, i.e., irreplaceable and unique persons (cf. Habermas 2003, 28). Insofar as a foreign domination of the inner nature damages other persons in their claim of being irreplaceable and unique, another request for the indisposability of the human nature for others arises. This time, however, the reasoning is primarily moral insofar as it would represent a violation of the equal autonomy of all persons. In the German Constitution it is also protected by article 1.1, the legal rule that human dignity is *untouchable* (cf. Lohmann 2004).

The situation is different when manipulation occurs in the pre-personal stadium of a human. And it is exactly this case that we are confronted with in eugenic manipulations. Here, moral arguments don't grasp directly, but *indirectly* (Lohmann 2003). Because and insofar as all of us have an interest to preserve the moral conditions of a mutual ascription of responsibility, we all bind ourselves to the request to respect an indisposability of human nature in these cases, as well. The reasoning here is *species-ethical* in the above sense in that it expresses interest in preserving universal modern morality and equal respect for all humans by all humans, like it is presupposed by the human rights concept.

However, this is no longer a particular and plural *ethical* question, but one with a stronger claim. Because the universal morality of respect at the same time represents the founding moral concept of human rights, there does not seem to exist any alternative to it at the level of morality. Naturally, there are other culturally defined morality concepts, but those are unable to provide the foundation for a universal and egalitarian consideration towards all humans in a satisfying way. Thus, if one wants a moral regime that corresponds with the universal validity of human rights, one does, at this point, not have any alternative choice. However, one still has the choice to decide in favor of this universal morality at all. It is exactly this more fundamental and embracing option of choice that the term *species-ethical* refers to, in referring, at the same time, to the natural preconditions of a universal morality.

11.5 "Species-Ethics": Post-metaphysical and Inter-culturally Balancing?

It seems to be decisive in terms of our question that the character of the whole argument is neither pure juridical in the sense of human rights nor immediately moral or ethical in Habermas' sense. Habermas claims his *species-ethical* argument to be of a higher status than simply that of a (plural) evaluative *ethical* argumentation. Therefore, it is not about a "common" balancing of goals and preferences that are to be assessed socially differently in which the greatest good for all can be decisive. Habermas' argument is, thus, at least according to his self-understanding, not to be called utilitarian.

Rather, it is about a complex conflict of balancing which seems to answer a contemporary question—e.g., "Should PGD be allowed?"—differently in the light of a later, anticipated scenario of conflicts than at the current moment. If this was only about the question of whether PGD should be allowed, it seems to me that Habermas would have to allow PGD for the similar moral reasons that he allows, or at least accepts, negative eugenics, which prevents severe handicaps by means of genetic manipulations. I suppose a similar argument can be found in favor of the question of whether therapeutic cloning should be allowed. This moral permission does, however, appear problematic in light of a later conflict. The later conflict includes our desire for a functioning universal morality of respect and the (anticipated) use that (some) parents and children will take from a liberal, positive eugenic and the chance for healing that we hope to achieve by therapeutic cloning. Leaving aside all restrictions and assumptions on both sides, it is a conflict between our interest for moral principles and possibly contrasting anticipated uses. From the perspective of a universal morality of respect, the necessary preconditions for morality are given a weight similar to that of following the moral principles itself. As, within morality, principles are to be given priority over uses, it seems logical that the necessary preconditions of morality are equal to the moral principles and, accordingly, it is logical to prioritize granting their protection over possible uses. Our interest for morality feeds from the weight that morality itself claims. Nevertheless, as Habermas explicitly points out, "an assessment of morality as a whole is itself not a moral judgement, but an ethical one, a judgement which is part of the ethics of the species" (Habermas 2003, 73). The *species-ethical* judgment is an evaluative judgment that expresses (as Habermas polemically puts it) that "we prefer to take the status of a member of a community that requires an equal respect and solidary responsibility for all" to that which would mean "life in a moral vacuum". Now, this evaluative judgment is problematic in three different regards:

1. It is unclear whether the alternative relative to which preferability is being expressed has to be that unambiguously negative.
2. It is unclear whether the feared alternative does necessarily occur when the genetic manipulations that are at question take place. With regard to this, we deal with empirical probabilities and accuracies of prognoses concerning future technical developments—typical problems of arguments of a "slippery slope".
3. We deal with the weight of the *species-ethical* evaluative judgment itself which, as such formally, with regard to diameter and range, accords to the evaluative judgments that concern a comprehensive conception of life. However, due to the moral content, which claims a universal validity even within differing *Weltanschauungen*, this *species-ethical* evaluative judgment appears *content-wise* with a claim for universality.

This doubling and tension, however, has so far been somewhat neglected in Habermas' explications. Has the universal morality of respect, according to Habermas, so far been founded on hopefully "friendly" forms of life (which, under post-metaphysical conditions, cannot be theoretically and philosophically further defined); the theoretician is now confronted with the task of delivering a formal and universal theory of exactly these contents, including an evaluation of the mankind's very own natural history.

In my opinion, the status of this "species-ethical" evaluative judgment is parallel to the inter-culturally balanced evaluation of human dignity as the new normative basis of human rights. Like in the historical situation after the Second World War, confronting all humans with "crimes against humanity," the new interpretation of "human dignity" was invented and constructed relative to the alternative—that those crimes should never be possible. The negative alternative to the new evaluative principle for human rights was an absolute negative. But the argument for the new basis on which to justify the universal claims of human rights is not an absolute argument, but an interpretation of historical experiences with unbelievable cruelties and barbarism (Lohmann 2011b). It claims something like a historically-relative argument for universalism and egalitarism (cf. Lohmann 2010a).

Whether this does not, after all, require a redefinition of the borders between metaphysics and post-metaphysics is what makes these questions so philosophically intriguing (see Habermas 2012). With comprehensive content, every *species-ethical* judgment ultimately poses a metaphysical claim. This would also match with the implicit tendency towards metaphysics that seems to be attached to the concept of "indisposability". But this claim could no longer be redeemed in the classical way of metaphysics, namely by simply making absolute assumptions concerning human nature. Alternatively, an intercultural trial of understanding offers itself. It would treat the universal claim of a *species-ethical* evaluation as an empirically open question that can be redeemed by means of a mutual convincement through good arguments (Lohmann 2012a).

References

Birnbacher, D. 2002. Habermas' ehrgeiziges Beweisziel – erreicht oder verfehlt? *Deutsche Zeitschrift für Philosophie* 50: 121–126.

Buchanan, A., D.W. Brock, N. Daniels, and D. Wikler. 2000. *From chance to choice*. Cambridge, MA: Cambridge University Press.

Habermas, J. 1958. Anthropologie. In *Philosophie*, ed. A. Diemer and I. Frenzel. Frankfurt am Main: Fischer.

Habermas, J. 2002. Replik auf Einwände. *Deutsche Zeitschrift für Philosophie* 50: 296.

Habermas, J. 2003. *The future of human nature*. Cambridge: Blackwell.

Habermas, J. 2005. Ich selber bin ja ein Stück Natur" – Adorno über die Naturverflochtenheit der Vernunft. Überlegungen zum Verhältnis von Freiheit und Unverfügbarkeit. In *Dialektik der Freiheit. Frankfurter Adorno-Konferenz*, ed. A. Honneth, 13–40. Frankfurt am Main: Suhrkamp.

Habermas, J. 2012. *Nachmetaphysisches Denken II*. Berlin: Suhrkamp.

Hager, F.P. 1984. Art. Natur. In *Historisches Wörterbuch der Philosophie (6)*, ed. J. Ritter et al. Basel: Schwabe.

Lohmann, G. 2003. On the relation between moral, juridical and evaluative justifications of Pre-implantation Genetic Diagnosis (PGD). *Ethical Perspectives, Journal of the European Ethics Network* 10(3–4, Leuven): 196–203.

Lohmann, G. 2004. Unantastbare Menschenwürde und unverfügbare menschliche Natur. In *Menschenwürde. La Dignité de l' etre humain*, Studia Philosophica Vol. 63. Jahrbuch der schweizerischen philosophischen Gesellschaft, ed. E. Angehrn and B. Baertsch, 55–75. Basel: Schwabe.

Lohmann, G. 2008. Evaluations of the inner and outer nature of humans within biotechnology. A request towards Chinese colleagues. In *The ethics of today's science and technology. A German-Chinese approach*, ed. Wenchao Li and H. Poser, 224–231. Berlin: Lit Verlag.

Lohmann, G. 2010a. Kulturelle Besonderung und Universalisierung der Menschenrechte. In *Universelle Menschenrechte und partikulare Moral*, ed. G. Ernst and S. Sellmaier, 33–47. Stuttgart: Kohlhammer.

Lohmann, G. 2010b. Zur moralischen, juridischen und politischen Dimension der Menschenrechte. In *Recht und Moral*, ed. H.J. Sandkühler, 135–150. Hamburg: Felix Meiner Verlag.

Lohmann, G. 2011a. 'Menschenwürde' – formale und inhaltliche Bestimmungen. In *Menschenwürde und moderne Medizintechnik*, ed. J.C. Joerden, E. Hilgendorf, N. Petrillo, and F. Thiele, 151–160. Baden-Baden: Nomos.

Lohmann, G. 2011b. Menschenwürde als, soziale Imagination'. Über den geschichtlichen Sinn der Deklaration der Menschenrechte und Menschenwürde nach 1945. In *Facetten der Menschenwürde*, ed. N. Knoepffler, P. Kunzmann, and M. O'Malley, 54–74. Freiburg/München: Alber Verlag.

Lohmann, G. 2012a. Interkulturalismus und 'cross-culture'. In *Menschenrechte. Ein interdisziplinäres Handbuch*, ed. A. Pollmann and G. Lohmann, 210–215. Stuttgart/Weimar: J.B. Metzler.

Lohmann, G. 2012b. Umweltzerstörung. In *Menschenrechte. Ein interdisziplinäres Handbuch*, ed. A. Pollmann and G. Lohmann, 438–443. Stuttgart/Weimar: J.B. Metzler.

Lohmann, G. 2013. Individuelle Menschenrechte und Pflichten zugunsten von Gemeinschaften. In *Transkulturalität der Menschenrechte: Arabische, chinesische und europäische Perspektiven*, ed. P. Brunozzi, S. Dhouib, and W. Pfannkuche, 147–170. Freiburg/München: Alber Verlag.

Plessner, H. 1981. Die Stufen des Organischen. In *Gesammelte Schriften*, Bd. IV, ed. H. Plessner. Frankfurt am Main: Suhrkamp.

Ritter, J., et al. (eds.). 1984. *Historisches Wörterbuch der Philosophie*, vol. 6. Basel: Schwabe.

Siep, L. 2002. Moral und Gattungswesen. *Deutsche Zeitschrift für Philosophie* 50: 111–120.

Chapter 12
Species, Potentiality and Their Manipulation

Markus Rothhaar

12.1 Introduction

Among all the arguments put forward in the ongoing debate on the moral status of human embryos, the arguments of potentiality and species membership seem to be the most widespread and popular ones, at least in continental Europe. In Germany both entered the legal definition of a human embryo as given by the Embryo Protection Act of 1990, which defines a human embryo as "the human egg cell, fertilised and capable of developing, from the time of fusion of the nuclei, and further, each totipotent cell removed from an embryo that is assumed to be able to divide and to develop into an individual under the appropriate conditions for that" (The Human Embryo Protection Act, 1990, §8,1).[1] This reference to species membership and potentiality as criteria for being a bearer of human rights and human dignity is a recurring feature of the German juridical discourse. The most important example is certainly the German Constitutional Court, which stated in its—admittedly inconsistent—first decision on abortion in 1975:

> Wherever there is human life, it owns human dignity; it is not important if the bearer is conscious of this dignity and able to preserve it. The potential capabilities, inherent to human life from the very beginning, are sufficient to establish human dignity. (BVerfGE 39, 1)

It is obvious now that the arguments of species membership and of potentiality in ethics directly point to the relationship between nature and normativity. This

[1]The official English translation, provided by the German Ministry for Foreign Affairs can be found here: http://www.auswaertigesamt.de/cae/servlet/contentblob/480804/publicationFile/5162/EmbryofProtectionAct.pdf.

M. Rothhaar (✉)
Department of Philosophisch-Pädagogische Fakultät, Catholic University of Eichstätt-Ingolstadt, Ostenstr. 26, 85072 Eichstätt, Germany
e-mail: markus.rothhaar@ku.de

M. Albers et al. (eds.), *Human Rights and Human Nature*, Ius Gentium: Comparative Perspectives on Law and Justice 35, DOI 10.1007/978-94-017-8672-0_12,

is not only true for species membership, which unquestionably is a natural trait, but also for potentiality, since the potential of developing e.g. self-consciousness is something that is rooted in the natural endowment of a human embryo or fetus. Proponents of both arguments therefore have to justify a type of normative theory that allows for nature and natural properties to have at least some normative dimensions. Such a justification—which is already a difficult task in itself—is further complicated by the fact that modern biotechnology has developed some techniques which make it possible to manipulate species membership and developmental potentiality of early embryos. Hence, in order to defend these arguments, one would not only have to give a rational account of the relationship between nature and norms, but would also have to show why the possibility of these manipulations does not interfere with this account. In the following, I shall try to present an approach which may satisfy both conditions. The basis for this approach will be provided by Immanuel Kant's reflections on human procreation, as well as Robert Spaemann's theory of personhood.

Let us consider the argument of species membership. Whenever this argument is invoked in bioethical debates, the objection is made that it constitutes a naturalistic fallacy. This objection is obviously true for versions of the argument which put forward that being a member of the human species already constitutes a sufficient reason to be endowed with human rights and human dignity, since no norms or rights follow from the mere fact of being a member of a certain biological species. However, the versions of the argument that can be taken seriously—e.g. the ones of Robert Spaemann (1989), Wolfgang Wieland (2003) and, in a certain way, Immanuel Kant—never proposed anything like that. Their claim is, rather, that species membership constitutes only the *criterion* of the moral status, but not its *reason*. According to this differentiation between reason and criterion, the *reason* for being treated as bearer of human rights and human dignity is personhood or subjectivity. The *criterion*, however, for identifying a being as a person or a moral subject would be membership to the human species. Thus, every member of the human species would have to be considered a person during the whole of its existence. It is now evident that such a theory avoids the trap of a naturalistic fallacy, since being a member of the human species does not provide the reason for recognition as somebody endowed with human rights and human dignity, but instead provides only a criterion identifying an entity as a bearer of these rights.

If this is a good description of the theories of Spaemann, Wieland, and some other authors, another objection arises, which might be expressed as follows: if being a person or a moral subject is the very reason for having human dignity and human rights, then why should one introduce a criterion different from that reason? Why don't we simply take the reason for the criterion and state that being a person is also the criterion for being recognized as a bearer of rights and dignity? Although this objection seems prima facie reasonable, it relies on a hidden premise which can easily be proved false. This premise assumes that it is clear what it means to be a person and that it can somehow be recognized empirically, whether an entity is a person or not. It is, however, completely opaque as to how something non-empirical, like being a person or a moral subject, could be identified via empirical data. If there is anything like "recognizing a person as a person" this whole act must take place on a level of experience that is different from empirical cognition. If this is true, then every theory of moral status, even those that do not refer to the moral status of the unborn, is requested to provide not only a reason for the moral status, but also a criterion for identifying those entities that fulfill the requirements of this reason. Hence the difference between reason and criterion of the moral status is not limited to one special theory, but is instead a general feature of all theories of the moral status.

Even if we look at the theory of moral status that can be considered the very antipode of species-related theories, namely the classic preference utilitarianism of Peter Singer (1979), Hoerster (2003) and others, the difference between reason and criterion of moral status arises. According to preference utilitarianism, the reason for "having a right to X" is the fact of "having an interest in X". For example, the reason for having a right to life would be the fact that certain beings are aware of their finiteness and therefore have an interest in continued living. Thus, if a preference utilitarian is asked, what is the criterion for having a right to life according to his theory, he would probably answer that criterion and reason are the same: "having an interest in living". This answer is, however, just a pseudo answer because it is utterly unclear as to what "having an interest" means. Who has an "interest in living"? Only someone who, at this moment, thinks that he would like to live on? If this is the answer, then most people would not have any rights during most of their life (one would not even have to refer to sleep or vegetative states). If, however, "having an interest" does not mean being in some mental state of "being interested in something", then what does it mean at all? Is it the fundamental ability to be in such a mental state; the external attribution of "having an interest" or perhaps even a reflexive, a priori insight that being alive is the essential condition of being a person (in which case Singer's approach would be quite close to Kant's)? It is evident that each of these possible answers leads to different criteria for awarding the moral status connected with "having interests". This example illustrates that the difference between reason and criterion is a universal feature of *any* theory of moral status. Any of these theories has to provide us with not only a valid *reason* for a given moral status, but also with a rationally justified criterion for identifying the beings that fall under this status. Johann Gottlieb Fichte was among the first authors to become aware of this problem. Fichte's description of the problem may in a certain way be considered its "classic" and still prevailing description:

> A vexing question for philosophy, which, as far as I know, it has not yet anywhere resolved, is this: how do we come to transfer the concept of rationality on to some objects in the sensible world but not on to others; what is the characteristic difference between these two classes of objects? Kant says: act so that the maxim of your will can be the principle of a universal legislation. But then who is to be included in the kingdom governed by such legislation and thus share the protection it affords? I am supposed to treat certain beings such that I can will that they, in turn, treat me in accordance with the same maxim. Yet every day I act upon animals and inanimate objects without ever seriously posing the question raised above. Now someone will say to me: it is obvious that we are speaking only of beings that are capable of representing laws, and therefore only of rational beings. With this, I admit, I have replaced the first indeterminate concept with another, but I certainly do not have an answer to my question. For then how do I know which particular object is a rational being? (Fichte 2000, 75)

12.2 The Moral Status in Kant's Practical Philosophy

In the same year Fichte wrote these considerations, the question as to which phenomenal criteria must be applied to identify the bearers of human rights and human dignity also arose in Kant's Doctrine of Rights. The fact that this happens in Kant and Fichte is not astonishing since Kant and Fichte have drawn the most radical possible conclusions from the naturalistic fallacy. According to their conclusions, human beings are characterized by having a phenomenal and a noumenal aspect. In this dichotomy freedom, autonomy and normativity belong completely to the noumenal side of human existence, while the phenomenal side is determined entirely by the empirical laws of nature. However, as soon as one makes this assumption, the question of which empirically perceivable entities are free and autonomous rational beings becomes unsolvable within the means of theoretical philosophy. It has to be noted here that this is true not only for the unborn, but for virtually every human being. Yet, we usually only become aware of the question in the doubtful case, like, for example, in the case of human embryos or fetuses. But how does Kant answer the question if the problem is theoretically unsolvable. First, we have to note that for Kant it seems relatively indisputable that unborn human beings are considered to be bearers of human rights and human dignity. Evidence for this is given in Kant's reflections on suicide which he begins with the remark that a suicide may not only be a crime against one's own person, but in some circumstances also against the person of another, as when a pregnant woman commits suicide (Kant, I.: 1996b, 546). This is even clearer in his Doctrine of Law, in which he rather casually establishes the moment of conception as the beginning of personhood and legal subjectivity:

> For the offspring is a person, and it is impossible to form a concept of the production of a being endowed with freedom through a physical operation. So from a practical point of view it is a quite correct and even necessary idea to regard the act of procreation as one by which we have brought a person into the world without his consent and on our own initiative [...] They [the parents] cannot destroy their child as if he were something they had made (since a being endowed with freedom cannot be a product of this kind) or as if he were their property, nor can they even just abandon him to chance, since they have brought not merely a worldly being but a citizen of the world into a condition which cannot now be indifferent to them even just according to concepts of right (Kant 1996b, 429f.).

With this passage, Kant leaves no doubt as to his view: human beings are to be accorded the status of *homines noumenales* with human dignity from the moment of conception.[2] The reason Kant gives is not as trivial as it might seem given

[2]Volker Gerhardt argues against this by interpreting the passage as stating that the embryo is a person before birth only if the parents would acknowledge him as a person in an act of freedom that corresponds to the act of love that lead to conception. Admittedly, this interpretation suffers from the disadvantage that it projects onto the text precisely the opposite of what is clearly stated. Moreover, it attributes to Kant a very un-Kantian, very romantic concept of freedom. The result is more an act of hermeneutical violence than a hermeneutical act of love enacted in freedom (cf. Gerhardt 2004, 122–27).

the occasional nature of the remark. His tone does not succeed in masking the fact that here Kant is stumbling over a fundamental problem of his transcendental anthropology, the problem of how the two sides of phenomenal and noumenal human beings relate. It appears under the guise of the question as to how one can at all conceive the beginning of a noumenal creature's existence in the phenomenal world. Kant's answer is as simple as it is rigorous: one simply cannot conceive such a beginning theoretically. This answer is rigorous because conception is in fact a "physical operation", that is, a natural event in the phenomenal world which is subject to natural laws, causality and the forms of intuition of space and time. But *homo noumenon*, by virtue of which human beings alone possess dignity, stands outside the phenomenal world. It is, therefore, impossible in principle to comprehend it as having originated in a spatio-temporal event structured by the category of causality. In particular, its genesis cannot be comprehended as a genesis because every genesis, every event, presupposes a temporal before and after.[3]

On the other hand, human beings in the phenomenal world clearly have not always existed, but come to be and pass away. This contradiction can no longer find a theoretical solution within the framework of Kant's philosophy. Instead, the two sides are bridged by a "practical idea"—a procedure Kant frequently adopts. In the case before us, conception is defined as the beginning of the noumenal being's existence. But this definition, and Kant puts great store by this, is not the result of an arbitrary decision of the individual, but instead a "correct and necessary one": a definition that is compelling for every rational being and therefore morally obligatory. The real difficulty here lies in the fact that Kant offers no further justification for this very claim. In my view, it is nevertheless possible to at least partially reconstruct such a justification.

Indeed, it would seem that the very absence of prior conditions and the development they imply allows conception to appear as the "necessary" starting point of existence. Since the noumenal human being stands outside the form of intuition of time and is not subject to the causality of nature, its coming to be cannot be conceived as a development over time. Every hiatus which is preceded by biological or social development over time, from embryo to fetus, or from unborn to born human being, for example, is thus ruled out. Since the validity of human dignity is a transcendental quality that cannot depend on empirical conditions—such as brain activity, extra-uteral viability, etc.—, all of which, by definition, belong to the phenomenal world, the starting point must be as unconditional and as free of presuppositions as possible; the condition of the validity of human dignity must, if there is to be one at all, be of a non-empirical sort.

The only natural event that fulfills both criteria is conception. Conception is, first of all and despite a certain temporal duration, an event that is not temporally

[3]Cf. Kant's explanation in the footnote to the above quoted §28 of the Doctrine of Right (Kant 1996b, 429 f.), in which he goes so far as to say that since the noumenal side of human beings cannot be subjected to the form of intuition of time, it is not even possible to think of the homo noumenon as created by God!

preceded by a natural development. It is an event marked by necessary discontinuity, namely between the non-existence and the existence of a human organism. Just such discontinuity and timelessness is required by practical reason's analogical inference for the absolute beginning of noumenal existence.

Second, starting with conception means that no empirical states of affairs such as brain activity, or extra-uteral viability can serve as conditions for the validity of human dignity. Only being human, as such, counts. At this point one might object that belonging to the species "human being" is obviously an empirical property. But in the framework of Kant's philosophy, such an objection loses its plausibility as it only concerns the phenomenal side of being human. But being human is distinguished from other empirical conditions precisely by the fact that humans possess a phenomenal and a noumenal side. Even if one was not convinced by this argument, starting with merely being human would still be "correct and necessary" given Kant's premises, since being human would still be the most general, most empty and poorest in presuppositions of all possible empirical properties. To that extent, simply being human fulfills the second of the postulates of practical reason mentioned above, according to which if human dignity is to be valid, then either no empirical conditions are admitted, or only the one with the least content.

Both postulates are based on something one might call a "transcendental-pragmatic analogical inference" from the noumenal to the phenomenal world and from the theoretical to the practical point of view. Since the existence of a noumenal human being cannot be conceived if from a theoretical point of view as either temporal or as conditioned, only such a natural event can be selected as the starting point from a practical point of view that is the most proximate analogy to being empirically unconditioned and temporally discontinuous. And that is what conception is. Given this background it is unsurprising that a whole series of well-known philosophers and constitutional scholars in Germany who take their bearings from Kant can be found on the pro-life side of the discussion about the status of unborn children. Examples include Gerold Prauss (2001), Reinhard Brandt (2003), Wolfgang Wieland (2003) and Otfried Höffe (2002, 70–72).

12.3 Species Membership as a Criterion of the Moral Status

From all the above mentioned authors we can find arguments that show striking resemblances to Kant's reflections on the criterion for the moral status. Probably the most influential of these arguments has been put forward by Robert Spaemann. Spaemann departs from an analysis of the concept of "human rights" by contrasting the usual criteria for moral status which have been proposed in bioethical debates. According to Spaemann, the main problem with all of these criteria is that they consider the status of a person as something that is awarded only under the condition of having certain properties. Since there is, however, an unavoidable arbitrariness in the postulation of these properties, the idea of human rights itself would be undermined because everybody could be excluded from the community of human rights subjects at any time, i.e. as soon as he doesn't fulfill certain requirements set by others:

> To the status of a person it is essential, that it is not bestowed upon by someone, but that each person enters the community of persons by her own right. If human rights are 'bestowed upon' or 'conceded', as Hoerster writes, they do not exist at all. For in this case the question of who is granted these rights and who is not is a question of who has the power of definition. Society becomes a 'closed shop', which can arbitrarily co-opt or exclude on the grounds of criteria that are established by a majority [. . .] (Spaemann 2001b, 78f.).

And in an earlier book, Spaemann had already exposed this argument:

> If there exists anything like human rights at all, they can only exist under the condition that nobody is entitled to judge, whether someone is a subject of these rights. For the concept of 'human right' means that a human being is a member of the human community not by cooptation on the grounds of certain properties, but that she enters it in virtue of his own right. In virtue of her own right however can only mean: in virtue of her biological membership to the species of homo sapiens (Spaemann 1989, 220).

If we take a closer look at this argument we see that it is actually a twofold argument. Its first aspect says that the concept of human rights implies the indisposability (in German: "Unverfügbarkeit") of those rights of a subject to any other subjects. This indisposability would be invalidated if any person were given the power to define criteria that others have to fulfill in order to be granted the status of a human rights subject. Such a procedure would introduce an asymmetry of rights and power into interpersonal relationships that is incompatible with the concept of human rights. If the argument is reconstructed in this way, it seems to be circular because such an asymmetry is, of course, only incompatible with the concept of human rights if it takes place between persons. Hence the argument seems to presuppose what it is meant to prove: that unborn human beings are persons or at least have to be treated like persons. The objection of circularity can only be refuted if we reformulate the argument into a kind of a "tutioristic" argument. This reformulation would say that we don't know whether unborn human beings have the status of persons because any proposed criterion for the moral status is in the sense "arbitrary", in that it relies on certain metaphysical assumptions (e.g. assumptions concerning the mind-body-problem) which can be disputed with good reasons. For this reason we would have to rely on the criterion that requires the least metaphysical assumptions possible and is therefore the least arbitrary. Biological membership of the human species must be the "least arbitrary criterion" since any other criterion would require more comprehensive assumptions. This reformulation leads directly to the second aspect of the argument, the aspect that is indicated by the expression that each person enters the human community "in virtue of her own right". This phrase expresses the idea that the concept of human rights requires that membership in the community of human rights subjects has to be thought of as something that does not depend on any criteria at all because, in that case, it would never be an unconditional admission in its own right. Human rights, however, only makes sense if they are accorded unconditionally and in virtue of what their bearer is in and of itself. On the other hand, human rights would not make any sense if they were granted to anything that exists; therefore, there has to be some criterion, although it is, in a certain way, forbidden to postulate. Thus, resorting to mere "biological membership of the human species" would be the answer to an—albeit necessary—contradiction in the concept of human rights: the contradiction that this concept simultaneously requires and forbids postulating any conditions for being a member of the community of human rights subjects. And this is because biological membership in the human species would be the only criterion among all possible criteria that comes close to a non-criterion. On the one hand, it is the least comprehensive criterion and, on the other hand, it refers to something that is by definition beyond human disposability: Nature. The concept of nature, thus, does not enter into the argument in the way of a naïve reference to some simple biological facts, but instead as the result of a highly sophisticated reflection on the indisposability of human rights and human dignity, bridging the gap between the practical necessity to set criteria for the membership to the community of human rights subjects and the theoretical prohibition to do so. But what again is the criterion which allows us to determine whether a living being is a member of the human species? Is the criterion a morphological one, as suggested, for example, by Johann Gottlieb Fichte (2000, 75f.)? Is it "the human genome"? Or is it something like "being born by a human being"? And what happens if biological species membership itself enters the realm of human disposability? What happens, in other words, if the boundaries between the biological species can be blurred by human intervention?

12.4 Species Membership and Potentiality

The questions expressed above are not merely theoretical ones, as proven by the possibility to create genetic chimeras and genetic mosaics. At present, the problem of cytoplasmic hybrids, so-called "cybrids" (cf. for example Chen et al. 2003), that are created by introducing a human cell nucleus into an animal cytoplast, which still contains mitochondrial animal DNA, is especially controversial.[4] Given all these methods used to blur the borders between the species, anybody who advocates the species argument must answer two questions: first, which criteria he applies for species membership, e.g. genetical, morphological, genealogical or social; second, what would be the moral status of genetic "cybrids", chimeras or mosaics.

To answer these questions it is important to note that neither Kant nor Spaemann consider the status of a person to be something that is exclusive to the human species. If there would be another species (e.g. the Neanderthals, if they had survived until today, or some alien species that might be discovered one day) whose members can be addressed as persons at some stage of their development, then all of their members would need to be considered to be persons as well. Kant explicitly men-

[4]For a good overview on the biological and moral issues cf. HFEA (2007).

tions the possibility of "other rational species" several times, notably in the third part of his precritical work from 1755, "[a]llgemeine Naturgeschichte und Theorie des Himmels oder Versuch von der Verfassung und dem mechanischen Ursprunge des ganzen Weltgebäudes, nach Newtonischen Grundsätzen abgehandelt" (Kant 1902) and in the later "Anthropologie in pragmatischer Hinsicht" from 1798 (Kant 1907). Further, Spaemann writes:

> And we must assume, that there might exist beings on other planets, who expose those traits on the grounds of which we consider someone as member of the community of persons. It is even conceivable, that there are some beings on earth that we until now consider as animals, but that we should consider as persons. I don't think that it is very probable, but it is not inconceivable that we might finally come to consider dolphins or primates as persons [...] (Spaemann 2003, 45f.).

Hence, for both authors it is true that although every member of the human species is a person, not all persons are necessarily members of the human species. But if that is the case, then what are the criteria to identify the species, whose members are persons?

Kant does not give any answer to this question, and Spaemann's answers are not very explicit. On the one hand Spaemann quotes (Spaemann 2001a) David Wiggins' account of a "person" as "any animal, the physical make-up of whose species constitutes the species' typical members thinking intelligent beings, with reason and reflection, and typically enables them to consider themselves as themselves, the same thinking things, at different times and different places" (Wiggins 1980, 188). On the other hand, Spaemann frequently refers to genealogy as the constitutive characteristics of a species (cf. Spaemann 1996, 2003). According to this reference, any single member of the human species comes into being by procreation. Since now for persons, procreation is not just a biological event but an interpersonal act, "human species" differs from any other species insofar that it is not a merely biological phenomenon but also an interpersonal, social phenomenon. "Genealogy" in this sense links the spheres of natural life with the sphere of social life and is therefore unique to human beings. The problem with this argument, however, is that today not only is the membership of the biological species at the technical disposal of man, but procreation is even more so. For example, if a human being was created through artificial cloning techniques instead of through an interpersonal act of procreation, thereby still showing every sign of personhood, we would certainly consider her to be a person, regardless of how she was created.

For this reason, Spaemann's argument of genealogy is unconvincing. Rather we should stick to a modified version of the argument of potentiality. This version departs from the idea that the personhood of another person can usually be experienced through communicative acts at some point in her development. This, however, does not mean that she is only a person if she is experienced as a person by other persons. Rather, we have to recognize anyone who can be experienced as a person at some point in her life as someone, who has already been a person during her whole lifespan. This recognition necessity ("Anerkennung") results from the reason that Spaemann has already exposed: if we would not recognize someone who can be experienced as a person at some point of her life as someone who is a person during her whole lifespan, then we would postulate certain stages of development, e.g. single properties as conditions for recognition as a person. This, however, would contradict the concept of human rights and human dignity. Hence we are obligated to recognize everyone who can be experienced as a person *at some point* in her life as a person *during all of her lifetime*. Since now members of the human species can typically be experienced as persons at some point in their lifetime, all members of the human species—no matter if "species" is defined genetically, morphologically or genealogically—have to be recognized as persons from the beginning until the end of their lives. The same would apply to any other species whose members can typically be experienced as persons at some point in their development. It would also apply to any other beings who can be experienced as persons at some moment in their lifetime or of whom we have good reasons to suppose that they could be experienced as persons if we would let them grow up. Thus cybrids, genetic mosaics or chimeras have to be recognized as persons as soon as good reasons exist to suppose that they could be experienced as persons at some stage in their development.

This proposal might, however, be confronted by the fact that not only species membership but also the potential of early embryos to develop can be manipulated by biotechnical means, maybe even to a greater extent (cf. Denker 2006; Meissner and Jaenisch 2006; Nagy et al. 1993). Given recent findings on induced pluripotent stem cells (iPS cells), it almost seems that the development potential of embryos can be directed towards any direction desired. Yet all of these artificially induced transitions from "something" to "someone" do not pose any greater problem than the transition from the "something" of a human ovule and a human sperm to the "someone" of a human embryo, which happens in procreation. If, however, some embryonic entity would be created, that is genetically human, but that does not have any potential to develop and thus does not contain even the most basic characteristics of personhood, it would certainly be difficult to address it as an embryo. There may still be good reasons to neither destroy such an entity nor to create it beforehand, but personhood could presumably not count as such a reason.

References

Act for Protection of Embryos (The Embryo Protection Act). 1990. *Gesetz zum Schutz von Embryonen (Embryonenschutzgesetz – ESchG).*

Brandt, R. 2003. Natürlich sind Embryonen Personen. In *Frankfurter Allgemeine Zeitung*, 19 February 2012.

Bundesverfassungsgerichtsentscheidung (BVerfG) 39,1, *Schwangerschaftsabbruch I*, (25 Feb 1975).

Chen, Y., et al. 2003. Embryonic stem cells generated by nuclear transfer of human somatic nuclei into rabbit oocytes. *Cell Research* 13(4): 251–263.

Denker, H.-W. 2006. Potentiality of embryonic stem cells: An ethical problem even with alternative stem cell sources. *Journal of Medical Ethics* 32: 665–671.

Fichte, J.G. 2000. *Foundations of Natural Right*, ed. by Neuhouser, F., trans. by M. Baur, Cambridge: Cambridge University Press.

Gerhardt, V. 2004. *Die angeborene Würde des Menschen*. Berlin: ParErga.
Hoerster, N. 2003. *Ethik und Interesse*. Stuttgart: Reclam.
Höffe, O. 2002. *Medizin ohne Ethik?* Frankfurt am Main: Suhrkamp.
Human Fertilization and Embryology Authority (HFEA). 2007. *Hybrids and chimeras*. London.
Kant, I. 1902. *Allgemeine Naturgeschichte und Theorie des Himmels oder Versuch von der Verfassung und dem mechanischen Ursprunge des ganzen Weltgebäudes, nach Newtonischen Grundsätzen abgehandelt, Akademie-Ausgabe*, Vol. I (Vorkritische Schriften I), 215–368. Berlin: Preussische Akademie der Wissenschaften.
Kant, I. 1907. *Anthropologie in pragmatischer Hinsicht, Akademie-Ausgabe*, vol. VII, ed. G. Reimer, 117–334. Berlin: Preussische Akademie der Wissenschaften.
Kant, I. 1996a. Groundwork of the metaphysics of morals. In I. Kant, *Practical Philosophy*, ed. and trans. by M.J. Gregor. Cambridge: Cambridge University Press.
Kant, I. 1996b. Metaphysics of morals. In: I. Kant, *Practical Philosophy*, ed. and trans. by M.J. Gregor, Cambridge: Cambridge University Press.
Meissner, A., and R. Jaenisch. 2006. Generation of nuclear transfer-derived pluripotent ES cells from cloned Cdx2-deficient blastocysts. *Nature* 439(7073): 212–215.
Nagy, A., J. Rossant, et al. 1993. Derivation of completely cell culture-derived mice from early-passage embryonic stem cells. *Proceedings of the National Academy of Sciences of the USA* 90: S. 8424–S. 8428.
Prauss, G. 2001. Geprägte Form, doch zweckbewusst zerstückelt. In *Frankfurter Allgemeine Zeitung*.
Singer, P. 1979. *Practical ethics*. Cambridge: Cambridge University Press.
Spaemann, R. 1989. *Glück und Wohlwollen: Versuch über Ethik*. Stuttgart: Klett-Cotta.
Spaemann, R. 1996. *Personen. Versuche über den Unterschied zwischen 'etwas' und 'jemand'*. Stuttgart: Klett-Cotta.
Spaemann, R. 2001a. Gezeugt, nicht gemacht. In *Biopolitik. Die Positionen*, ed. C. Geyer, 41–50. Frankfurt am Main: Suhrkamp.
Spaemann, R. 2001b. Wer jemand ist, ist es immer. In *Biopolitik. Die Positionen*, ed. C. Geyer, 73–81. Frankfurt am Main: Suhrkamp.
Spaemann, R. 2003. Sind alle Menschen Personen? In *Menschenleben – Menschenwürde*, ed. W. Schweidler, H.A. Neumann, and E. Brysch, 45–50. Münster/Hamburg/London: LIT-Verlag.
Wieland, W. 2003. Pro Potentialitätsargument. In *Der moralische Status menschlicher Embryonen*, ed. G. Damschen and D. Schönecker, 149–167. Berlin/New York: De Gruyter.
Wiggins, D. 1980. *Sameness and substance*. Oxford: Blackwell.

Chapter 13
Collection of Human Tissue Samples in Biobanks: Challenges to Human Rights and Human Nature

Bianka S. Dörr

13.1 Introduction

Collections of human tissue samples and body parts have been made almost as long as people have aimed to understand the human body and its diseases. These early collections varied in form and size and were mainly built up by and located in pathology institutes; the materials conserved were used for, among others, research and education purposes.

Since the mid-1990s—in particular, alongside the successful sequencing of the human genome that greatly increased progress in the field of molecular genetics and biotechnological analytical processes—a new category of human tissue samples and data collections is proliferating, namely biobanks for research. Biobanks are private or public institutions that gather and store samples of human tissue (cells, tissue, blood, organs) or constituents thereof (serum, DNA) from patients or healthy volunteers in order to correlate them with donor-related data (genes, phenotype, diseases, lifestyle, genealogical, medical, environmental information) and make them accessible for scientific research purposes. Thus, the main feature of a biobank is the combination of material and information. By linking samples to data, it is hoped that exploring the complex interface between genes, lifestyle, and environment will shed light on the causes of numerous diseases. The aim in the long term is, therefore, to develop more effective strategies for the prevention, diagnosis, and treatment of a wide range of serious and life-threatening diseases, as for example cancer, metabolic disorders, stroke, depression, cardiovascular or nervous diseases.

In recent years, biobanks have become major strategic and powerful tools for undertaking medical-scientific research. Nowadays, almost all European countries

B.S. Dörr (✉)
Faculty of Law, University of Lucerne, Frohburgstr. 3, 6002 Lucerne, Switzerland
e-mail: bianka.doerr@unilu.ch

M. Albers et al. (eds.), *Human Rights and Human Nature*, Ius Gentium: Comparative Perspectives on Law and Justice 35, DOI 10.1007/978-94-017-8672-0_13,

are maintaining or launching (large-scale) human tissue collections in multiple shapes, some of which strive for a substantial number of samples and genetic data. A prominent example of a large population-based biobank in Europe is the UK Biobank that—within five years—successfully recruited 500,000 people aged 40–69 years from across the country to participate. UK Biobank aims to improve the prevention, diagnosis, and treatment of a wide range of serious and life-threatening diseases, including cancer, diabetes, arthritis, eye disorders, depression, and forms of dementia (www.ukbiobank.ac.uk).

However, the use of human tissue and donor-related data for research and biomedical applications raises important legal and ethical problems. In this chapter, I focus on the concept of informed consent, an issue that has advanced to become one of the most contested topics in biobanking. Much has been written in recent years to address the challenges connected to research biobanks and informed consent, in particular with regard to various consent models discussed by legal and ethics scholars, but the issues are still far from being solved. In a second part, I will discuss the role of informed consent within utilitarian and human rights approaches and its implications for human nature.

13.2 The Concept of Informed Consent—A Brief Overview of Its History and Development

The concept of informed consent is at the center of biomedical law and ethics, and has been for over 60 years. In 1947, the *Nuremberg Code*, a set of research ethics principles, was released. The code states that the voluntary consent of the human subject is "absolutely essential" for participation in research studies (The Nuremberg Code 1947, no. 1). This implies that competent individuals must be able to exercise, absent any intervention, free choice in their decision to take part or reject participation in any research project. In order to make this decision, competent individuals should have sufficient knowledge and comprehension of the elements of the research project, in particular, the nature, duration, purpose, method, means, quality control, risks, and benefits of the proposed research study (The Nuremberg Code 1947, no. 1).

Only a year after the Nuremberg Code, in 1948, the *Universal Declaration of Human Rights* (UDHR) was adopted by the UN General Assembly; this Declaration represents the universal recognition that basic rights and fundamental freedoms are inherent to all human beings, and all individuals are equal in dignity and rights (United Nations 1948, article 1) and supported by respect for liberty and security of the person (United Nations 1948, article 3).

In 1954, the World Medical Association (WMA) released the *Principles for Those in Research and Experimentation*, a collection of ethical principles that further supported informed consent by including the option that the participant's next of kin or legal representative may, in exceptional cases, provide sufficient consent for sick and incapacitated persons to participate in an experiment (WMA 1955, no. 3–5).

Ten years later in 1964, the WMA adopted the *Declaration of Helsinki*, regarded as a cornerstone document of human research ethics, it contains an important set of guidelines for physicians, further highlighting the requirement of obtaining informed consent from research subjects. As to content, the Declaration advanced the ten principles of the Nuremberg Code but more specifically addressed clinical and non-clinical biomedical research. A remarkable deviation from the Nuremberg Code, however, was the Declaration's easing of the conditions of consent; whereas *Nuremberg* considered consent to be "absolutely essential", the Declaration only required doctors to obtain consent "if at all possible", and, in cases where a proxy consent was procurable, research was allowed without consent (WMA 1964, article II.1). The Declaration of Helsinki was revised last in 2008.

With the adoption of the *International Covenant on Civil and Political Rights* (ICCPR) by the UN General Assembly in 1966, informed consent found its first adoption into a legally binding document, with recognition as an absolute right. This multilateral treaty commits parties to respect the civil and political rights of individuals, including such rights as self-determination, physical integrity, individual liberty, security of the person, and procedural rights (i.e., the right to due process and a fair trial). Article 7 ICCPR explicitly rules that no one shall be subjected to medical or scientific experimentation without his/her free consent. Through this prohibition, the ICCPR expresses the fundamental human value that is held to be protected in scientific experimentation.

The *International Ethical Guidelines for Biomedical Research Involving Human Subjects*, issued by the Council for International Organizations of Medical Sciences (CIOMS) in association with the World Health Organization (WHO) in 1993 (also known as "CIOMS Guidelines") and lastly updated in 2002, places a particular focus on biomedical research ethics. The revised text consists of a description of general ethical principles and 21 guidelines, including commentary. Regarding informed consent, the CIOMS Guidelines provide concise requirements for investigators and sponsors to be observed when obtaining informed consent from prospective research subjects for biomedical research (CIOMS 1993/2002, Guideline 4–6). According to guideline 4, informed consent is a decision to participate in research, taken from a competent individual who has received and adequately understands the necessary information and, after consideration, has reached a decision. In particular, guideline 5 specifies 26 essential pieces of information that have to be provided to prospective research subjects before requesting consent, such as the subject's right to refuse and withdraw from research, the purpose of research, the procedures to be carried out, foreseeable risks, potential pain or discomfort, and the expected duration of the individual's participation or direct benefits. In addition, guideline 6 lists obligations and duties for sponsors and investigators to observe when obtaining informed consent.

The Council of Europe's *Convention for the Protection of Human Rights and Dignity of the Human Being with Regard to the Application of Biology and Medicine* from April 4th, 1997 (i.e., "Convention on Human Rights and Biomedicine", or

"Oviedo Convention") creates a model legal document that sets forth the fundamental rights to be observed in core areas of medicine and biology in order to create more specific international minimum standards for the protection of individuals against new developments in these fields and their possible misuse. Thus, the Oviedo Convention aims to provide a common framework for the protection of human rights and human dignity in the mentioned areas; it sets out only the most important principles, whereas additional standards and more detailed questions are dealt with in additional protocols. Articles 5 through 9 of the Oviedo Convention deal with informed consent and affirm, among other things, as a general rule, that no intervention in the health field may be carried out without the person concerned being given free and informed consent to it. Intervention is to be understood in its broadest sense and covers all medical acts, in particular those performed for disease prevention, diagnosis, treatment, rehabilitation, or research purposes (Council of Europe 1996, para. 34). Patient's consent is considered to be given free and informed if it is based on objective information as to the purpose, nature, consequences and risks of the intervention (Council of Europe 1996, para. 35). Freedom of consent implies that consent may be withdrawn at any time and that the decision to withdraw shall be respected once s/he has been fully informed of the consequences. The special significance of the Oviedo Convention lies in the fact that its definition and requirements of informed consent are—for the first time—mandatory as a general principle for any biomedical intervention (not only scientific research) in an international binding instrument (Andorno 2005, 133–143, 138).

The *Universal Declaration on Bioethics and Human Rights*, adopted by the United Nations Educational, Scientific and Cultural Organization (UNESCO) in October 2005, is designed as a universal framework of principles and procedures that aims, among other things, to promote respect for human dignity and protect human rights by ensuring respect for the life of human beings and fundamental freedoms, consistent with international human rights law. This UNESCO Declaration, which is not legally binding, places great emphasis on the concept of informed consent. In particular, article 6 (1) states that any preventive, diagnostic, or therapeutic medical intervention is only to be carried out with the prior, free, and informed consent of the person concerned, based on adequate information. The consent should, where appropriate, be express and may be withdrawn by the person at any time and for any reason, without disadvantage or prejudice. Further, scientific research should only be carried out with the prior, free, express, and informed consent of the person concerned. The information should be adequate, provided in a comprehensible form, and should include modalities for withdrawal of consent (UNESCO 2005, article 6 (2)).

Two things become evident from this analysis: First, the informed consent clauses used reveal the underlying assumption that any decision made concerns the values and life of a single competent and rational acting individual. Second, it is obvious that the mentioned declarations, conventions, and guidelines established over the past decades since World War II have adopted a clear human rights approach and the principle of informed consent enjoys the status of a universal human right. Thus, it plays the significant role in international biomedical law and research ethics of protecting individuals from harm when participating as research subjects in biomedicine. There are several reasons for this: On the one hand, biomedical advancements are closely linked with basic human rights as they, for example, in research or health care activities, often have direct implications for physical integrity or the right to life (Andorno 2008, 59–60; 2009, 235). On the other hand, human rights are regarded as universal norms and, as such, they enable the verbalisation of universal standards in the biomedical field (Andorno 2008, 60). Further, human rights are understood as an articulation of common values that transcend cultural differences (Andorno 2009, 235). Every human being, regardless of his/her nationality, ethnic origin, sex, age, religion, or language, possesses these fundamental rights just by the fact of being human. Another aspect that calls for a human rights framework in the biomedical field is that there are only very few mechanisms or instruments available other than human rights that could serve as the "global normative foundation" for biomedicine (Thomasma 2001, 300; Andorno 2009, 235; Meslin and Garba 2011, 454–460).

13.3 The Concept of Informed Consent and Its Controversy in the Biobanking Context

Apart from the above mentioned documents, numerous other national and international documents, declarations, guidelines, and policy statements dealing with the legal and ethical aspects of human genetic databases and/or research biobanks have been published. The principle of informed consent is basically discussed in almost all documents, however, in various forms and can, as such, be regarded as a standard requirement in biobanking research. Just to mention a few examples: In 2003, UNESCO released the *International Declaration on Human Genetic Data* and provides in article 8 (a) that "prior, free, informed and express consent [...] should be obtained for the collection of human genetic data, human proteomic data or biological samples, [...] and for their subsequent processing, use and storage, whether carried out by public or private institutions". In 2006, the Council of Europe released its *Recommendation Rec (2006)4 of the Committee of Ministers to Member States on Research on Biological Materials of Human Origin* and foresees—without defining the notion consent in detail—that appropriate consent has to be given for obtaining, collecting, and using biological materials for research (Arts. 10, 12, 13, 14, 15, 21 and 22). In 2009, the OECD adopted the *Guidelines on Human Biobanks and Genetic Research Databases*. This document provides a set of guidelines with regard to the establishment, management, governance, operation, access, use, and discontinuation of human biobanks and genetic research databases and regulates, in great detail, the requirements for informed consent.

As previously stated, the concept of informed consent is included in almost all relevant documents governing biobanks and it is unquestioned that some form of written informed consent is required at least once in the process of collecting samples and data or storing them in biobanks and genetic databases (Boggio et al. 2005, III, 2–8; Salvaterra et al. 2008, 309; Cambon-Thomsen 2004, 869; Elger 2010, 141; Hofmann et al. 2009, 3ff). To some extent, these documents reveal the universally accepted components of informed consent. However, some important issues remain highly contested in the biobanking context, such as, for example, the adequate level of information that needs to be given to a research participant at the time the sample is collected or the permissibility of the use of irreversibly anonymized tissue without informed consent for genetic research projects (Kegley 2004, 832; Elger 2010, 142). Further, it is highly controversial as to whether individuals must give a blanket consent in order to allow their samples and/or data being used for any further research purposes or whether they must be re-contacted each time their samples and/or data are to be included in new research projects (Boggio et al. 2005, III, 3–5; Andorno 2007, 35–36; Salvaterra et al. 2008, 309; Elger 2010, 142).

As a result, lawyers and ethicists have given much thought to how these issues within the biobank context could be solved and explored whether "less informed" consent models could be considered legally and ethically sound (Macilotti 2012, 271–272). The proposed models range from traditional informed consent to blanket consent, from partially restricted to broad consent, from implied to presumed consent, and even to a waiver of consent (Macilotti 2012, 271–272; Salvaterra et al. 2008, 307; Hansson et al. 2006; Elger and Mauron 2003, 269–296; Elger 2010, 142–156; Cambon-Thomsen 2004, 869; Harmon 2009, 31; Hansson et al. 2006, 267–269; Allen and McNamara 2011, 158; for the waiver model: Manson and O'Neill 2007, 20–21). Each model has been equipped with its own theoretical basis and features; however, in practice, it is often very difficult to differentiate between the various models (Macilotti 2012, 272; Harmon 2009, 31). As described earlier, traditional informed consent encompasses only specific and known uses of samples and data. Blanket consent, however, is only given once (preferably at the time when the tissue samples and data are collected) but covers any future use of the material and/or data at any time (Kegley 2004, 833; Elger 2010, 142–145). In situations like this, the consent is clearly missing the informed component. Broad consent, by contrast, allows a broad but relatively specified radius of activities (Caulfield and Kaye 2009, 85–86), whereas presumed consent assumes one's agreement for the unlimited use of his/her samples and/or data unless the individual opts out (Kegley 2004, 833). If the samples and data are to be used in specific immediate research and future studies associated with them, then a partially restricted consent is available (Salvaterra et al. 2008, 309).

In order to justify the various departures from the traditional informed consent model, many interesting arguments have been put forward. Chadwick and Berg argued:

> "It has become clear that some principles of biomedical ethics, such as individual informed consent, that have been highly regarded for some time, might not be ideally equipped to deal with the issues that arise in large-scale population genetic research" (Chadwick and Berg 2001, 318).

Others regard it as legitimate to diminish the requirements for participation in research because of the low risk to the research persons and the high potential outcome and benefit for a greater audience (Hansson et al. 2006, 266–267). Likewise, the argument has been brought that informed consent is impractical (O'Neill 2003, 4–7; Manson and O'Neill 2007). Another set of arguments put forward emphasizes that the traditional concept of informed consent overprotects the human subjects (Wilcox et al. 1999, 362), hinders research (Hansson et al. 2006, 266), and endangers the quality of research (Wilcox et al. 1999, 362; Hansson et al. 2006, 266), and by doing so, also hazards the fundamental public benefit of this efforts (Wilcox et al. 1999, 362; Wadman 2000, 115). Moreover, the argument has been made that biobank research conceptually differs from other research activities involving humans that requesting renewed consent would be regarded as invasion of privacy (Eriksson 2003, 182–184; see also the view of Santosuosso 2013, 105ff). Correspondingly, it has even been argued that specific consent restricts a person's autonomy, while broad and future consent enhances autonomy as long as the risks of research are low and sufficiently well controlled (Hansson et al. 2006, 267).

Beyond that, several empirical studies have been conducted aiming to find out what people think about informed consent (for an interesting overview see: Ring and Kettis Lindblad 2003, 198ff; Hofmann et al. 2009, 3ff., 12). According to the study of Wendler and Emanuel that looked at individuals' attitudes when consent for research on stored samples is to be obtained, most people indicated that their informed consent was not necessary for research (Wendler and Emanuel 2002, 1461). However, the respondents of the study either wanted themselves to be informed or their physicians being informed about research results of uncertain clinical significance (Wendler and Emanuel 2002, 1461). A study by Ring and Kettis Lindblad found that 46 % of the respondents did see "no need to be asked for informed consent in relation to new projects if the research ethics committee had approved the project" (Ring and Kettis Lindblad 2003, 203); however, 18 % requested "the strictest informed consent procedure" (Ring and Kettis Lindblad 2003, 203). A recent study from Australia showed that people regarded the giving of consent "as a valued act of self-determination" (Allen and McNamara 2011, 161–163, 165); however, the study did not find anything in the data that suggests "that broader unspecified consent would have been any less satisfactory" (Allen and McNamara 2011, 161–163, 165).

Further, legal and ethics scholars have enriched the discussions on modified consent approaches by including public good rationales, including solidarity and equity, transcending the individualistic focus of autonomy-based principles (Caulfield and Kaye 2009, 91). To say it in the words of Chadwick and Berg:

> "The question arises as to whether existing ethical frameworks are sufficient to mediate between the competing interests at stake. One approach is to strengthen mechanisms for obtaining informed consent and for protecting confidentiality. However, there is increasing interest in other ethical frameworks, involving solidarity – participation in research for the common good – and the sharing of the benefits of research" (Chadwick and Berg 2001, 318).

Knoppers and Chadwick find clear words in promoting solidarity:

"The increasing prominence of solidarity in the ethical debate might be seen as one aspect of a 'communitarian turn' in ethics, moving away from the paramount position of individualism and autonomy. [...] However, today the challenge to the primacy of the choice model tends to be framed not in the language of eugenics, but in terms of responsibility and solidarity" (Knoppers and Chadwick 2005, 76 with further references).

Also Wagstaff has proclaimed that

"[...] the whole enterprise is based on the principle of solidarity, where the gain from these voluntary donations are disseminated and used for the public good" (Wagstaff 2011, 29).

The three quotes clearly show that the arguments for not seeking informed consent is based on the benefits to society from the research; the individual from whom the sample was taken does not play a role because the social good that will result from the research is considered superior to individual rights (Caulfield and Kaye 2009, 89 with further references).

It is evident from the (more or less convincing) arguments justifying the departure from the traditional informed consent concept that there is a major lack of consensus between legal and ethics scholars on how to address and deal with these questions. This controversy is far from being solved—if at all feasible. The aim of this paper is not to add another consent model to the already confusing discussion or to even solve this major controversy, instead I want to draw the reader's attention to a new, much less discussed aspect, namely that of the implications a departure from the traditional informed consent scheme would have on human rights and human nature.

13.4 Implications for Human Rights and Human Nature

Informed consent sits right in the middle between individual rights and common good. The very fact that so many scholars see the need to impose lower requirements on informed consent in the biobanking context compared to the requirements found in a traditional informed consent model (as outlined above) is not only surprising but also raises questions about human rights and human nature. Human rights are fundamental rights that are inherent to all human beings simply by being human, regardless of nationality, ethnicity, color, religion, language, sex, or place of residence, among others. These rights are all interrelated, interdependent, and indivisible and their existence is, in principle, recognized by almost all countries today. The principle of universality of human rights—although the subject of much political debate and argument—is regarded as a cornerstone of international human rights law.

The concept of informed consent enjoys the status of a universal human right and rests on the principle of individual autonomy, or, derived therefrom, the right to self-determination—an idea that enjoys a leading position predominately in Western societies. In the medical and health care context, the right to self-determination empowers everyone to decide what will be done with his/her own body, and, thus, his/her bodily integrity is protected against intervention by others. As such, any such intervention requires prior and informed consent of the person concerned as many of these rights are—at least to some extent—protected and exercised through consent practices (Kegley 2004, 833; Harmon 2009, 27–29). However, it is quite obvious that practices involving either lesser-informed or blanket consents do not fit well within an ethic of human rights.

Although the relevant documents analyzed show a strong human rights approach and set clear requirements for informed consent, many scholars advocate for a different view in the biobanking context. In particular, the call for solidarity and equity by participating in research for the common good derived from biobank research and the sharing of benefits is an expression of utilitarian philosophy. A central feature of the utilitarian approach is the assessment of each action in terms of its consequences or outcomes. It aims to maximize utility and minimize disutility in order to optimize welfare (human health and happiness) (Brownsword 2007, 15). In other words, utilitarians seek to achieve the greatest good for the largest number of people while creating the smallest detriment or preventing the least amount of distress; an individual's rights are not given particular emphasis under this philosophy (Brownsword 2007, 15–16).

On the contrary, a human rights proponent does not share this view; instead, s/he will insist that public good, however good, must be pursued in a way that entirely respects individual rights (Brownsword 2007, 15). In particular, there is no reason why the rights and entitlements of individuals should be disregarded for the sake of a larger good. Each individual ought to be taken seriously and treated with respect (Brownsword 2007, 16). One has also bear in mind how much effort stakeholders have taken over decades in order to empower patients and human research subjects and to convert histories of revelations of research malfeasance and medical paternalism into a relationship of mutual trust and dialogue with consent at its centre. This was achieved under the heading of human dignity and is still a work in progress (Harmon 2009, 28). Taking into account all these reasons, human rights should not simply be ignored or restricted in this particular area of tension; instead, a clear commitment to a human rights approach should be adopted, one that values and respects the individual as the sample donor and asks for his/her informed consent in cases where his/her bodily material will be used for current and future research.

By advancing to the implications this has on human nature, the first conclusion is that the establishment of major biobanks and the call for lesser-informed or blanket consents is undermining the traditional understanding of human beings as autonomous individuals. Once consent to participate in a biobank research project has been provided, and the samples and data are collected, the human being as an individual not only dissolves into a set of data but also vanishes as a person as soon as the anonymization of his/her samples and data is complete. Consequently, determining at a later point whose samples and data have actually been collected and stored may become more and more difficult—if not impossible. As genetic information never solely concerns the individual but also genetically related family members or members of a particular community, the lines between the individual and the general public, between the person, the family, the population, and even the species are increasingly permeable. Moreover, it is more and more difficult to reconcile traditional principles such as informed consent with large-scale biobank projects as it is in the interest of the latter to pursue new projects with unknown goals for future research.

13.5 Conclusion

Setting up biobanks and genetic databases constitutes new ways of governing life and raises some major challenges. In particular, the consent problem is a very complex one. It is not only about several varieties of consent forms circulated in a myriad of publications as a departure from the standard requirements of traditional informed consent, but also about the appropriate balancing of individual and collective rights, values, and interests. As informed consent sits at the heart of biomedical law and health care ethics, one has to pony up convincing positions and arguments in order to justify subordinating settled standards.

References

Allen, J., and B. McNamara. 2011. Reconsidering the value of consent in biobank research. *Bioethics* 25(3): 155–166.

Andorno, R. 2005. The Oviedo convention: A European legal framework at the intersection of human rights and health law. *Journal of International Biotechnology Law* 2: 133–143.

Andorno, R. 2007. Population genetic databases: A new challenge to human rights. In *Ethics and law of intellectual property: Current problems in politics, science and technology*, ed. C. Lenk et al., 27–45. Aldershot: Ashgate.

Andorno, R. 2008. Warum braucht eine globale Bioethik die Menschenrechte? In *Gibt es eine universale Bioethik?* ed. N. Biller-Andorno, P. Schaber, and A. Schulz-Baldes, 59–72. Paderborn: Mentis.

Andorno, R. 2009. Human dignity and human rights as a common ground for a global bioethics. *Journal of Medicine and Philosophy* 34: 223–240.

Boggio, A. et al. 2005. *Comparing guidelines on biobanks: Emerging consensus and unresolved controversies*. Geneva. http://www.academia.edu/303963/Comparing_Guidelines_on_Biobanks_Emerging_Consensus_and_Unresolved_Controversies.

Brownsword, R. 2007. Biobank governance: Property, privacy and consent. In *Ethics and law of intellectual property: Current problems in politics, science and technology*, ed. C. Lenk et al., 11–25. Aldershot: Ashgate.

Cambon-Thomsen, A. 2004. The social and ethical issues of post-genomic human biobanks. *Nature Reviews Genetics* 5: 866–873.

Caulfield, T., and J. Kaye. 2009. Broad consent in biobanking: Reflections on seemingly insurmountable dilemmas. *Medical Law International* 10: 85–100.

Chadwick, R., and K. Berg. 2001. Solidarity and equity: New ethical frameworks for genetic databases. *Nature Reviews Genetics* 2: 318–321.

Council for International Organizations of Medical Sciences (CIOMS). 1993/2002. *International ethical guidelines for biomedical research involving human subjects*. Geneva. http://www.cioms.ch.
Council of Europe. 1996. *Convention for the protection of human rights and dignity of the human being with regard to the application of biology and medicine: Convention on human rights and biomedicine, explanatory report*, ETS No. 164. http://conventions.coe.int/Treaty/EN/Reports/Html/164.htm.
Council of Europe. 1997. *Convention for the protection of human rights and dignity of the human being with regard to the application of biology and medicine: Convention on human rights and biomedicine*. Oviedo. http://conventions.coe.int/Treaty/en/Treaties/Html/164.htm.
Council of Europe. 2006. *Recommendation rec (2006)4 of the committee of ministers to member states on research on biological materials of human origin*. Strasbourg. https://wcd.coe.int/ViewDoc.jsp?id=977859.
Elger, Bernice. 2010. *Ethical issues of human genetic databases: A challenge to classical health research ethics?* Aldershot: Ashgate.
Elger, B., and A. Mauron. 2003. A presumed-consent model for regulating informed consent of genetic research involving DNA banking. In *Populations and genetics: Legal and socio-ethical perspectives*, ed. B.M. Knoppers, 269–296. Leiden: Brill Academic Publishers.
Eriksson, S. 2003. Mapping the debate on informed consent. In *Biobanks as Resources for Health*, ed. M.G. Hansson, M. Levin, 165–190. Uppsala: Universitetstryckeriet.
Hansson, M.G., et al. 2006. Should donors be allowed to give broad consent to future biobank research? *The Lancet Oncology* 7: 266–269.
Harmon, S.H.E. 2009. Semantic, pedantic or paradigm shift? Recruitment, retention and property in modern population biobanking. *European Journal of Health Law* 16: 27–43.
Hofmann, B., et al. 2009. Consent to biobank research: One size fits all? In *The ethics of research biobanking*, ed. J.H. Solbakk et al., 3–23. Dordrecht/Heidelberg/London/New York: Springer.
Kegley, J.A.K. 2004. Challenges to informed consent: New developments in biomedical research and healthcare may mark the end of the traditional concept of informed consent. *EMBO Reports* 5(9): 832–836.
Knoppers, B.M., and R. Chadwick. 2005. Human genetic research: Emerging trends in ethics. *Nature Reviews Genetics* 6: 75–79.
Macilotti, M. 2012. Reshaping informed consent in the biobanking context. *European Journal of Health Law* 19: 271–288.
Manson, N., and O. O'Neill. 2007. *Rethinking informed consent in bioethics*. Cambridge: Cambridge University Press.
Meslin, E.M., and I. Garba. 2011. Biobanking and public health: Is a human rights approach the tie that binds? *Human Genetics* 130: 451–463.
OECD. 2009. *Guidelines on human biobanks and genetic research databases*. http://www.oecd.org/sti/biotech/44054609.pdf.
O'Neill, O. 2003. Some limits of informed consent. *Journal of Medical Ethics* 29: 4–7.
Ring, L., and A. Kettis Lindblad. 2003. Public and patient perception of biobanks and informed consent. In *Biobanks as resources for health* ed. M.G. Hansson, M. Levin, 197–206. Uppsala: Universitetstryckeriet.
Salvaterra, E., et al. 2008. Banking together: A unified model of informed consent for biobanking. *EMBO Reports* 9(4): 307–313.
Santosuosso, A. 2013. Should privacy be abolished in genetics and biobanking? In *Comparative issues in the governance of research biobanks. Property, privacy, intellectual property, and the role of technology*, ed. G. Pascuzzi et al., 105–130. Heidelberg/New York/Dordrecht/London: Springer.
The Nuremberg Code. 1947. http://www.hhs.gov/ohrp/archive/nurcode.html.
Thomasma, D. 2001. Proposing a new agenda: Bioethics and international human rights. *Cambridge Quarterly of Healthcare Ethics* 10: 299–310.
UNESCO. 2003. *International declaration on human genetic data*. Paris. http://www.unesco.org/new/en/social-and-human-sciences/themes/bioethics/human-genetic-data/.

UNESCO. 2005. *Universal declaration on bioethics and human rights*. Paris. http://www.unesco.org/new/en/social-and-human-sciences/themes/bioethics/bioethics-and-human-rights/.
United Nations. 1948. *The universal declaration of human rights*. New York. http://www.un.org/en/documents/udhr/index.shtml.
United Nations. 1966. *International covenant on civil and political rights*. New York. http://www.hrweb.org/legal/cpr.html.
Wadman, M. 2000. Geneticists oppose consent ruling. *Nature* 404(6774): 114–115.
Wagstaff, A. 2011. International biobanking regulations: the promise and the pitfalls. *Cancer World* 42: 22–29.
Wendler, D., and E. Emanuel. 2002. The debate over research on stored biological samples: What do sources think? *Archives of Internal Medicine* 162: 1457–1462.
Wilcox, A.J., et al. 1999. Genetic determinism and the overprotection of human subjects. *Nature Genetics* 21: 362.
World Medical Association (WMA). 1955. Principles for those in research and experimentation. *World Medical Journal* 2: 14–15.
World Medical Association (WMA). 1964. *Declaration of Helsinki: Ethical principles for medical research involving human subjects*. http://www.wma.net/en/30publications/10policies/b3/.

Chapter 14
The Body, the Law, and the Market: Public Policy Implications in a Liberal State

Ingrid Schneider

14.1 Introduction

The human body is often seen as the epitome of human nature. However, the human body also enjoys a special moral and legal status, as it is intimately linked to individuality and the human person. Over the last decades, we have witnessed a deep transformation in the availability and circulation of human body parts. The body's substance—its "flesh and blood"—and not just its potential for work is being utilized for scientific, diagnostic, therapeutic, and industrial purposes. Nowadays, there is hardly any part of the human body that cannot, in some way or another, be used in biomedicine. This new view of the body as a source, a mere resource, for biomedical purposes has raised a panoply of medical, ethical and political questions. La Mettrie's (1784) famous "L'homme Machine" (Vartanian 1960) has thus transmuted into "L'homme Resource" in contemporary times. In this contribution, the interaction of the law, the market and public policy in the governance of body parts will be analyzed. Special attention will be devoted to public policy implications in a liberal state in which autonomy and self-determination form constitutional foundations. How is the donation and circulation of human body parts to be regulated? What are the norms, and how are new opportunities and demands deriving from advanced biomedical practices to be incorporated? What is the role of the state, and what can and should be left to the laws of the market? How does the law conceptualize these new developments? Can and should demands for a regulated market in human organs be met, will it comply with the rule of law, and can it be reconciled with demands for a just and equitable social policy?

I. Schneider (✉)
FSP BIOGUM (Centre for Biotechnology, Society, and the Environment), Universitaet Hamburg, Lottestr. 55, 22529 Hamburg, Germany
e-mail: Ingrid.Schneider@uni-hamburg.de

M. Albers et al. (eds.), *Human Rights and Human Nature*, Ius Gentium: Comparative Perspectives on Law and Justice 35, DOI 10.1007/978-94-017-8672-0_14,

To respond to these questions, this contribution proceeds as follows: The first section of this chapter lays out the legal and philosophical groundwork. First, the universal norm of non-commercialization of the human body will be explored as it is enshrined in legal documents, declarations and professional codes of ethics. Second, the major philosophical lines of tradition with respect to the relationship of humans towards their bodies will be reconstructed. As will be argued in the third step, philosophical foundations and deontological reasoning for non-commodification of the human body are predicated on Immanuel Kant's philosophy of morals. The non-instrumentalisation paradigm imposes duties to oneself and upon society, and seems to restrict inclusion of body parts in civil contracts and for the laws of the marketplace. The fourth section explores how the continental European law deals with the duality of the human body as both a person and a "material thing".

The second section of the article puts these constitutional norms and deontological principles to the test. Such norms have come under fire as liberal philosophers, economists, and medical professionals have proposed a regulated market for organs in order to overcome organ shortages. Therefore, in the second section, the models and arguments brought forth to justify financial incentives for organ procurement are subjected to logical and empirical scrutiny. This contribution provides a critique of such proposed market solutions.

14.2 The Norm of Non-commercialisation of the Human Body

The non-commercialisation of the body is an internationally accepted norm for which universal validation extends beyond cultural and social differences. Broad consensus about this norm has been reached, and it is enshrined in numerous international documents, declarations, and recommendations. The Charter of Fundamental Rights of the European Union (2000/C 364/01) proclaims in Article 3:

1. Everyone has the right to respect for his or her physical and mental integrity.
2. In the fields of medicine and biology, the following must be respected in particular: (...)
   - the prohibition on making the human body and its parts as such a source of financial gain.

Also, several statements of the World Medical Assembly expressly condemn any monetary acts by which human organs, ova, sperm, or embryos are offered for purchase or sale.[1] All financial incentives to donate an organ are rejected:

> "The World Medical Assembly condemns the purchase and sale of human organs for transplantation". (Statement on Living Organ Trade of the 37th World Medical Assembly

[1] In this contribution, the differences between regenerative, non-regenerative, and reproductive bodily substances cannot be tackled. See for reproductive tissues Schneider (2003b, 2009), and for a taxonomy Schneider (2008, 26–50).

> (WMA) in 1985.)[2] "Financial incentives for providing or obtaining organs and tissues for transplantation can be coercive and should be prohibited" (Statement on Human Organ and Tissue Donation and Transplantation, 52nd WMA in 2000).[3] It also called on national medical associations "to urge their governments to enact legislation prohibiting commercial transactions in human ova, sperm and embryos" (Resolution on the Non-Commercialisation of Human Reproductive Material, adopted by the 54th WMA in 2003).[4]

The World Health Organization (WHO) also calls for a prohibition on any kind of compensation for the donation of human body parts. In their Guiding Principles on Organ Transplantation from 1990, the fifth principle states:

> The human body and its parts cannot be the subject of commercial transactions. Accordingly, giving or receiving payment (including any other compensation or reward) for organs should be prohibited.[5]

Similarly, Article 21 of the Council of Europe's Convention on Human Rights and Biomedicine (2002) contains a non-profit rule for human body parts "as such":

> "The human body and its parts shall not, as such, give rise to financial gain or comparable advantage." This provision is specified further in Article 21 and 22 of the second additional protocol, whereby Article 22 once again explicitly stipulates: "Organ and tissue trafficking shall be prohibited."[6]

Several resolutions of the European Parliament (1993, 2005) and of the Council of Europe's Committee of Ministers (1978, 2001, 2003, 2004) condemn all organ-trafficking and financial incentives for the procurement of organs and recommend an express ban be adopted in the penal code.[7] These regulatory documents and recommendations are in broad agreement as to the inadmissibility of any relinquishment of organs for monetary gain or any other form of payment.

The Strasbourg-based Council of Europe is currently preparing a convention against trafficking in human organs. If adopted by the Committee of Ministers of the 47 member states, it would be the first legally binding international instrument devoted to combating organ trafficking. It may also include a road map for an additional protocol against the trafficking of human tissues and cells.[8]

However, all of these aforementioned supranational documents, whether as professional codes of practice or as standards of international law, have the character of "soft law" insofar as international sanctions are absent, and degrees of binding

---

[2]http://www1.umn.edu/humanrts/instree/organtrade.html

[3]http://www.portaledibioetica.it/documenti/001328/001328.htm

[4]http://www.wma.net/en/30publications/10policies/r1/

[5]http://www.who.int/ethics/topics/transplantation_guiding_principles/en/index1.html

[6]Additional Protocol to the Convention on Human Rights and Biomedicine concerning Transplantation of Organs and Tissues of Human Origin, Strasbourg, 24.1.2002, http://conventions.coe.int/Treaty/en/Treaties/Html/186.htm

[7]See references in: Breyer (2006).

[8]Council of Europe: Towards a Council of Europe convention to combat trafficking in organs, tissues and cells of human origin. Doc. 13082, 2012, available at: http://assembly.coe.int/ASP/Doc/XrefViewHTML.asp?FileID=19236&Language=E (last accessed 26 March 2013).

effect vary according to national enforcement powers. At the same time, however, these principles have constitutional character and quality given that they must be understood as fundamental principles and values. As such, they enjoy high ranks both in European and in international legal hierarchy and thereby extend to ordinary law. The prohibition of organ trafficking is found in a number of national legal systems (e.g., the criminal provisions of the German Transplantation Law, effective 1997).

14.3 Philosophical Foundations of the Exceptional Position Held by the Human Body

These norms must be considered in light of the exceptional position held by the human body. On the one hand, the human body is prerequisite for personal existence and represents a precondition for being a human subject. On the other hand, it is possible for a human being to have command over his or her own body and to dispose of it as s/he sees fit—within certain limits—as if it were an object or a mere "thing".

In simplified terms, two distinct philosophical lines of tradition with respect to the personal relationship and attitude to one's own body can be discerned:

- In the **Cartesian-Instrumental Tradition**, the body is understood to be an instrument of the person's will. One strives for the highest possible control over the body, both in its shape as well as in its functions. Bodily substances can be regarded as biological goods, resulting in broad availability as well as potential transferability of such body parts.
- In the **Kantian and Judaeo-Christian Lines of Tradition**, the body has intrinsic moral significance and possesses a cultural symbolic quality that is essential for personal identity. The intactness of the body and its integrity are high-level principles and values. The predominant model of relationship to the body is stewardship (Murray 1987). Attentiveness to and custody of one's own physical being is viewed as a responsibility to oneself (or toward a transcendental institution), but it is not considered as ownership over the body (George 2001).

Both types of bodily relationships impact issues of accountability in the use of bodily substances on the part of the medical profession or the pharmaceutical industry. The former category stipulates, first and foremost, that appropriate mechanisms for securing the autonomy of those giving bodily substances must be met, and, if necessary, appropriate monetary compensation must be considered. In contrast, aspects of inalienability are of significance in the latter category. These range from giving body parts away as a donation or gift to absolute inalienability which means unavailability altogether for any third party.[9] The informed consent of the person

[9] For "market-inalienability" see Radin (1987), for introducing an additional category of "absolute inalienability" see Schneider (2003b).

in question is a necessary but still not sufficient provision to morally justify the removal and transfer of bodily substances. Both philosophical models imply that a line should be drawn between legitimate and illegitimate uses in order to avoid abusive and exploitative practices, but these lines are defined differently. In the first category, the spectre of possible uses of human body parts consists mostly of individual choice and may include experimental, cosmetic, and industrial use. In the second category, use is mostly restricted to therapeutic purposes; respective biomedical practices must be responsible and must consider the welfare of the general public.

14.4 Kantian Maxims and Duties

The philosophical reasoning for prohibiting the commercialisation of organ donation is grounded in the deontological tradition. It relies primarily on three topoi from Kant's doctrine of virtues and duties; Kant famously contrasted dignity with price, as human beings have an intrinsic worth, i.e., *dignity,* which makes them valuable "above all price". Therefore, a human being should never be reduced to his or her "superficial value" or usefulness, as it has an unconditional value that has no price (Kant 1793, AA VI: 435–436).

Closely related to this line of reasoning is the question of self purpose or, accordingly, the non-instrumentalisation maxim deriving from the second formulation of the categorical imperative, which stipulates: "Act in such a way that you treat humanity, whether in your own person or in the person of another, always at the same time as an end and never simply as a means" (Kant 1785, AA IV: 429). Organ donation motivated by financial reasons is insofar understood as self-utilisation and exploitation.

A third force in this reasoning is the "duties to oneself", under which Kant (1793, AA VI: 423) includes, amongst others, the prohibition of self-mutilation. Although these responsibilities to oneself are sometimes handled rather polemically in current philosophical discourse, the ethical standard not to damage oneself continues to be an important reminder of the duties that arise from human dignity (Steigleder 2002, 268; Chadwick 1989, 129–40).

14.5 The Body in Law: Property Rights and Personality Rights

The intermediary position of the body between human person and thing is also reflected in the law. On the one hand, as a whole, the body is part of the person and thus, in terms of the old Roman law, a "res extra commercium". On the other hand, separate body parts are treated—though in a restricted manner—as things and, therefore, can potentially be subjected to property rights and market mechanisms.

The problems that arise from this dual position are solved in legal doctrine by the idea of an "overlap" of personality and property rights (Schünemann 1985; Halàsz 2004; Schneider 2002). However, this legal doctrine creates still further tensions and frictions with regard to individual rights over detached body parts: How and to what extent can a person make ex-ante dispositions over parts of his body? Furthermore, it begs the question, how far a person may make use of her own body as an object without thus undermining her position as a legal subject, and hence her free will and freedom of choice.

When pondering and weighing rights over the body, both personality rights and property rights should generally be distinguished as different legal approaches in continental European Law; as a legal interest (*Rechtsgut*), personality rights protect the physical integrity of the body, health and self-determination. Above all else, personality rights offer protection against unauthorised violations of privacy, be it from the state or from a third party. Property rights, in contrast, imply a right to use and exploit for monetary gain. Property rights protect against wilful damage, enable a contractual transfer to a third party, and, in principal, also include *in personam* access. Privacy laws derived from personality rights are thus designated as defense rights which protect oneself from intrusion by others. Property rights, however, pertain first and foremost to positive rights, i.e., authority of disposition. For the cases in question, they primarily authorise the extraction of body parts; however, for the most part, this does not extend towards determining who can be the beneficiary of a human organ or the recipient of another bodily substance.

In the following section, we will turn towards the debate over whether financial incentives or outright sale of human organs should be allowed. Thus, the status of human body parts in legal practice and in public policy will be critically discussed by a case study on organ donation.

14.6 Push for the Commercialisation of Organ Donation

For the past four decades, there has been a push to commercialise organ donation[10] initiated by transplant physicians and supported by legal professionals, economists, and philosophers. The utilitarian argument focuses on the shortage of organs. This is predicated on the premise that financial incentives will increase the availability and supply of organs. Furthermore, it is postulated that this would mean a win-win-situation for both the monetarily compensated organ donor and for the organ recipient, at least in the case of kidney transplantation. Moreover, it is viewed as a cost-savings for society at large because dialysis treatment is more expensive than kidney transplantation.

The central argument is built around the autonomy of the organ seller. In this context, autonomy is meant as a comprehensive right to command over one's

[10]An overview of such initiatives is given in Schneider (2003a).

own body, including the "right to inflict injury to oneself". Body parts are treated as property. A further essential feature of this argument is based on the liberal paradigm. Protagonists of liberalism are against what is alleged to be state interference, branding it a form of obsolete paternalism and excessive over-regulation. They proclaim instead the "freedom" to sell one's organs (Wilkinson 2003).

Paradoxically, however, a strong etatism can be found in such a suggestion. Proponents call for a *regulated* organ market of which the central element is the restriction of organ circulation to the national realm; another condition is minimum standards for willing donors, e.g., they should be of a legal age to give consent, but also that there is no *in personam* obligation to sell organs. In addition, a fixed flat rate has been recommended for the sale and purchase of an organ. The allocation of organs should be organized and monitored by a central governmental agency, and distribution or assignment of organs should be made regardless of a person's ability to pay (Breyer 2002, 111–123). In the European context, this means that the national health service or health insurance should cover the cost of organ purchase. Despite the apparent call for liberalisation and de-regulation of organ transplantation, the model of a "regulated market for organs" is thus connected to greater re-regulation: Oversight by governmental administration and intervention is linked with a marked optimistic view of regulation, which assumes a strong federal government is in place along with effective and efficient bureaucratic processes in order to avert abuse.

14.7 Incentive Models for Post-mortem Organ Donation

Using financial incentives to increase the supply of organs for transplantation applies to organ donation after brain death (i.e., post-mortem donation) as well as to donations from living donors. For the post-mortem incentive models, it must be analytically distinguished who is to be addressed and who can dispose of the organs in question:

Ex-ante declarations are directed toward the presumptive donors; several models discuss that willingness to donate organs would be increased through a reduction in health insurance costs or by tax-deductions.

However, the feasibility of this model fails already when economic effectiveness is considered because dissociated brain death seldom occurs. Given that there is an average donor potential of 30–60 persons per year in a population of one million inhabitants (Breyer et al. 2006, 42), it is hardly worth compensating any willingness to donate organs in the first place. If a person's health insurance costs are only marginally reduced when they complete a donor card, then it is unlikely that the motivation to sign up will significantly increase. Furthermore, a person's willingness to donate organs usually increases when the person himself is in need of an organ, most often the case with those sick in their advanced years. These individuals tend to be unsuitable donors.

Ex-post incentive models, in contrast, are directed toward the next of kin. For example, perhaps willingness to donate organs increases if burial expenses are assumed.

However, this model runs into fundamental ethical problems; such an inquiry is tied to monetary value and would place additional emotional burden on the next of kin. Immediately following the loss of a beloved one, relatives would be subject to acute pressure to make a decision whilst amidst an already exceedingly difficult situation. Under such a premise, asking for consent to organ donation with a link to monetary offers may be disrespectful and a breach of the reverence and obligations incumbent on the next of kin. Bioethicist Thomas Murray highlights that the intimate social and moral connectedness to a recently deceased relative is transformed by the infiltration of pecuniary aspects:

> The relationship is transmuted (...) to something more like the relationship that owners of old cars have with their vehicles, which are now being stripped of still-useful parts in a salvage yard: impersonal, with the hope that a little utility and a little money can be extracted from the now lifeless hulk (Murray 1996, 117).

As of yet, all pushes for monetary incentives for organ donation post-mortem have failed politically in spite of several legislative initiatives having been put forward in the USA and some other countries.

14.8 Compensation for Living Donations

Economists have suggested that kidney and partial kidney living donations could potentially create a supply equal to the demand. Living donation is an elective procedure, a supposed benefit of which is the achievement of a better medical outcome for the organ recipient. However, it is contended whether the risks for the paid donor—the danger of complications, serious injury, disability, and even death, in extreme cases—can be deemed acceptable (Feuerstein 1995; Schneider 2004).

One of the most significant arguments in favour of the sale of organs equates the purchase of body parts with the purchase of manpower. Those who intuitively sense and assert a fundamental difference between the two are often countered on the grounds that the distinction is merely incremental. After all, fire fighters, miners, race car drivers, and mercenary soldiers all place their health and even their lives on the line as part of their job description (Schutzeichel 2002).

A noteworthy counter to this argument is that manpower is only bought for a limited time. The worker remains the owner of his/her own body and can terminate the employment contract at any time. In contrast, the sale of a non-regenerative organ is a one-time irreversible act. The removal of an organ is invasive and necessitates harming the body, harm that is not to be simply accepted as an indirect consequence (ERM 2005, 69).

Semantic strategies seeking to gradually remove the taboos surrounding the trafficking of organs can be summed up with the phrase "rewarded gift", a term coined by the Indian medical doctor C. T. Patel in 1987.[11] With the aim of generating acceptance for new organ sale options, a discursive prettification is taking place. While shying away from "aggressive" terminology such as organ trafficking and rampant commercialisation, monetary rewards are declared "compensation" for pain and suffering. However, these linguistic strategies misjudge the fundamental change taking place from the introduction of monetary means into the transplantation system. Before examining this transition in more detail, attention will be paid to the empirical evidence for which a "win-win-situation" is propagated.

14.9 Improving the Living Situation of Organ Sellers? Empirical Studies

To some extent, some proponents even fashion the sale of organs as a form of welfare or development aid. They argue that it would be cynical and hypocritical to deny impoverished individuals this opportunity to improve their living condition. The prohibition of selling their organs would force people to live in conditions that they themselves assess as worse than selling a kidney (Wilkinson 2003, 120).

Conversely, one may argue that there is no empirical evidence to support that people are better off, able to be freed from social plight, or can ensure a sustainable livelihood with the start capital obtained as a result of selling an organ. In fact, a number of empirical studies show that by no means organ sale leads to a sustainable improvement of life.

In 2002, the Journal of the American Medical Association published the results of a survey of 305 kidney sellers in Chennai, India. The survey participants were, on average, 6 years postoperative. The survey yielded the following results:

- 71 % of kidney sellers were women, many of whom were urged, to some extent, by their husbands to contemplate kidney sale.
- 96 % of participants sold their kidneys to pay off extensive debt of their family.
- On average, each person received US-$ 1,070 per kidney.
- Three-quarters of the respondents remained in debt following payment, and the number of participants living below the poverty line actually increased.
- Approximately 86 % reported deterioration in their health after the kidney removal.
- Nearly four out of five participants (79 %) would not recommend that others sell a kidney.

The authors concluded that, "among paid donors in India, selling a kidney does not lead to a long-term economic benefit and may be associated with a decline in health" (Goyal et al. 2002, 1589–1593).

[11] According to Patel in Daar (1991, 182). See also Oberender and Rudolf (2003).

These results have been reinforced by qualitative studies: Thirty kidney sellers interviewed in a slum in Chennai, India found themselves once again in crippling poverty and overdebted within a few years of their donation. In an area known as the "kidney zone", organ vendors intensified their search for donors while creditors, at the same time, even aggressively enforced debt collection. Furthermore, relatives of affluent kidney patients shied away from contemplating altruistic kidney donation and instead pointed to the opportunity that a kidney could be bought from another source, outside the family (Cohen 1999). Similar results were shown in a study from Pakistan, which also indicated a stigmatisation of families and social rejection of whole communities living in poverty and known as kidney resources, and, likewise, the study reported a decline in altruistic donations after brain death in parallel to the increase in kidney sales (Moazam et al. 2009).

Iran is the only nation worldwide that has a regular purchasing system for the procurement of kidneys, institutionalised by the Iranian state since the end of the 1980s. Approximately 1,500 paid kidney transplantations are performed each year. In addition to a state controlled payment of approximately 900 euros, a kidney seller receives 1 year of free healthcare and, as a general rule, a negotiable sum from the recipient after the operation (Griffin 2007; ERM 2005, 26–27). A survey of 300 Iranian kidney vendors found that two thirds of donors struggled with the negative impacts of the donation on their workplace, and a majority suffered from severe psychosocial deterioration, such as depression, anxiety, and isolation. None of the respondents escaped poverty and debt. Four out of five could not take advantage of post-operative treatment. Eighty-five percent of the respondents would not sell their kidneys again if given the chance (Zarghooshi 2001).[12]

Even if there is doubt as to whether these results from India and Iran can be carried over to industrial nations, in general, a one-time monetary payment (e.g., winning the lottery) generally does not lead to a sustainable improvement in living conditions. Instead, other factors such as a solid education and secure employment are necessary for the stabilisation of precarious living conditions.

[12]The result, as summed up by the author states: "Poverty prevented 79 % of vendors from attending follow-up visits, and vending caused negative effects on employment in 65 %. Of the families 68 % strongly disagreed with vending, which caused rejection of 43 % and increased marital conflicts in 73 % of vendors, including 21 % who divorced. There were 70 % of vendors isolated from society, and 71 % had severe de novo postoperative depression and 60 % anxiety. Vending caused somewhat (20 %) to very (66 %) negative financial effects. It also had negative effects on the physical abilities in 60 % of vendors, and 80 % were dissatisfied with postoperative physical stamina, which was decreased mostly by depression. Of the vendors 37 % concealed the truth of kidney sale from anyone, 14 % disclosed it only to spouses, 43 % to first generation relatives and 94 % were unwilling to be known as donors. The mental preoccupation with kidney loss was usually (30 %) to always (57 %) present and interfered negatively with vendor life, and 62 % reported negative effects on sense of being useful. Effects on general health were somewhat (22 %) to very (58 %) negative. When thinking about vending, the majority cited negative feelings. They responded that if they had another chance 85 % would definitely not vend again, and 76 % strongly discouraged potential vendors from "repeating their error". Half the vendors were ready to lose greater than 10 years of life and 76–100 % of properties to regain kidneys" (Zarghooshi 2001, 1790–1799).

14.10 Systemic Change: From Gift Exchange to Regulated Market

According to prevailing social and cultural norms, gifting organs is subject to social modes of gift exchange[13] as a moral economy: Organs are given voluntarily. The donation occurs with the expectation that the organ will be accepted, appreciated, and put to therapeutic use. A donation that proceeds free of charge is based on the principle of generalised reciprocity, according to which, in principle, every person can be an organ donor and also a recipient. It is attached to an ideal of communitarisation and morally—but not juridically—implies a responsibility to each other. Organs circulate in a "sphere" outside of market mechanisms.

In contrast, the commercialisation approach strives for a market model in which organs are considered property, and thus open to contractual mechanisms. However, proponents of this approach never expressly call out for the sale and purchase of organs to become fully subjected to the laws of the marketplace, as this would allow for full freedom of contract between organ vendor and buyer and pricing based on supply and demand. Instead, they advocate strong restrictions on the freedom of contract as well as comprehensive state regulation. Also, they call for a central regulatory agency to be established. The allocation of organs should not be dependent on a recipient's ability to pay. Because leaving organ allocation subject only to the purchasing power of those desperate for an organ would result in taking organs from the poor to give to the rich, irrespective of medical necessity and urgency.[14]

Pertaining to the relationship between these two different exchange models, gift-giving versus (regulated) market, in a given society, it is questionable whether a co-existence would be possible over a longer period of time. This respective philosophical dispute already took place in the 1970s and is known as the famous Titmuss-Arrow Controversy (Titmuss 1971; Arrow 1972). Subsequently, a multitude of theoretical and empirical research findings have strongly pointed to the likelihood of *crowding-out effects* that lead to the morally-motivated gift being superseded and fully replaced by the market-based incentive system (Benkler 2004). As a result, it is to be expected that organ supply would either result in a zero sum game (more purchased organs but fewer altruistically donated organs), or an even greater reduction in the overall number of organ donations would occur (Gold et al. 2001, 19).

---

[13]For the anthropological concept of gift exchange applied to transplantation see: Fox and Swazey (1992), Kalitzkus (2003), Motakef (2011), and Sanal (2011). For objections to the model of gift exchange see: Waldby and Mitchell (2006).

[14]See Reich (2000). The model of a state regulated organ market raises numerous unresolved questions, for instance, as to the relative arbitrary pricing of any kind, and the question of "reasonable and fair" prices, the handling of price variations (e.g. in various federal states or in EU countries) and to the enforceability as well as effectiveness of regulatory and monitoring procedures.

14.11 Consequentialist Objections to the Commercialisation of Living Donations

In particular, consequentialist philosophical reasoning allows for a rejection and condemnation of organ sales since it calls attention to the socio-economic selectivity in donor recruitment: Let us assume that a legal and equitable organ allocation could be secured through pooling, i.e., organ purchases organised by the state would establish a (virtual) pool of available organs; subsequently, organs would be allocated according to urgency, waiting time, and efficiency of expected results. Even assuming that such a pool model would work with respect to the distribution of collected organs, there would still be a social bias with respect to those willing to provide/sell an organ; procuring organs would be one-sided and at the expense of the economically underprivileged population. This would only exacerbate already existing social inequalities. Economically vulnerable groups would be expected to give up their right to physical integrity. The fundamental civic norm of equality between citizens with respect to the right to physical integrity would thereby be abandoned. Therefore, introducing a regulated organ market would fundamentally reshape, dislocate, and shift political axes with respect to the relationship between freedom and equality; equality norms would be sacrificed for the sake of liberty. However, the question arises, *whose* freedom is being defended?

Permitting the sale of organs will lead to a social imbalance in favour of the organ recipients. The proclamation of material self-exploitation puts a price on what was until now considered priceless. The barrier of the skin is permeated, and the body's interior thus becomes a measurable and commensurable entity of economic worth. What formerly was untouchable and sacrosanct has thus become accessible. In other words, it is the physical integrity of a social underclass which is brought into question and which is declared an organ-bank for society at large. Hence, the function of the debate is to create financial access to what was up until now not for sale: The vital parts of another human being's body. By allowing—however regulated—a system for the sale of organs, members of vulnerable social groups are made to bear the burden of the imposed decision to inflict bodily harm for the sake of monetary gain.

14.12 International Dimension

All models advocating a "regulated organ market" stipulate that access and supply to such a market solution must be restricted to the national territory. This is a prerequisite which seems to be unrealistic in a globalised world given that the transplantation system is already organised on a transnational basis. In Germany, for instance, living donations are legally restricted to family members and those with close personal ties (German Transplantation Act §8 (1)), but there are no restrictions upon the nationality of donors. A restriction of living donors to German citizens or persons with permanent residency in Germany would hardly be legally permissible (because of anti-discrimination laws). With regard to financially motivated organ sales in times of economic difficulty, restrictions that would exclude those receiving social welfare benefits or heavily indebted individuals from organ donation would be easily susceptible to manipulation and thus could not be effectively enforced. Regulations always create new requirements for monitoring as well. Paradoxically, regulative "obstacles" set in place to preserve the "genuine autonomy" of organ vendors will most probably promote a tendency to turn to organ sellers in other countries, where regulatory requirements are weaker. Thus, medical tourism to countries with more "favourable" conditions for supplying organs would increase.

Moreover, the assumption that a regulated organ market would dry out a "black market in organs" (Oberender and Thomas 2003) is utterly implausible. It is more likely that a regulated organ market would increase medical tourism as patients would seek out the most inexpensive organ (Schlitt 2002; Joralemon 2001). Consequently, a gradual removal of taboos clears the way for organ trafficking, and, on the whole, international inequalities are exacerbated even further. An international regulatory race to the bottom, undercutting other countries' ethical and legal standards, could even be the consequence (Schneider 2013). In view of the fact that, according to UN estimates, more than a billion people worldwide must survive on less than \$1.00 per day, a rapid "price decline" in the international organ market would be foreseeable. The mere fact that more than 1.3 billion people have no access to clean drinking water begs the urgent question, what are the priorities being set in healthcare policy: Which lives are being saved? Who can afford to improve their quality of life through transplantation? These questions concerning international social injustice must be brought back into the organ transplantation debate (Scheper-Hughes 2002).

14.13 Consequences of a Systemic Change from Gift to Market

The gradual removal of taboos with regard to the payment for organ donation evokes a qualitative shift, namely a demoralisation of the relationship to oneself and to others. When a person has command over his/her own body parts as if they are objects, this leads to an altered relationship to oneself. The organ as an object is thus separated from the physical being and identity of a person. The relationship to others is also altered; the body part of another thus becomes a commodity over which one can lay claim (Taupitz 2007; Potthast et al. 2010).

Therefore, the improved quality of life of dialysis patients through kidney transplantation or a life saving partial liver transplantation occurs at the expense of social cohesion, and further erodes solidarity with the socially underprivileged who rely on community support. From a socio-political perspective, this means that the right to physical integrity of the economically weak is forfeited for and on behalf of autonomy. In other words, it is rarely, if ever, the high school teacher who opts to sell a kidney in order to finance a journey to the Seychelles instead of spending holidays at the local seaside; it is instead the unemployed or those who are in extreme debt who have to make the difficult choice. It will be the individuals that find themselves in hopeless situations who may come to see their organ as their "last financial asset".

To postulate autonomy in this case is to glorify such acts of desperation as rational acts of freedom and self-determination. It remains noteworthy that this demand for the right to self-determination does not stem from the poor or disenfranchised. Nowhere in the world have citizens founded a civil society organisation in order to fight for the right to sell their organs. This absence of "self-articulation" and self-organisation of willing organ vendors is significant as it points to the implicit partisanship of scholarly pushes for the establishment of an organ market. Although these pushes are said to advocate the "autonomy" of the compensated organ donors, they are often more concerned with the vested interest of presumptive organ recipients. Accordingly, it is those who identify with the organ *purchase* that spur on such strategies. It is always with regard to the body of *another* person that "autonomy" is used to claim a right of access. One's own body is never included in this construct. Primarily, the *freedom* of the organ recipient is being defended at the expense of social *equality* with regard to the inviolability of bodily integrity. The sense of solidarity with those who are sick and "in need of organs" is at the expense of social solidarity with economically vulnerable social groups.

14.14 Constitutional Law and Redistribution Arguments

At times, the debate gets narrowed down to juridical arguments of constitutional law. It is in this context that it is frequently argued that the sale of organs remains a voluntary decision of self-determination (in accordance with Article 2 (1) of the German Basic Law). In a liberal constitutional state, self-inflicted injury and self-instrumentalisation must be accepted. Protecting an individual "from him/herself" is not the function of the state, and should not be enforced, least of all by means of penal law.

There is nothing wrong with this, but one should also bear in mind that the jurisprudence of Germany's Federal Constitutional Court allows for an intervention into the liberties of fundamental rights under certain limited circumstances in order to prevent the affected parties from inflicting major harm on themselves. Laws, for example, requiring the use of protective helmets and seat belts, prohibiting doping, or banning (excessive) hashish consumption also serve to protect citizens "from themselves" and are constitutionally permitted (König 1999). Therefore, lawmakers and state authorities enjoy a margin of appreciation as to the implementation of such self-protective measures ordered from top-down. They must also consider that the prerequisites for freedom and for capacity to act are both physically and socially mediated.

Furthermore, the decision to be made is explicitly a political one: A regulated organ trade threatens the individuals' freedom and self-determination by economic mechanisms. There is a social responsibility to avert such situations through social policy measures. It is a humanitarian and socio-ethical imperative to offer support to those in either economic or physical hardship. Permitting organ sales, on the other hand, allows for some to profit on the despair of others.

As of yet, a citizen's body is exempt from redistribution for the purpose of fairness. Redistribution, for example, occurs through taxes. The fact that one has two healthy eyes does not obligate oneself to donate a cornea to a blind person (Herrmann 2011).

Defendants of explicit prohibition on organ trade have been accused of paternalism. However, it has been hardly noted that those proponents arguing in favour of a regulated organ market implicitly subscribe to the ideas of solidarity and social responsibility (*Sozialpflichtigkeit*) with respect to any human body without first providing a plausible reason. This attitude could also be denounced as paternalism or even as "biosocialism". If and to what extent the state is allowed to have command over the body of its citizens remains a question to be socially and politically negotiated. However, a "right" to the organ of another cannot exist.[15]

Therefore, such protagonists opting for the "liberalisation" of transplantation law usually ask: "Is the state allowed to prohibit the sale of organs as an encroachment on the individual's freedom of contract?" Framing the question like this is erroneous. The question is rather, "Do we want to live in a society, in which human lives are saved through organ transplants carried out *in this way,* in which the quality of life of kidney patients is improved *in this way*—namely, by accepting to endanger the lives of others?" The cost to society for a potential increase in the organ supply is high. Furthermore, it remains to be seen whether or not the assumption that a substantial increase in organ procurement can be reached by financial incentives is justified.

14.15 Conclusion: Unintended Side Effects and Impact on Transplant Medicine as a Whole

Market-based approaches operate according to a logic that assumes economic efficiency and rationality. With this fixation on incentives to increase organ supply, unintended side effects are brushed off.

[15] Walzer (1983) observed that besides the body, there are also other "things" that should not be allowed to be bought with money. That is if the foundation for plurality and equality in a democratic state governed by the rule of law should not be undermined. Among these "things" he mentions, for example, the right to vote, judicial judgments, public offices, polygamy, education and certain fundamental social securities. See similarly: Sandel (2012).

However, it is to be assumed that the introduction of financial incentives will over time lead to

- a shift from post-mortem donations after brain death to living donations; and
- a crowding-out effect that leads to a shift from uncompensated living donations between family members toward the sale of organs between strangers.

With regard to the second crowding-out effect, it is noteworthy that an uncompensated living donation in the hitherto existing system is limited to family members and closest friends. It thus functions inasmuch as an "egalitarian" happening of fate because sickness may befall one's parents, siblings, life partner or children. Concern for a specific other, whether close friend or family member, is generally more pronounced than the concern felt for an abstract, distant other. However, this is altered through the introduction of money into the system, based on the consideration: "Why should I put a family member at risk when I can buy an organ from a stranger?" (Cohen 1999, 161; Cohen 2001).

Money thus liberates the wealthy from the anguish of having to ask a beloved one to endure the risks of organ donation. Hence, the donor's sacrifice is made less palpable to the recipient who does not want to see the scar on his next of kin. In practice, this means a shift of health risks to those occupying a lower status in the social hierarchy.

With regard to the former crowding-out effect, the undermining of post-mortem donations would strengthen a trend that has already gained footing in many Western nations: The number of post-mortem organ donations has stagnated while the number of living donations has increased.[16] Certainly, however, permitting organ procurement by means of financial compensation will brings to light conflicts within the transplantation community itself, as cardiac surgeons remain reliant on post-mortem organ donations to perform heart transplants.

A legalisation of organ sale, even if it remains nationally confined, would irreparably damage the credibility and social trust in transplant medicine. The debate therefore shows little sensibility for the social/moral tightrope upon which transplant medicine walks.

In conclusion, the debate over the market-based regulation of organ sales is determined by an excessive conceptualisation of individual rights of disposal over one's own body. However, as has been demonstrated, this postulate of autonomy masks a one-sided support of the vested interests of potential organ recipients. At the same time, it involves implicit social obligations and the imperative of redistribution with regard to vital body parts of human beings. From the perspective of public policy, pushing such debates and demands for introducing a "regulated market in organs" requires that the economically vulnerable groups surrender their physical integrity. As a matter of fact, these measures promote the social divide between

[16]In Germany, for instance, the rate of kidney transplantations by living kidney donors increased from 15.6 % in 2000 to 27.9 % in 2011 (German Foundation for Organ Transplantation DSO, http://www.dso.de/medien-und-presse/pressebilder-und-grafiken.html, last accessed 26 March 2013).

haves and have-nots. The financial models proposed will likely fail to measure up even to their own requirements of effective organ procurement. "Monetary incentives" and a "regulated organ market" are thus ineffective tools for abating organ shortage.

From a sober and pragmatic perspective, it has been shown that, despite several decades of intense debates among professionals, a regulated market for human organs cannot be implemented. As the article has argued, there are good reasons—from deontological, consequentialist, and social policy perspectives—to maintain the norm of non-commercialisation and to preserve the special status of the human body.

References

Arrow, K.J. 1972. Gifts and exchanges. *Philosophy & Public Affairs* 1(4): 343–362.

Benkler, Y. 2004. Peer production of survivable critical infrastructures. http://www.dourish.com/classes/readings/Benkler-CriticalInfrastructures.pdf. Accessed 27 Mar 2013.

Breyer, F. 2002. Möglichkeiten und Grenzen des Marktes im Gesundheitswesen. Das Transplantationsgesetz aus ökonomischer Sicht. *Zeitschrift für Medizinische Ethik* 48: 111–123.

Breyer, F., W.d. van Daele, M. Engelhard, G. Gubernatis, and H. Kliemt (eds.). 2006. *Organmangel: Ist der Tod auf der Warteliste unvermeidbar?* Berlin/Heidelberg: Springer.

Chadwick, R.F. 1989. The market for bodily parts: Kant and duties to oneself. *Journal of Applied Philosophy* 6(2): 129–140.

Cohen, L. 1999. Where it hurts: Indian material for an ethics of organ transplantation. *Daedalus* 128(4): 135–165.

Cohen, L. 2001. The other kidney: Biopolitics beyond recognition. *Body & Society* 7(2–3): 9–29.

Daar, A.S. 1991. Rewarded gifting and rampant commercialism in perspective: Is there a difference? In *Organ replacement therapy: Ethics, justice commerce*, ed. W. Land and J.B. Dossetor, 181–189. Berlin/Heidelberg: Springer.

ERM (Enquete Commission Ethics and Law). 2005. Parliamentary Study Commission, Zwischenbericht der Enquete-Kommission "Ethik und Recht der modernen Medizin": Organlebendspende. *Deutscher Bundestag. Drucksache 15/5050.*

Feuerstein, G. 1995. *Das Transplantationssystem Dynamik, Konflikte und ethisch-moralische Grenzgänge*. Weinheim/München: Juventa-Verlag.

Fox, R.C., and J.P. Swazey. 1992. *Spare parts: Organ replacement in American Society*. New York: Oxford University Press.

George, A. 2001. *Property in the human body and its parts reflections on self-determination in liberal society*. San Domenico: Department of Law, European University Institute (EUI).

Gold, S.M., K.H. Schulz, and U. Koch. 2001. *Der Organspendeprozess Ursachen des Organmangels und mögliche Lösungsansätze: Inhaltliche und methodenkritische Analyse vorliegender Studien; eine Expertise im Auftrag der Bundeszentrale für gesundheitliche Aufklärung (BZgA), Forschung und Praxis der Gesundheitsförderung* 13. Köln: BZgA.

Goyal, M., R.L. Mehta, L.J. Schneiderman, and A.R. Sehgal. 2002. Economic and health consequences of selling a kidney in India. *Journal of the American Medical Association* 288(13): 1589–1593.

Griffin, A. 2007. Kidneys on demand. *British Medical Journal* 334: 502–505.

Halàsz, C. 2004. *Das Recht auf bio-materielle Selbstbestimmung*. Berlin/Mannheim: Springer.

Herrmann, B. 2011. *Der menschliche Körper zwischen Vermarktung und Unverfügbarkeit, Freiburg im Breisgau*. Tübingen: Alber.

Joralemon, D. 2001. Shifting ethics: Debating the incentive question in organ transplantation. *Journal of Medical Ethics* 27(1): 30–35.
Kalitzkus, V. 2003. *Leben durch den Tod*. Frankfurt a. M./Göttingen: Campus-Verlag.
Kant, I. 1785. *Grundlegung zur Metaphysik der Sitten*: AA IV. http://www.geocities.com/nythamar/gms.html.
Kant, I. 1793. *Die Metaphysik der Sitten*: AA VI. http://www.korpora.org/kant/aa06/.
König, P. 1999. *Strafbarer Organhandel*. Frankfurt a. M./München: Lang.
Moazam, F., R.M. Zaman, and A.M. Jafarey. 2009. Conversations with kidney vendors in Pakistan: An ethnographic study. *Hastings Center Report* 39(3): 29–44.
Motakef, M. 2011. *Körper Gabe*. Bielefeld/München: Transcript-Verlag.
Murray, T.H. 1987. Gifts of the body and the needs of strangers. *The Hastings Center Report* 17(2): 30–38.
Murray, T.H. 1996. Organ vendors, families, and the gift of life. In *Organ transplantation: Meanings and realities*, ed. S.J. Youngner, R.C. Fox, and L.J. O'Connell, 101–125. Madison: University of Wisconsin Press.
Oberender, P.O., and T. Rudolf. 2003. Das belohnte Geschenk: monetäre Anreize auf dem Markt für Organtransplantate. *Wirtschaftswissenschaftliche Diskussionspapiere der Universität Bayreuth*, ed. Universität Bayreuth. Bayreuth: Lehrstuhl für Volkswirtschaftslehre III.
Potthast, T., B. Herrmann, and U. Müller. 2010. *Wem gehört der menschliche Körper? Ethische, rechtliche und soziale Aspekte der Kommerzialisierung des menschlichen Körpers und seiner Teile*. Paderborn: Mentis-Verlag.
Radin, M.J. 1987. Market-inalienability. *Harvard Law Review* 100(8): 1849–1937.
Reich, K. 2000. *Organspendeverträge: Geldzahlungen als Anreiz zur Organspende in den USA und in der Bundesrepublik Deutschland*. Hamburg: LIT.
Sanal, A. 2011. *New organs within us: Transplants and the moral economy*. Durham: Duke University Press.
Sandel, M.J. 2012. *What money can't buy: The moral limits of markets*. New York: Farrar, Straus and Giroux.
Scheper-Hughes, N. 2002. The ends of the body – Commodity fetishism and the global traffic in organs. *SAIS Review* 22(1): 61–80.
Schlitt, H.J. 2002. Paid non-related living organ donation: Horn of Plenty or Pandora's box? *The Lancet* 359: 906–907.
Schneider, I. 2002. Körper und Eigentum: Grenzverhandlungen zwischen Personen, Sachen und Subjekten. In *Konfiguration des Menschen: Biowissenschaften als Arena der Geschlechterpolitik*, ed. E. Kuhlman and R. Kollek, 41–59. Opladen: Leske + Budrich.
Schneider, I. 2003a. Ein Markt für Organe? Die Debatte um ökonomische Anreize zur Organspende. In *Organtransplantation, Organgewinnung und -verteilung*, ed. F.S. Oduncu, U. Schroth, and W. Vossenkuhl, 189–208. Perspektiven/Göttingen: Vandenhoeck & Ruprecht.
Schneider, I. 2003b. Gesellschaftliche Umgangsweisen mit Keimzellen: Regulation zwischen Gabe, Verkauf und Unveräußerlichkeit. In *Verkörperte Technik – entkörperte Frau: Biopolitik und Geschlecht*, ed. S. Graumann and I. Schneider, 41–65. Frankfurt a. M./New York: Campus-Verlag.
Schneider, I. 2004. Lebendspende: Kommerzialisierung des Unbezahlbaren? *Gen-ethischer Informationsdienst* 163: 33–40.
Schneider, I. 2008. Die soziale und rechtliche Regulation des Transfers von Körpersubstanzen: Kategorien, Klassifikationen und Normbildungsprozesse. In *Kultur und Bioethik: Eigentum am eigenen Körper*, ed. C. Steineck and O. Döring, 26–50. Baden-Baden: Nomos Verlag.
Schneider, I. 2009. Indirect commodification of ova donation for assisted reproduction and for human cloning research – proposals for supranational regulation. In *Altruism reconsidered: Exploring new approaches to property in human tissue*, ed. M. Steinmann, P. Sýkora, and U. Wiesing, 209–241. Farnham/Burlington: Ashgate.

Schneider, I. 2013. Race to the bottom or race to the top? Governing medical tourism in a globalized world. In *Bodies across borders: The global circulation of body parts, medical tourists and professionals,* ed. B. Parry, B. Greenhough, T. Brown, and I. Dyck. Farnham/Burlington: Ashgate Publishing (Forthcoming).

Schünemann, H. 1985. *Die Rechte am menschlichen Körper*. Frankfurt a. M./Göttingen: Lang.

Schutzeichel, C.I. 2002. *Geschenk oder Ware? Das begehrte Gut Organ. Nierentransplantation in einem hochregulierten Markt*. Münster/Bochum: LIT.

Steigleder, K. 2002. *Kants Moralphilosophie: Die Selbstbezüglichkeit reiner praktischer Vernunft*. Stuttgart/Weimar: Metzler.

Taupitz, J. (ed.). 2007. *Kommerzialisierung des menschlichen Körpers*. Berlin/Heidelberg: Springer.

Titmuss, R.M. 1971. *The gift relationship: From human blood to social policy*. New York: Pantheon Books.

Vartanian, A. 1960. *La Mettrie's L'Homme machine: A study in the origins of an idea*. Princeton: Princeton University Press.

Waldby, C., and R. Mitchell. 2006. *Tissue economies: Blood, organs, and cell lines in late capitalism*. Durham: Duke University Press.

Walzer, M. 1983. *Spheres of justice: A defense of pluralism and equality*. New York: Basic Books.

Wilkinson, S. 2003. *Bodies for sale: Ethics and exploitation in the human body trade*. London: Routledge.

Zarghooshi, J. 2001. Quality of life of Iranian kidney 'donors'. *The Journal of Urology* 166(5): 1790–1799.

Chapter 15
Should Society Guarantee Individuals the Right to Maintain "Normal Functioning"?: A Genetic Minimalist Approach in a Globalized World

Tetsu Sakurai

15.1 Introduction

Many arguments have been put forward against so-called "bioliberalism", which contends that prospective parents are morally entitled to use their reproductive freedom even to the extent of choosing their children's genetic traits through use of genetic technology. In my book, *Liberal Eugenics and Justice*, I have divided the existing counterarguments into three types, namely technological, political and philosophical, and then pointed out that none of these counterarguments is altogether decisive (Sakurai 2007, 133–76). Since the book was written in Japanese, let me briefly summarise its content here.

Liberal Eugenics and Justice attempts a philosophical inquiry into the logic of liberal eugenics, also known as bioliberalism. It examines the merits and implications of liberal eugenics and considers whether the principles of modern liberalism can persuasively impose any restrictions on liberal eugenics. Having traced the history of modern eugenics from Thomas Robert Malthus through Francis Galton to Hermann Muller in the book's first half, I found that there is a remarkable continuity from the Darwinian theory of evolution to contemporary bioliberalism. First, I noted that all the theorists dealt with here more or less agree that the process of natural selection is not functioning in contemporary society as it did in primitive times. Second, I discovered that these Darwinian theorists are generally optimistic about the prospect of using social policies and/or advancing technologies in order to carry out large-scale eugenics and to accomplish a grandiose artificial evolution of humankind.

T. Sakurai (✉)
Graduate School of Intercultural Studies, Kobe University, 1-2-1 Tsurukabuto, Nada-ku, 657-8501 Kobe, Japan
e-mail: sakurait@kobe-u.ac.jp

M. Albers et al. (eds.), *Human Rights and Human Nature*, Ius Gentium: Comparative Perspectives on Law and Justice 35, DOI 10.1007/978-94-017-8672-0_15,

After noting that such liberal philosophers as John Rawls, Ronald Dworkin and Robert Nozick are all sympathetic to the idea of genetic improvement of biological human nature, I concluded that liberalism not merely provides a solid moral foundation for liberal eugenics but also imposes "intrinsic" ethical restrictions on liberal eugenic practices. The Millean harm principle, for example, while justifying some forms of human genetic modification for the sake of our offspring, imposes ethical restrictions on the procreative liberty of the prospective parents. The harm principle will not allow prospective parents to infringe on the freedom of their offspring to form their own life plan through the use of genetic or other forms of intervention, even when it becomes technically possible in the future to improve the offspring's innate abilities by means of genetic engineering techniques.

Based on this analysis, I argued that the liberal harm principle presumably approves of physical genetic enhancement that gives prospective children "all-purpose" advantages such as the enhancement of the immune system and the slowing down of aging, because it does not infringe on the right of such children to an "open future" but rather facilitates their effective pursuit of any conception of the good. On the other hand, I believe that the same liberal values are bound to inhibit the use of *moral* enhancement where it is, for example, a question of genetically curbing a penchant for violent crime, since such genetic enhancement would lead us to think that the causes of social problems lie in the genetic makeup of individuals whilst also discouraging criticism of political institutions. Given that rebellious aggression can often bring about social reform, the genetic modification of moral traits such as aggressiveness and rebelliousness leads to the danger of undermining the freedom of prospective children to form and pursue their own political ideals in a critical spirit. Indeed, if such moral enhancement leads to the inhibition of an offspring's potential beliefs, as I think it does, then it is bound to prevent him or her not only from challenging or changing the political institutions and its administration but even from achieving any form of self-government. It is doubtful whether our offspring's "open future" can ever be assured as long as their political autonomy is undermined by eugenic practices. Even liberal eugenics would not tolerate such genetic intervention, for it is designed to guarantee the right of prospective children to an "open future" (Sakurai 2007).

One of the arguments against bioliberalism which I have hitherto not treated upon in sufficient detail is the argument that social disparity with respect to access to advanced technology might widen the economic gap between rich and poor, especially in light of the fact that the technology of human genetic intervention is bound to be very expensive during the initial phase of its practical use. It is argued that genetic technology will give its already rich beneficiaries enhanced health and physical capacities, resulting in double inequalities both in wealth and innate endowments. In my book, I dismissed this idea by stating that, although the practical uses of genetic engineering would initially benefit only the affluent in much the same way as automobiles and personal computers have done, there would come a time when, thanks to technological progress and a commensurate decrease in costs, all social strata would be able to enjoy its benefits (Sakurai 2007, 151–52).[1]

Although I still stand by this argument, I have to admit that my focus was, at the time, restricted to the problem of distributive justice within a society, whereby I paid little attention to the much-publicized fact that socioeconomic inequalities both within and between societies have been growing worldwide in recent decades (Pogge 2008; Brock 2011; Beitz and Goodin 2009; Singer 2004; Shapiro and Brilmayer 1999; Rawls 1999; Beitz 1999). There are, then, some questions that need to be asked: How can we speak of the moral acceptability of cutting-edge biomedicine, when we know that a large number of people in developing countries suffer from extreme poverty and lack even essential medicines and safe water, whilst governments of affluent countries are chiefly concerned with matters of employment and economic growth in their own countries? How are we to evaluate bioliberalism's claims from the perspective of global justice when we are, for example, trying to accomplish a just distribution of benefits well beyond the boundaries of individual countries?

This article examines the moral implications of the development of genetic technology in the laboratories of affluent societies in the light of the fundamental inequalities that exist on a global scale. Such an examination is, of course, based on the assumption that the "inequality of health care" in the global context is unjust and that it is necessary to ascertain who is responsible for this injustice and who should set about putting an end to it.

15.2 Chapman's Criticism of Human Germline Engineering

It was Audrey Chapman's article, "The Implications of Inheritable Genetic Modifications for Justice", that first prompted me to re-examine the moral implications of human genetic engineering from the perspective of global justice. According to Chapman, the only justifiable occasion for using inheritable genetic modification for purposes of medical treatment is when no other forms of treatment are available, for example when both parents are homozygous for a monogenic defect and desire to have a genetically related child without that defect. Except in this kind of extremely rare case, parents can avoid the birth of a child with a serious genetic disorder by using embryo selection. Since both statutory health insurance systems and private health insurance policies rarely cover anything considered nontherapeutic, enhancement technology will probably be developed within the private sector and is likely to be fairly expensive. This, she believes, would mean that germline engineering would only be possible for the biological enhancement of a narrow,

[1]Ronald Dworkin also says that "as in the case of more orthodox genetic medicine, the techniques available for a time only to the very rich often produce discoveries of much more general value for everyone" (Dworkin 2000, 440).

wealthy segment of society. What she envisages is a gloomy scenario in which a biologically stratified society would ultimately appear on the scene, because the wealthy classes would repeatedly try to hand down their innate superior capacities through genetic enhancement technology over many generations (Chapman 2003, 135–43).

One of the major challenges for Chapman's egalitarian theory of justice is to establish what health care services should be guaranteed to all members of a society. One possible means of forestalling her gloomy scenario, given the technological safety of the inheritable genetic intervention, would be to aim at an equal distribution of medical resources by giving every individual the moral claim-right to genetic technology through various health insurance systems. Even though some scholars have already advocated this idea, Chapman is not quite as optimistic as they seem to be about the feasibility of such an egalitarian scheme. As she says:

> So does a just distribution of societal resources require that IGM [inheritable genetic modification], if and when it becomes available, be part of a package of universally available health care benefits provided by either the state or private insurers? A final determination would depend on a number of factors, including efficacy, cost, and the alternatives available, but it is difficult to envision a scenario in which a commitment to justice would require funding these interventions. (Chapman 2003, 149)

Obviously, an egalitarian distribution of advanced medical technology would be feasible depending on the cost of the technology used and on the public policies concerning the distribution of limited health care resources in each society. In addition, the choice of a distributive principle with reference to medical resources should surely be considered an important issue not just within a society but also in the international community. However, it is noteworthy that many developing countries still fail to provide their people with even basic medical services. Moreover, ongoing economic globalization and megacompetition seem to be widening the economic gap between the rich and poor countries and exacerbating health inequalities on a global scale. If, as Chapman believes, expensive genetic enhancement technology is used in rich countries, it will inevitably aggravate the whole situation by creating "doubly-strong competitors", to use Parens's well-turned phrase (Parens 2000, 122). As Chapman asserts from her essentially egalitarian point of view, we need to prohibit the development and application of human germline engineering since the technology would very likely "make current injustices and inequalities worse and far more difficult to rectify" (Chapman 2003, 152).

Although this is one typical example of egalitarian criticism of bioliberalism, it is necessary to mention two important considerations here. First, germline genetic engineering is not the only biomedical technology causing new inequalities in our mental or physical capacities—a point I shall be discussing in the next section. Second, if the increased use of this technological development makes its cost sufficiently low (which is not likely to happen overnight), then the promotion, as against the prohibition, of universal access to genetic technology will, as I hope to make clear below, emerge as a new ethical challenge in the future.

15.3 Three Responses to Chapman's Criticism

There are three possible rejoinders to Chapman's arguments against germline engineering. First, like other egalitarian critics of genetic technology, Chapman is inclined to presuppose the permanence of the present socioeconomic inequalities, whether within a society or between societies. It is, however, impossible to predict with any certainty whether such inequalities will persist indefinitely in the future. Admittedly, today's world is pervaded by glaring political and socioeconomic inequalities between countries on account of constant civil or racial conflicts, uneven distribution of natural resources, the dominance of capitalism, the arms race and so on. These huge disparities have helped to widen the gap between rich and poor, and caused further inequalities between peoples with respect to health. Nevertheless, if we look into the future, the day may come when such forms of worldwide disparity will have disappeared from the face of the earth.

The molecular biologist Lee Silver has predicted that an analysis table of the human genome will be completed by the mid-twenty-first century (Silver 1998, 245). Yet human genetic engineering has in fact made comparatively little progress in the meantime, and many commentators believe that it has a long way to go before it can be used safely. Given this historical background, there is a possibility that, thanks to global political progress, the wealth gap between the rich and poor countries will be drastically reduced by the time we have realized a safe modification of the human genome. Thus, in a situation where substantial equality between societies had been practically established, Chapman's criticism would certainly be out of place.

The second possible rejoinder focuses on the socioeconomic gradient that the practical use of any advanced medical technology necessarily produces in any society. If the extant wealth disparity both within and between countries continues, it is quite possible, as Chapman fears, that there will emerge a social class unable to enjoy the benefits of medical technology. Indeed, Chapman is deeply concerned that, if access to genetic therapy and enhancement remains unequal, the genetic technology will continue to be used only by the richer classes and thus further widen the wealth gap within and between nations.

However, I should point out that unequal access to such new technology can apply to any sort of technology as long as wealth disparity persists. Given the technical problems of human genetic modification, other technologies such as regenerative medical techniques and cybernetic organisms may be put to use sooner than germline engineering. Since these technologies are expected to be expensive in their early stages, it will be necessary to find ways and means of working out principles of justice to address the distribution of those technologies. Thus, it will be not so much a technological problem as a problem of public policy in the society providing the benefits of the technology.

For instance, many people all over the world are even now anxiously waiting to receive the benefits of regenerative medicine. Since it will perforce cost a good deal to implement the technology of regenerative medicine in its early phase, there arise important questions such as: Who should enjoy the benefits of this technology? What should be the order of priority for those receiving the benefits of this technology? Who should bear the cost? These problems of distributive justice are not unique to regenerative medicine, but rather common challenges for any society that has a new, costly technology as well as many members of the society waiting to benefit from it. Technological development leading to this sort of distribution problem does not mean that the technology itself is at fault.

When we have access to an advanced medical technology such as regenerative medical techniques and cybernetic organisms, we certainly have access to physical capacities which were never before available. Given that these new technologies help us to more than make up for defects to which ordinary human beings are sometimes subject, they should be regarded not so much as a means of therapy but as an enhancement of our biological nature. If our society should differentiate into a social class that has easy access to advanced technologies and a class that does not, a significant inequality in health and capacities is bound to emerge between these classes. This would also be true with reference to where human genetic engineering is at stake.[2]

As we have seen, when we address the question as to whether such advanced technologies as genetic engineering can be provided in an egalitarian manner, we need to examine not just the cost of implementing the technology but also the public policy and institutions of each society concerning the distribution of its benefits. In terms of my second rejoinder, then, it is a mistake to judge the appropriateness of a new technology only by its cost and without taking into consideration the public policy's alternative possibilities.

15.4 The Right of Access to Genetic Intervention

Although my second rejoinder seems convincing enough to me, I should now like to consider the possibility of another argument for bioliberalism. This approach presents "the right to health care" as an integral part of basic human rights and tries to include in the right to health care access not only to traditional public health and first-aid treatment but also to a certain level of genetic technology.

[2]Allen Buchanan also points out that the problem of injustices in the distribution of valuable biomedical enhancement technologies is only one aspect of the larger problem of injustices resulting from the inadequate diffusion of technologies. From this perspective, it is much more important and effective for the reduction of distributive injustices to promote the development and diffusion of valuable innovations in an egalitarian manner than to prohibit it, since only the former strategy could remove existing unjust disadvantages as the diffusion progresses (Buchanan 2011, ch. 8).

In other words, this approach seeks to secure for people egalitarian access to certain benefits of biomedicine, assuming as it does the evolvability of biotechnology and the variability of wealth disparities on a global scale. This is obviously a very challenging task. Nevertheless, provided that we refrain from limiting our perspective to the near future or from hastily extrapolating the future from existing politico-economic inequality, such an approach seems to me a realizable choice.

This approach requires us to make two conceptual extensions. First, we need to extend the concept of "the right of access to health care" to include access to advanced, even potentially practicable, biomedicine that is at present not included. This extension involves not only the economic problem of increasing medical costs but also the technical problem of developing more and more new technologies such as human genetic engineering. Since these problems can probably only be solved in the distant future, the concept also requires a temporal extension. If this counterargument is, then, futuristic by definition, it should not be condemned for being so. Since Chapman's standpoint, which presupposes that genetic enhancement technology will be in widespread use only among the affluent, is in itself futuristic, any objection to it cannot but be futuristic.

Second, we need to extend still further the concept of the right to access "genetic technology as health care" so that it should be considered not only as a domestic legal right but also as a universal human right which is guaranteed to all individuals, which means that the spatial scope of the right to health care would also need to be expanded on a global scale. As is well known, Article 25 of the Universal Declaration of Human Rights stipulates that "[e]veryone has the right to a standard of living adequate for the health and well-being of himself and of his family, including food, clothing, housing and medical care and necessary social services". "The right of everyone to the enjoyment of the highest attainable standard of physical and mental health" is also affirmed in Article 12, paragraph 1, of the International Covenant on Economic, Social and Cultural Rights. What matters is who is responsible for specifying the nature of the right to health care through certain sorts of procedures and who should have the responsibility to guarantee that right concerning which we have so far seen a wide range of viewpoints.[3]

15.5 The Right of Access to Preimplantation Genetic Diagnosis

Here it is worthwhile to mention Leonard Fleck's claim, namely that considerations of justice *warrant*, if not require, public funding for access to preimplantation genetic diagnosis (PGD) in some circumstances. In the 1990s, Fleck argued that considerations of justice required the development and dissemination of some emerging genetic technologies, given the existence of universal health insurance

[3]See, for example, Buchanan 2009, ch. 9, Daniels 2008, ch. 13, and Miller 2007, ch. 10.

coverage. Thus, in distinguishing between therapy and enhancement in genetic engineering, Fleck insists—arguing from the perspective of health care justice—that genetic engineering for the purpose of eliminating deleterious genes has a stronger claim to research funds than the very expensive life-prolonging technologies (Fleck 1994, 142–50). He follows Norman Daniels in believing that the social obligation to ensure "fair equality of opportunity" is at the heart of health care justice, and that our duty to maintain normal functioning is the corollary of this obligation (Daniels 1985). Fleck also agrees with Daniels in linking "needs that have a just claim to medical resources with the degree to which fulfilment of such needs allows an individual to access a normal opportunity range in that society" (Fleck 1994, 143). He even acknowledges that, given the future lower cost of genetic technology and the universal health care system being in place, genetic enhancement engineering may displace other possible therapeutic interventions and be included as part of that package (Fleck 1994, 142).

However, in his later article Fleck dismisses as "too futuristic" the question concerning the possibilities of germline genetic engineering and, instead, focuses on the serious issues raised by our financial capacity to use PGD. Since the cost of achieving a successful pregnancy via PGD amounts to as much as $40,000, if no financial support is provided to equalize access to this form of technology for members of all socioeconomic classes, a disproportionate number of children with serious genetic disorders are likely to be born to less well off parents. Asking if we can confidently assert that such an outcome is merely an "unfortunate" consequence, Fleck goes further by suggesting that this is a moral wrong we are obligated to redress as a matter of justice (Fleck 2009, 302–03).

Since there is a limit to the medical resources available to members of a society it is, financially speaking, important to distinguish between PGD that needs social funding and PGD that does not. Instead of adopting a conventional dichotomy between therapy and enhancement, Fleck distinguishes between genetic disorders that would adversely affect the length and quality of a child's life from the early stages, and those genetic disorders that emerge in midlife or later, adversely affecting the length and quality of an adult's life. He claims that considerations of health care justice warrant public funding for access to PGD for any couple who have been informed through genetic tests that they are at risk of having a child with the former type of disorders.[4]

Why does Fleck say public funding is warranted for access to PGD only in the former case? First of all, in the light of Daniels' "fair equality of opportunity" account of health care justice, while children with the latter type of disorder will have had many of the opportunities that validate a human life, those with the former type of disorders are severely deprived of opportunities to live a normal human

[4]He cites as examples of the former category cystic fibrosis, Duchenne's muscular dystrophy, Canavan's disease, fragile X syndrome, hemophilia, Tay Sachs, autism, and Lesch-Nyhan syndrome. As examples of the latter, he gives breast cancer linked to the BRCA1 gene, somewhat earlier onset forms of Alzheimer's disease linked to APOE, and Huntington's (Fleck 2009, 303).

life. Perhaps the most important reason for his assertion is that we have only limited resources to meet the virtually unlimited health care needs for members of a society. If society prevented the former type of serious disorders by funding PGD, the cost of delivering PGD could be more than offset by the savings achieved by not having to provide very expensive medical and social interventions for children with these genetic disorders. In the case of the latter type of disorders, even if we used PGD to eliminate them, the cost saved by the prevention of these diseases would be very little, given the considerable expenses incurred in the treatment of a whole range of other medical problems to which everyone is subject (Fleck 2009, 303–05).

Critics of public funding for access to PGD might argue that embryo selection based on PGD, unlike normal medical practice or germline genetic manipulation, involves the replacement of one future possible person afflicted with a serious genetic disorder by a different future possible person free of that disorder. What they see as a major moral problem is Fleck's proposal of a policy which uses social resources for this purpose, rather than a policy of allowing the rich to implement embryo selection at their own expense (Fleck 2009, 308–09).

According to Fleck, this objection rests on the flawed premise that "future possible persons" in a petri dish have some sort of strong moral right to become actual persons. On the contrary, "[f]uture possible persons have no actual interest" (Fleck 2009, 309). He goes on to say:

> If multiple embryos have been brought into existence for a morally permissible purpose, such as avoiding the actual birth of children with serious genetic disorders that would adversely affect both the length and quality of their lives, then both the parents and the larger society are morally free to choose which of those embryos will have an opportunity to become actual persons. (Fleck 2009, 309)

Obviously, such a view is plausible if no moral significance is attached to the biological moment of fertilization and the embryo formation thereafter. Then, logically speaking, there must be a morally significant moment when a human embryo turns into an actual person and claims its own moral right to existence; a moment that Fleck has apparently failed to identify.

Even though Fleck's argument regarding "persons" may be somewhat logically unsound, it needs to be said that we cannot define the moment of fertilization as the beginning of an actual person. As Lee Silver persuasively states, fertilization itself is not an "instant" but a series of slow, complex processes in which a sperm gradually moves into and fuses with an egg, so that you cannot pinpoint the split second when the future possible person becomes an "actual person" (Silver 1998, 43–45). Even if you focus on the short-term process of fertilization, it is still a most difficult task to point to the instant during which a potential person turns into an actual one. In other words, whether you focus on the short-term process of fertilization or on the whole course of embryogenesis, it is equally and most difficult to point to the crucial moment when an actual person appears on the scene. In this sense, despite Fleck's silence on this awkward problem, he is in a no less advantageous position than those who dare to date the beginning of a person as happening at the moment of fertilization.

15.6 Moral Foundation of Public Funding for PGD

How does Fleck justify his claim that access to PGD should be made available with public funding? This he does in two specific ways. First, he invokes the moral consideration of "fair equality of opportunities", namely, John Rawls' idea which Daniels has applied to health care justice; he maintains that we have, as a society, an obligation of justice to provide an opportunity to access aid devices, therapy and rehabilitation, as well as any medical technology needed in order to restore those who have suffered serious damage by disease or injury. However, children born with serious genetic disorders such as Duchenne's muscular dystrophy or Canavan's disease are destined to suffer from severe physical and mental impairment, as well as being condemned to a short life. They are, therefore, deprived of even the possibility of recovering "fair equality of opportunity" by way of rehabilitation or medical devices. We healthy people may regard those with impairments acquired through illness or injury as merely unfortunate. However, from the perspective of health care justice, Fleck considers it irresponsible of us, now that genetic diagnosis technology is in practical use, to adopt the same attitude towards children with serious genetic disorders. Accordingly, if we take seriously the responsibility to guarantee "fair equality of opportunities" to all members of society, we ought to provide the social resources needed for PGD in order that more people may come to enjoy "the normal opportunity range" in life (Fleck 2009, 310–11). Still, it is worth noting that the actual person who, in order to escape a serious genetic disorder, was born through embryo selection based on PGD, is not the same as the possible person who may have been born with a genetic disorder as a consequence of a spontaneous pregnancy.

Second, the right to claim public funding for PGD presupposes the parents' reproductive freedom. When potential parents know that they are at risk of having a child with a severe disorder such as cystic fibrosis or Canavan's disease, and that PGD is an available option, Fleck believes that society should not deny those parents access to such medical technology. At the same time, if there is a publicly funded programme in place that makes this option available, those parents should also be given the right to reject this option on, say, religious grounds. They should, furthermore, not be held in any way morally responsible for "harming" a child by allowing it to be born with a serious genetic disorder, especially given that Fleck's liberal pluralism argues for the social funding for PGD based on respect for the procreative liberty of potential parents (Fleck 2009, 312).

Finally, we need to consider Fleck's claim that such public funding would be "*warranted* by consideration of health care justice, but not required" (Fleck 2009, 310, italics original). This claim is also related to his liberal pluralism. He admits that the "fair equality of justice" strand of justice does not always show us the way to what is morally requisite. On occasion, however, the utilitarian strand, or the strongly egalitarian strand, or even the libertarian strand, does. Moreover, as Fleck says, when we see "the entire expanding universe of medical needs" (such as protease inhibitors for HIV, prescription drugs for the elderly, neonatal ICU for extremely premature infants, etc.), we recognize that "there are some considerations of justice (one or another of our strands) that would provide moral support for meeting any one of these needs", but rarely will that support be decisive and uncontroversial (Fleck 2009, 311).

Fleck also relates such value pluralism to the need for medical care rationing, since we have only limited resources to fall back on in order to meet health needs. Even though priorities must be established among these competing needs, it will be impossible for various strands of health care justice to yield such a list of justice-authorized priorities. Indeed, we ought to resort to "processes of rational democratic deliberation" in order to achieve a reasonable ordering among our health care priorities. Since this ordering can only be "just enough," we are bound to say that "funding PGD for the circumstances we identified may be *warranted* by considerations of health care justice, but may not be required" (Fleck 2009, 311, italics original).

15.7 The Implications of the Right of Access to PGD

Fleck's argument for the right of access to PGD seems to me to be valid, suggesting as it does that we can, for example, legitimately claim the right to an advanced reproductive treatment according to our medical needs. It is worth noting that Chapman also argues that the rationale for providing access to preimplantation diagnosis if it is effective is far stronger than that for subsidizing germline genetic intervention (Chapman 2003, 149).

However, even if public funding for (limited) access to PGD is guaranteed in any society, it is obviously impossible for every society to guarantee such access to its members because of the lack of both financial and medical resources. That is why it is important to remember that establishing access to PGD in some developed countries implies that other kinds of health care disparity could emerge between rich and poor societies.

Fleck's egalitarian argument is important because it is, at least in principle, applicable to other cutting-edge medical technologies. Its importance is obvious when we think about the prospect of human iPS cells in regenerative medical techniques. In Japan, despite the revision of the Act on Organ Transplantation in 2009 that helped facilitate more organ transplants, we are still a long way from solving the problem of organ shortages. The most promising new method being developed at present is inducing the differentiation of iPS cells made from the patient's somatic cells in order to perform organ transplants. However, when this is technically possible in the near future, it will still involve a great deal of expense to carry out a transplant, especially in its earlier stages. What matters is who should bear the cost. If patients themselves were to bear the cost of such advanced medical technology, we would see the emergence not only of a new wealthy class capable of living a long and healthy life as never before possible, but also of a poor class severely deprived of such medical privileges. On the other hand, if we adopt an egalitarian policy of publicly funding access to regenerative medicine in the case of serious illnesses, as Fleck does with regard to PGD, we can expect a considerable improvement in the health levels of people in all social strata. Nevertheless, just as a social policy of guaranteeing access to PGD might lead to a new health disparity on a global level, so public funding for a broad range of regenerative medicines might create further health disparities between rich and poor countries.

What I should like to point out here is that, once regenerative medicine enables people to live longer and healthier lives than they have ever done, this should not be called mere therapy but rather enhancement of human biology. It is, of course, natural that, as we get older, we become senescent, suffer from age-related diseases like cancer and cardiac illness and eventually die. But if we can benefit from regenerative medical techniques, we are bound to change, or rather enhance, human biology with their aid. One might argue that, unlike germline engineering, regenerative medicine does not modify our genotype. However, from a moral perspective, it does not make any difference whether we implement physical enhancement by altering our genotype or by regenerating our own organs. Both regenerative medicine and genetic engineering have their own technical difficulties and risks, but if they are successful and are safe they can bring in completely new benefits to human biology. In light of their remarkable impact on human health and longevity, regenerative medical techniques are functionally comparable and, therefore, morally equivalent to human genetic engineering.

15.8 Genetic Decent Minimum?

How far can we expand the range of public funding for access to advanced medical technologies? Even if the cost of cutting-edge medical services gradually decreases with technological development, medical resources will still be limited, which will, in turn, mean rationing such resources. Fleck and Chapman are both in support of public funding for access to PGD, which at present costs about $40,000. Would it be ridiculous to suggest socially funded access to, say, human genetic engineering, which is expected to be much more expensive? It may be at the moment, but will it necessarily be so forever?

Daniels claims that what is morally special about health care is its impact on opportunity, since it serves to maintain "normal functioning". Disease and disability impair the normal functioning of the body and restrict the range of opportunities open to individuals. Health care contributes to protecting the equality of opportunity by helping people to recover and maintain their normal functioning, and thus enables them to participate in the political, social, and economic life of their societies. In other words, health care aims to protect an individual's fair share of "the normal range of opportunities (or a plan of life)" that reasonable people would choose in a given society (Daniels 2002, 6–7).

What is interesting here is that the maintenance of normal functioning by health care can be, theoretically, extended to include even the use of human genetic engineering. In his *Theory of Justice*, Rawls tries to mitigate the effect of inequality caused by "the natural lottery" by applying what he calls the Difference Principle (Rawls 1971). However, as the authors of *From Chance To Choice* indicate, if we can identify a safe and efficient method of eliminating disadvantageous genetic traits, then there seems to be nothing in Rawls' theory to hinder this course of action but, indeed, "much to support it".[5] Because he extends part of Rawls' Second Principle to health care, Daniels' principle of "fair equality of opportunity" finds it difficult to rule out the possibility that sometimes we may even be obliged to use genetic technologies to alter the distribution of innate endowment (Buchanan et al. 2000, 129).

It is important to note here that Buchanan and his fellow authors stop short of advocating "genetic equality", which aims both to eradicate all inequalities in natural assets and to redistribute otherwise normal but competitively disadvantageous capacities. Instead, they modestly suggest as an appropriate objective the attainment of a "genetic decent minimum", which means "a strong societal commitment to use advances in genetic intervention to prevent or ameliorate the most serious disabilities that limit individuals' opportunities across a wide range of cooperative frameworks" (Buchanan et al. 2000, 81–82).

Yet the idea of a genetic decent minimum is in some sense too much to expect, since it requires a public policy to provide all members of a society with a certain genetic endowment that will enable them to participate in all spheres of life as normal collaborators and competitors. In that respect, the idea of a genetic decent minimum presupposes that the parents' right of access to health care extends even to the genetic endowment of their offspring. This can be seen as another expansion of the domain of art into the nonhuman part of nature. As our genetic technology develops, the domain of nature is bound to be annexed to the domain of art and gradually diminish and the newly-won territory is colonized by ideas of law and justice (Buchanan et al. 2000, 83–84).

However, it must be pointed out that the idea of a "genetic decent minimum" could also conflict with the procreative liberty of parents referenced above. The public decision to ensure a "genetic decent minimum" necessitates that the state or any other public agency be given "a legitimate role as guardian of the genetic well-being of future generations" (Buchanan et al. 2000, 336). This allows the state not only to monitor the market in order to protect patients from unwise and hasty uses of genetic technology, but also to restrict some reproductive decisions

[5]Amartya Sen argues for his capability approach by saying, "if a person has a high income but is also very prone to persistent illness, or is handicapped by some serious physical disability, then the person need not necessarily be seen as being very advantaged, on the mere ground that her income is high" (Sen 2009, 234). He repeatedly emphasizes that a person with disability is subject to more restraints than a person without physical disadvantages (see also Sen 2009, 253). In this sense, at least on ethical grounds, it is likely that Sen would not even object to human genetic engineering for the prevention of serious genetic diseases, even though he claims that "only a fairly moderate proportion of the 600 million people living with disabilities were doomed to these conditions at conception, or even at birth" (Sen 2009, 259).

of prospective parents so that their decisions do not damage an "open future" for their children (Buchanan et al. 2000, 341–42). Thus, the responsibility of the state or some public agency as the guarantor of a genetic decent minimum involves exercising its authority to regulate and restrict the procreative liberty of parents and to legally require them to refrain from impetuously and irresponsibly choosing a leading-edge genetic technology or, alternatively, even to use certain genetic diagnosis technology.

Moreover, even if we guaranteed a "genetic decent minimum" in our society, it would then create another health disparity between rich and poor people on a global scale. While it is necessary to deploy considerable medical resources to secure a genetic decent minimum, there are many countries that cannot afford to meet all the expenses on their own. If we are to achieve a truly egalitarian distribution of health care and prevent the most serious genetic diseases on a global level, it is necessary to construe the right to a "genetic decent minimum" as one of the universal human rights and set out to guarantee the right as part of a "global minimum".

Obviously, the concept of a global minimum is closely related to the theory of basic human rights. David Miller, for example, regards basic human rights as being essential to his theory of global justice. He sets out his theory of basic human rights in order "to identify a list of rights that can specify a *global minimum* that people everywhere are entitled to as a matter of justice" (Miller 2007, 166, italics mine). When justifying basic human rights, he seeks a common foundation in basic human needs. Despite his nationalist bent, Miller does claim that, if a national community cannot, owing to some natural disaster, ensure for its citizens one of the basic human rights, e.g. the right to subsistence, then other nations should be obliged to come to its aid (Miller 2007, 166–67; De Schutter and Tinnevelt 2011).

The greatest appeal of the right to a "genetic decent minimum" derives from the fact that it can be on a list of universal human rights as advocated by Miller. It can claim protection not just within the borders of a country but well beyond its borders. It is true that, when applied on a global basis, it is confronted by a number of procedural questions that need to be answered: Who is supposed to specify the concrete content of a "genetic decent minimum"? What procedure does it need to follow? Who has the authority to interpret and guarantee the content? In addition, any attempt to provide a "genetic decent minimum" cannot escape the problem of rationing of limited resources.

However, this sort of distributive problem is one of the common challenges that any rights-based doctrine has to deal with (Daniels 2008, 314).[6] Certainly, the most

[6]Daniels makes it clear that his advocacy for a moral right to health and health care is different from a human rights approach that (wrongly) believes that searching for the philosophical foundations of human rights can better resolve conflicts among some of these human rights and among the means of realizing them (Daniels 2008, 315–18). In other words, a human rights approach needs a fair deliberative process to resolve reasonable disagreements about priorities in designing institutions, as Daniels' moral right approach does (Daniels 2008, ch. 4). That is why the determination of the specific content of the right to health, be it moral or human, necessarily involves reasonable choices by appropriate agents in a society. In this context Daniels justly points out that "the moral

serious dilemma faced by the doctrine of the right to a genetic decent minimum is that, on the one hand, we have to regulate and restrict the procreative liberty of prospective parents in a coercive manner in order to secure an open future for future generations while, on the other hand, providing social resources in a liberal pluralistic society to support the choices of couples who would either demand or reject a genetic decent minimum (Fleck 2009, 312–13). In this respect, the idea of a genetic decent minimum and respect for reproductive choice sharply conflict with one another. If a couple had a 50 % chance of having a baby with a serious genetic disorder that would not emerge until midlife at the earliest, and yet wanted to give birth to the child without using PGD or any other preventive measures, what would the doctrine of a genetic decent minimum do for the couple? What if the chance were 25 %? In either case, the doctrine of a genetic decent minimum would arguably interfere with the couple's decision to have that child.

Therefore, the idea of a genetic decent minimum also seems to contradict Fleck's opinion that public funding for PGD is warranted, but not required, from the viewpoint of justice. If the idea of a genetic decent minimum gives the state or some public agency “a legitimate role as guardian of the genetic well-being of future generations” (Buchanan et al. 2000, 336), as in fact it does, then the state has no alternative but to require couples at risk of having a child with a serious genetic disease to use PGD. This, however, is clearly not a policy that Fleck would advocate.

15.9 Two Kinds of Decent Minimum

Probably, no one would argue against the idea of a “decent minimum” or a “global minimum”, since it is clearly desirable that every human being be guaranteed a minimum level of welfare, whether as a matter of justice or on humanitarian grounds (Ignatieff 2001; Beitz and Goodin 2009). On the other hand, almost anyone might question or even protest against the idea of a “genetic decent minimum”. What is the specific difference between these two kinds of “decent minimum”? Is the adjective “genetic” so alarming? Obviously, only the doctrine of a “genetic decent minimum” involves, and even necessarily restricts, the procreative liberty of prospective parents. In this sense, it also goes against value pluralism, which claims to secure such parents a wide range of reproductive choices. It is important at the same time, however, to pay attention to those elements that both kinds of decent minimum have in common, and these are, to my mind, “the fulfilment of basic needs” and “the avoidance of harm”.

First, both these decent minimums are designed to meet the “basic needs” of their beneficiaries. Just as the fulfilment of a “global minimum” is indispensable for

---

right to health and health care that I defend has a specific content only relative to a specific society” (Daniels 2008, 317).

building a brighter and more optimistic future for people plagued by famine, civil war, or persistent poverty, so a "genetic decent minimum" provides our offspring with their basic "biological" needs that ensure the possibility of their "open future". We can consider a "genetic decent minimum" as fulfilling our "biological needs" in the same way that a supply of water, food, shelter, etc., i.e. a "decent minimum", meets our biologically crucial requirements. In this sense, the provision of basic needs for the vulnerable, both before birth and during their lifetime, is clearly essential for them to lead a decent, independent existence.

Second, the fulfilment of a decent minimum is absolutely necessary for all human beings who are anxious to avoid present or foreseeable harm. This can be seen more clearly in the case of a genetic decent minimum, because one of the main purposes of genetic intervention, including PGD, is the avoidance of "foreseeable harm" inflicted on a child with a severe genetic disorder which would adversely affect the length and quality of its life. Genetic intervention for other purposes, such as the creation of superhuman athletes, has nothing to do with the idea of a genetic decent minimum and, indeed, may be viewed with suspicion.

Henry Shue insists that the rights to three substances—subsistence, security, and liberty—are basic rights that should be honoured by all human beings. According to Shue, subsistence, or minimal economic security, means "what is needed for a decent chance at a reasonably healthy and active life of more or less normal length" (Shue 1996, 23). Interestingly, Shue considers that guarantees of subsistence and guarantees of security are equally essential factors in the actual exercise of all other rights. Accordingly, if a person lacks these basic rights, she "is readily open to coercion and intimidation through threats of the deprivation of one or the other [subsistence or security], and credible threats can paralyse the exercise of any other right" (Shue 1996, 26). It is easy to see that "coercion and intimidation" are closely related to the concept of harm, since they are done by threatening a person with some imminent harm. The parallel between security and subsistence clearly implies that avoidance of harm, as a main target of a "global minimum", is a principal aim in Shue's theory of rights (Shue 1996, 38–40).

In light of the factors common to these two kinds of decent minimum, the question then arises as to how we can safely embed an unpopular "genetic decent minimum" in a more readily-acceptable "decent minimum". There is little doubt that it will be impulsive and even harmful, especially for our offspring, to use human genetic engineering while the benefits of such technology remain unproven. And yet, I still wonder if the failure to provide a genetic decent minimum, when the technology is sufficiently developed, might not adversely affect the "open future" of our children-to-be and inflict "culpable harm" on them. Are we not morally required to prefer ensuring for our offspring a genetic decent minimum and their "open future" over the procreative liberty of their parents under certain conditions?

In fact, whether to choose a genetic decent minimum or procreative liberty is but a variation on the pro-choice and pro-life debate surrounding abortion. The procreative liberty of prospective parents should be respected, as long as there is no sufficient consensus about what should count as "responsible harm" to our offspring. This is clearly a logical conclusion from value pluralism, for which I have much respect. However, if we could ever reach some reasonable agreement on this matter through democratic deliberation with enough information and knowledge about genetic technology, I suspect there should be some cases in which we ought to give priority to the genetic decent minimum of offspring over the reproductive freedom of their parents.[7]

15.10 Conclusion

One might argue that a genetic decent minimum does not necessarily require that we use germline genetic engineering because embryo selection through PGD would be efficient enough for almost all prospective parents to have their own children born without serious genetic disorders. However, this argument presupposes that the content of a genetic decent minimum and our demand for it will ever remain unchanged. Consider, however, the current high death rate due to a variety of cancers and the future progress of cancer research. The day may come when we develop genetic enhancement technology that radically strengthens the human immune system against that terrible disease (Stock and Campbell 2000). In the same way, the cost of enhancement technology will eventually become affordable, when we may well regard it as part of a genetic decent minimum. Again, we should be careful, when considering the prospects of genetic technology, not to look too far ahead.

At the beginning of this article, the two values in conflict seemed to be the domination of human nature by genetic technology and a just distribution of the benefits of the technology. However, as I have tried to suggest, the most significant conflict of values from the perspective of health care justice turns out to be the one between the egalitarian idea of a genetic decent minimum and the procreative liberty of potential parents. Since these two concepts are both based on the validity and force of basic human rights, it will surely take some careful, unbiased deliberation before we can give one of these concepts the priority it may deserve over the other.

[7]This does not mean that I insist, for example, that a pregnant mother at risk of having a child with a serious genetic disease be denied her procreative liberty and required to terminate her pregnancy because of its possible failure to provide a genetic decent minimum. What is important here is that a genetic decent minimum is something society owes to its offspring, but not something prospective parents owe to the society or the state in which they live. Artificial termination of pregnancy necessarily puts an end to the very existence of a beneficiary of a genetic decent minimum, i.e. the foetus, and thereby makes physically impossible the fulfilment of the decent minimum itself. The idea of a genetic decent minimum can, as I understand it, "trump" the reproductive freedom of prospective parents under some circumstances, but cannot justifiably eliminate the existence of its rightful claimant.

References

Beitz, Ch.R. 1999. *Political theory and international relations*. Princeton: Princeton University Press.

Beitz, Ch.R., and R.E. Goodin (eds.). 2009. *Global basic rights*. Oxford: Oxford University Press.

Brock, G. 2011. What do we owe others as a matter of global justice and does national membership matter? In *Nationalism and global justice: David Miller and his critics*, ed. H. de Schutter and R. Tinnevelt, 62–77. London/New York: Routledge.

Buchanan, A. 2009. *Justice and health care*. Oxford: Oxford University Press.

Buchanan, A. 2011. *Beyond humanity? The ethics of biomedical enhancement*. Oxford: Oxford University Press.

Buchanan, A., D.W. Brock, N. Daniels, and D. Wilker. 2000. *From chance to choice: Genetics and justice*. Cambridge: Cambridge University Press.

Chapman, A. 2003. The implications of inheritable genetic modifications for justice. In *Designing our descendants*, ed. A. Chapman and C. Frankel, 130–155. Baltimore: The Johns Hopkins University Press.

Daniels, N. 1985. *Just health care*. Cambridge: Cambridge University Press.

Daniels, N. 2002. Justice, health, and health care. In *Medicine and social justice: Essays on the distribution of health care*, ed. R. Rhodes, M.P. Battin, and A. Silvers, 6–23. Oxford/New York: Oxford University Press.

Daniels, N. 2008. *Just health: Meeting health needs fairly*. New York: Cambridge University Press.

De Schutter, H., and R. Tinnevelt. 2011. *Nationalism and global justice: David Miller and his critics*. London/New York: Routledge.

Dworkin, R. 2000. *Sovereign virtue: The theory and practice of equality*. Cambridge, MA: Harvard University Press.

Fleck, L. 1994. Just genetics: A problem agenda. In *Justice and the human genome project*, ed. T. Murphy and M. Lappé, 133–152. Berkeley/Los Angeles: University of California Press.

Fleck, L. 2009. *Just caring: Health care rationing and democratic deliberation*. New York: Oxford University Press.

Ignatieff, Michael. 2001. *Human rights as politics and idolatry*. Princeton/Oxford: Princeton University Press.

Miller, D. 2007. *National responsibility and global justice*. Oxford: Oxford University Press.

Parens, E. 2000. Justice and the germline. In *Engineering the human germline: An exploration of the science and ethics of altering the genes we pass to our children*, ed. G. Stock and J. Campbell, 122–124. Oxford/New York: Oxford University Press.

Pogge, Th. 2008. *World poverty and human rights: Cosmopolitan responsibilities and reforms*, 2nd ed. Cambridge: Polity Press.

Rawls, J. 1971. *A theory of justice*. Cambridge, MA: Harvard University Press.

Rawls, J. 1999. *The law of peoples*. Cambridge, MA: Harvard University Press.

Sakurai, T. 2007. *Liberal eugenics and justice*. Kyoto: Nakanishiya (in Japanese).

Sen, A. 2009. *The idea of justice*. Cambridge, MA: Harvard University Press.

Shapiro, I., and L. Brilmayer (eds.). 1999. *Global justice*. New York: New York University Press.

Shue, H. 1996. *Basic rights: Subsistence, affluence, and U. S. Foreign Policy*. Princeton: Princeton University Press.

Silver, L.M. 1998. *Remaking Eden: How genetic engineering and cloning will transform the American family*. New York: Harper Perennial.

Singer, P. 2004. *One world: The ethics of globalization*, 2nd ed. New Haven: Yale University Press.

Stock, G., and J. Campbell (eds.). 2000. *Engineering the human germline: An exploration of the science and ethics of altering the genes we pass to our children*. New York: Oxford University Press.

Chapter 16
Enhancement, Human Nature, and Human Rights

Marion Albers

16.1 Introduction

Modern biotechnologies have a decisive influence on the ideas of human nature and human rights. They can affect the notion and the boundaries of the human as well as the concept of individual rights. Among the characteristics and consequences of advanced biotechnologies are radical transformations: The biological foundations of humankind are more and more accessible, can be modified in a targeted way and thus become the subject of decisions. Naturalistic self-descriptions are questioned and replaced by forms of description which are explicitly culturally constructed. This dimension of profundity corresponds to the basic nature of the controversies. Human nature, human dignity and human rights play as crucial a role in the disputes as does the question of universalizability of basic values and rights. While initially stem cell and embryo research, genetic diagnostics, cloning or patents on life were central issues, the debate on human enhancement is now the focus of attention. Enhancement links up with various biotechnologies and aims at increasing or improving human capacities, traits or moods. This ranges from cosmetic surgery, genetic intervention, smart pills, neuroprostheses, cyborgs or artificial life all the way to visions of "transhumanism" or "posthumanity." How enhancement can be described and defined is already a component of the debate and a problem prompting discussion of fundamental questions. Normative judgments are often based on recognized values and rights such as human dignity, autonomy or equality. A closer examination shows that the various lines of reasoning that are applied originate in entirely different as well as also often insufficiently elucidated approaches and preconditions. The debate on enhancement can benefit from being part of a wider discourse on human nature and human rights while also enriching this discourse.

M. Albers (✉)
Faculty of Law, University of Hamburg, Rothenbaumchaussee 33, 20148 Hamburg, Germany
e-mail: marion.albers@uni-hamburg.de

M. Albers et al. (eds.), *Human Rights and Human Nature*, Ius Gentium: Comparative Perspectives on Law and Justice 35, DOI 10.1007/978-94-017-8672-0_16,

This chapter treats the enhancement problem as a fundamental challenge and as a catalyst for reconsideration of the notion of human nature and human rights. Section 16.2 aims at analyzing and clarifying the concept of enhancement. In presenting fields and visions I will illustrate the range covered by this concept and elaborate on primary aspects. Conceptual dualities such as "therapy/enhancement," "normal/supranormal" and "natural/artificial" will provide further explanation. In the end, enhancement requires contextualization. Section 16.3 explores the controversy surrounding enhancement insofar as this controversy refers to human nature or central human rights. It is striking that the same normative standards are the basis of conflicting results. For the debate to lead somewhere, context, concepts and the status of discussions must be worked out more clearly in the future. Section 16.4 shows in how varied and complex a manner the relationship between human nature and human rights can be described and results in considerations on the role of human nature in the present-day human rights discourse. Against this background the fifth section deals with the interplay between enhancement, human nature and human rights and centers on some decisive aspects of the enhancement problem.

16.2 Enhancement: Concept and Contextualization

The concept of human enhancement and the practices and techniques involved are not immediately obvious, but are themselves the subject of a debate over describing and delimiting them. Defined in broad terms, "human enhancement" includes any activity which improves human bodies, minds, capacities, abilities, sociability, life span or general well-being (Harris 2007, 19 ff.; Bostrom and Sandberg 2009, see also Allhoff et al. 2011, 201 ff.; Menuz et al. 2013, 162). Physical exercise, a cup of coffee, school education, glasses or vaccinations are given as examples. Regardless of whether such an approach is perceived as involving an effort to achieve an impartial understanding or whether a strategy of de-problematization and legitimization is suspected (Wehling and Viehöver 2011, 14 f.), the matter cannot be left with this approach alone. To give the concept any explanatory and differentiating power at all it needs to be understood in the historical-social context in which it was formed. An appropriate concept must, in the first place, encompass the current discussion of human enhancement and also the potentially novel situation. Secondly, it must be as transparent as possible regarding its parameters, its underlying assumptions and its descriptive or normative elements.

An initial approach involves describing the fields which are discussed within this framework (Sect. 16.2.1). A further description is offered by conceptual dualities such as "therapy/enhancement," "normality/supranormality" and "naturalness/artificiality" (Sect. 16.2.2). This leads to a contextualization. Already from the outset, enhancement has to be understood as a concept which describes complex relationships that need to be fleshed out in a context-related manner, and which is inherently reflexive (Sect. 16.2.3).

16.2.1 Fields and Visions

The fields that are discussed with regard to human enhancement range from cosmetic surgery, genetic selection and intervention, smart drugs and neuroprostheses to the creation of chimeras, cyborgs and artificial life. Development in the various fields is characterized by heterogeneity and asynchronicity. In addition, the enhancement debate is carried on in society in general, in the political system as well as in various scientific disciplines and also in the form of futuristic visions or science fiction narratives (cf. Bostrom 2005; however, for a discussion of the dearth of appropriate academic futurology see Cornips and van Asselt 2013). Scientific contributions are being made in the natural sciences, the social sciences and the humanities or literary studies. With the specific perspectives of each discipline, they all have something to offer, but have their limits, too. Natural sciences reveal biotechnological possibilities, while the social sciences emphasize the social constitution of the development or realization of biotechniques. Futuristic visions or science fiction narratives communicate what is imaginable in the fantasies of a society, and can also promote technical ideas; however, they cannot be regarded as scenarios which are sure to become reality in the future.[1] Hence, appropriate analyses must keep the source and context of a particular line of thought in mind. Enhancement is about practices that are already being applied as well as about scenarios and visions of the future, and it is a matter of varying approaches. In the following, we can highlight some typical aspects considering central questions: What is meant by enhancement? How is enhancement delineated? What are the standards according to which something is regarded as an enhancement? How are enhancement practices carried out and what are the conditions of their establishment and dissemination?

Cosmetic surgery can serve as a "natural starting point" (Devereaux 2008, 160) to illustrate the enhancement issue. We can draw upon concrete established practices such as face lifts, Botox injections, rhinoplasty, chin implants or breast augmentation and liposuction. There are also some cases of people who strive to reinvent their appearance or, more accurately, themselves by resorting to cosmetic surgery to an extreme extent (see Toledano 2011). On the one hand, enhancement practices stand apart from what are merely cosmetic measures, such as hair coloring or removal. The most important differentiation criteria are the nature of the intervention in a person's bodily integrity using techniques which come from medicine – mostly from reconstructive surgery which repairs bodies torn by war, accident or disease –, the customary impossibility of reversibility, and the risks involved. However, as the examples of skin lightening or tattooing demonstrate, clear-cut differentiations are difficult. On the other hand, enhancement practices are distinguished from medically required measures. The central distinguishing

[1] Of course, we always deal with a future which is a specific part of each present – a present future – and constructed on the basis of, i.e., available knowledge, imaginations, value judgments or interests, see Grunwald 2013, 211 ff.

criterion here is differentiation between health and disease. Since this distinction is socially constructed, is based on a number of assumptions and is context-dependent (see Sect. 16.2.2.), it does not provide a clear-cut differentiation either. Even so, it can be used for identifying measures which enhance neither health nor function (see also Devereaux 2008, 169).[2] The standards to which enhancement practices are oriented and according to which something is judged to be an "improvement" are beauty ideals. These ideals are shaped by interplay among numerous individual and social factors. Images in the media and in advertising, role models and the cosmetic surgery industry have considerable influence on this (see, with different approaches, Davis 2003; Frazer 2003; Langer and Wimmer-Puchinger 2011; Stroop 2011; Gimlin 2012, 55 ff.). The standard for being "beautiful" is thus dependent on the culture and era, partly standardized, partly pluralistic and correspondingly varied and dynamic. The application of enhancement-practices in the area of cosmetic surgery shows national specificities (cf., comparing the US and the UK, Gimlin 2012, 26 ff.). In view of the relatively limited resources required in terms of technology and expertise, they are carried out mostly in a rather incremental and decentralized manner. This also results from the fact that the practices are less directly and instead indirectly regulated via general medical standards, via requirements for pharmaceuticals or via their exclusion from financing by medical insurance (see, for example, Damm 2011).

Genetic enhancement is based on developments in the decoding of the human genome and gene diagnosis which are dynamically advancing although still in their early stages. It presupposes close connections between personal characteristics or abilities and gene functions – an assumption that is, apart from explored monocausal relationships, highly problematic. We can distinguish between enhancement options in assisted reproduction, those involving living human beings and those involving the creation of artificial life. In the case of in vitro fertilization, germ cells or embryos can be analyzed and selected depending on their genetic constitution. In the future, genetic interventions are imaginable. In the case of living human beings, genetic interventions via somatic gene therapy, gene transfer techniques or germ line alterations are conceivable; though, technically, they are only in the beginning stages and not yet realizable. However, future scenarios foresee them, not least in view of relevant animal experiments (see, for example, Stock 2002). Transhumanist visions or science fiction narratives envisage the reshaping of humankind into a genetically modified post-human species as well as the creation and engineering of artificial life. Genetic enhancement can be distinguished from measures such as selection of partners by its technical character, the intentionality and directness involved and the particular objective desired by the procedures. And once again, the

[2]Expansion strategies take pains to justify cosmetic surgery by emphasizing that it is basically health-related in view of its effects on psychological well-being; promoting this well-being through measures which have no medical basis in themselves is, however, not the task of medicine (Devereaux 2008, 162 ff., 165 f.). Especially in the area of cosmetic surgery, we can observe vigorous discussion of professional ethics (see, for example Little 1998; Juengst 1998, 29 ff.; Ach 2006; Lanzerath 2011).

distinction between health and disease serves to distinguish enhancement practices from medically necessary measures. The standards according to which something is regarded as being an enhancement or improvement is, however, remarkably unclear when the model of healing illness can no longer be applied. What is "better"? A consensus on this is often assumed without any basis. In fact, the standards – improved vision or hearing, having greater intellectual faculties, looking beautiful, living longer – are characterized, as well as in cosmetic surgery, by an interplay among numerous individual and social factors. They depend on culture and era, are partly uniform, partly pluralistic, correspondingly varied and, above all, dynamic (simplifying Stock 2002, 116 ff.). The realization of genetic enhancement through biotechnological means has until now been limited by the fact that techniques are not yet adequately developed, a greater or lesser degree of medical professionalism is necessary and – differing from country to country – legal regulation restricts preliminary research on embryos or on living human beings or methods such as preimplantation genetic diagnostics.

Psychopharmacologic enhancement makes use of pharmaceuticals to improve cognitive abilities or emotional, motivational and/or conative states such as concentration, attention, alertness, memory and mood. Best known is the use of Ritalin® which was originally prescribed to treat ADHD (attention deficit hyperactivity disorder) but in the meantime serves as a "neuro-enhancer" aimed at an increase in concentration or attention. Other examples are Modafinil as a substance which has been approved for the treatment of sleep disorders and is being used to increase alertness or learning ability. Donepezil is a substance which has been developed for the treatment of dementia and is now being used as a cognitive enhancer to improve memory capacities. Fluctim® or Prozac®, originally developed as an antidepressant, has become a lifestyle drug for mood enhancement. Differentiation of psychopharmacologic enhancement from education, mental training and stimulant drugs such as caffeine, nicotine or alcohol mostly points to the development and application of the pharmaceuticals in a medical context or to their chemical effects, side effects and addictive potentials. Appropriate ways of differentiating, however, are disputed (cf. Bostrom and Sandberg 2009). The treatment of diseases on the one hand and enhancement on the other are distinguished as well. A distinction can be difficult, particularly in the case of mental states, as the examples of the diagnosis and treatment of ADHD or depression show (see also Sect. 16.2.2). The standard for assessing whether something is an enhancement derives from interplay among numerous individual and societal factors. Mostly, there is a focus on a particular function in a specified context, e.g., improving memory or alertness in exam or job situations. Again, the standard is subject to criticism: Alterations of this kind are not improvements but driven by the criteria of a performance-oriented competitive society and merely technocratic (see for instance Müller 2008a, 194 f.; Boldt and Maio 2009, 387 ff.). Carrying out psychopharmacologic enhancement is made easier by the fact that smart drugs can be prescribed "off label" and easily bought online, even if they are prohibited in a given country. As a consequence, enhancement practices have gradually increased and now seem to be quite widespread.

Last but not least, enhancement is discussed with regard to human-machine interfaces. Examples are high-tech artificial limbs, cochlear and retinal implants, neurosurgical intervention or neuroprostheses in various forms, and memory chips implanted in the brain (for an overview see Merkel et al. 2007, 117 ff.; Müller et al. 2009). Technical implants could improve sensory or cognitive functions. Deep brain stimulation, which is less or more invasive in the form of electronic stimulation or implantation of electrodes, is contemplated for enhancing associative memory or selectively inducing emotions with positive valence (cf. Synofzik and Schlaepfer 2008). Future scenarios envisage an increased use of bioelectronic and neural engineering systems in order to improve motor, sensory and cognitive traits and ultimately even create a symbiotic connection between the human biological system and various technical devices. The key word, although it is understood and used in a variety of different ways, is "cyborg" (cybernetic organism; see Kurzweil 2003; Clark 2003; Jones and Whitaker 2012, 259 ff.). The distinction between enhancement practices and other means such as glasses, mental training methods or computer networks is made with a view to the interconnections between human organisms and technical devices and to corresponding criteria of internality and externality. However, such differentiations rest upon a set of presuppositions – especially how a "human being" can be described; and they are thus subject to attacks (cf. Clark 2003; Bostrom and Sandberg 2009). Differentiation from medically necessary procedures can be made via the health/disease distinction (see, for instance, Merkel et al. 2007, 295 ff.; Bostrom and Sandberg 2009, 312). By way of illustration, retina implants are classified as enhancement if they outperform "normal" natural vision, for example, enable people to see like a honeybee or like a bat (Benford and Malartre 2007, 53 ff.; Jones and Whitaker 2012, 261 f.). Obviously, making a distinction involves numerous assumptions, although it is not impossible. The criteria for classifying an alteration as an enhancement are often left vague: to see better or to hear better – which means: to see a broader area or range of colors better and to hear more noises better –, to be able to remember more – or better still: to remember exactly what is needed at a given moment –, to be smarter, to feel good. Only a single isolated function is usually focused on when doing this. Upon closer examination, the relationship between the original state, the measure adopted, the final state and the assessment as enhancement involves many assumptions. We simply do not know exactly what it means to be able to see like a bat, or what consequences this will have for other functions. At present, carrying out enhancement applications in this area comes up against the obstacle that the procedures are at the beginning of their development and that in order to use them, a high level of expertise and technical resources are required. We largely have to do with prognoses, future scenarios and a great deal of science fiction.

This closer analysis of the fields in the context of which the problem of enhancement is discussed already shows the variety of forms of description that converge, the substantial role normative criteria play, and how complex the issue is. In the following sections, enhancement is more clearly defined with the help of several conceptual dualities.

16.2.2 Definitions and Conceptual Dualities

In addition to the approach based on the fields involved and examples, enhancement can be described by means of relevant dichotomies. Within their framework, the respective opposite concepts will facilitate understanding. Among the central dualities are "therapy – enhancement", "normality – supranormality", and "naturalness – artificiality". These dualities are not mutually exclusive, but complement each other, and are used in different contexts.

The distinction between therapy and enhancement is a basic and frequently used differentiation. This results from the fact that the enhancement debate derives from the question whether certain technologies that have been developed and established for therapeutic purposes should be used "beyond therapy" (President's Council on Bioethics 2003).[3] "Therapy" is understood as medical, surgical or psychological treatment aimed at restoring individuals with diseases or impairments to the original or to a so-defined normal physical and mental state of health or aimed at preventing diseases or disabilities. "Enhancement", by contrast, is used to describe efforts to increase or improve the condition, capacities and performance of healthy persons (cf. Merkel et al. 2007, 296 f.; Daniels 2000, 309). However, the usefulness of these classifications depends on clarification of the concept of "health" and "disease". Both concepts are historically and culturally variable, continuously evolving and context-dependent social constructions.[4] The way they are understood depends upon the area of application: "Health" as a political program which is reflected in the World Health Organization's definition of health as "a state of complete physical, mental and social well-being" (World Health Organization 1948) is understood differently from "health" as a criterion describing the medical necessity for treatment against the background of public or private insurance (see, for instance, Daniels 2000, 309 ff.). Scientific approaches have always been diverse and complex (Hofman 2001). Some approaches refer back to descriptions of functions and normality in the sense of an ideal type: "*Health* in a member of the reference class is *normal functional ability*, the readiness of each internal part to perform all its normal functions on typical occasions with at least typical efficiency" (Boorse 1977, 562; emphasis in the original; see also Schramme 2007). Normality is defined within historical and social processes correlating with the level of knowledge of the particular epoch (Lenk 2011, 68 f.). In objectivist approaches, this is in part

---

[3]This is a historical chronological description. From an analytic point of view, this distinction is not necessary; human-machine interfaces, for example, can arise independently in an information technology context.

[4]Originally, the understanding of "disease" as a social construction was motivated by an interest in preventing discrimination. Having said that, it would be expected that the problem of perfection should cease to exist and that there would be no necessity for any human enhancement. Surprisingly, in modern society an awareness of disease as a social construction and the quest for enhancement emerged simultaneously. An explanation might be that both approaches have in common the fact that they draw attention to the constructed and contingent nature of how humans are conceptualized.

left up to the biomedical sciences (Daniels 2000, 314 f.). Some considerations highlight values which are developed with a view to the natural functioning of the human organism and on which people would most probably agree, e.g., to be free from unbearable pain (Schramme 2011, 71 ff.). Other approaches accentuate more subjective self-evaluations and subjective feelings of well-being. The extent to which descriptions based on a differentiation between healthy and sick are capable of achieving consensus can also differ according to whether it is a matter of physical states or of psychosocial problems which are difficult to specify. In addition, physical or mental states have to be understood as a continuum, so that the distinction between health and disease necessarily involves gray areas. However, the fact that "health" and "disease" involve socially constructed, context-specific, multifactorial descriptions which are in need of concretization in no way means that a distinction between them is unnecessary. On the contrary, this distinction must be operationalized and applied within numerous contexts. This is also true for enhancement. To bring the area of interest and related problems into focus, "enhancement" should be grasped as a concept standing in opposition to treatment, even though grey zones have to be taken into account (see the critical considerations of Coenen et al. 2009, 17 ff.). The enhancement debate is thus to some extent affected by the debate about health and disease. This includes, among others, the debate on medicalization or "disease mongering". All in all, the distinction between "health" and "disease" is necessary for understanding enhancement; however, doing so is not sufficient. Further conceptual distinctions are needed, which are both to some degree connected with the distinction between health and disease and also independent dichotomies.

Among these further distinctions is the difference between "normality" and "supranormality". In its basic approach, normality is a relational concept based on specification of the reference points in terms of which something is "normal", the comparison group and the criteria according to which normality is measured. In our context, ideas of normality are often closely linked to descriptions of function in the same way as the functionality-oriented concept of health is based on these. Enhancement would involve activities which raise human capabilities beyond the species-typical level or the statistically normal range of functioning (cf. Allhoff et al. 2011, 203). The reference point is not an average value but a spectrum of typical functionalities characteristic of the species; the comparison group is the species; the criterion is what is typical. This approach implicitly assumes that the means by which an improvement is achieved are to be regarded as artificial – constellations in which people achieve outstanding performance through their own efforts, for instance in sports, would not transcend the species-typical level of functioning. Via artificiality of the means, constellations can also be considered in which people, without being sick, raise themselves to a level, for instance with smart drugs, which corresponds to the normal level.[5] However, the whole approach requires

[5]In order to make the constellations clearer, Grunwald describes this as "doping" to distinguish it from enhancement in the narrower sense, see Grunwald 2013, 204 f.

a description of the starting conditions, i.e., the state or the conditions that were assumed to exist when delineating normality. Notions of normality are based on a state achieved by a society, for instance, in terms of healthy living conditions which increase the maximum life span attainable. However, these notions do not include every imaginable method of improvement in performance. They thus merge with the idea of naturalness and of "natural limits." As a result, the concept of enhancement does not go without notions of normality (cf. also Lenk 2011, 82). These in their turn are full of presuppositions as well as rife with implications. Hence, the distinction between "normality" and "supranormality" cannot provide an exclusive framework for understanding enhancement.

The distinction between "naturalness" and "artificiality" is a central distinction as well. The meaning of nature and naturalness, however, is highly ambiguous. Their meaning depends on the corresponding opposite concepts and criteria (Bayertz 2003, 134; Birnbacher 2006, 6; Roughley 2011, 11 ff.; Witthøfft Nielsen 2011, 22).[6] An initial central and already traditional relevant distinction between approaches involves whether the human being is conceived of as part of nature while nature is distinguished from, for example, the supranatural, or whether "nature" is contrasted with human beings.[7] In the first case, nature is understood in a comprehensive sense. In the second case, human beings are distinguished from the rest of the world surrounding them, above all with the argument that for them it is less the laws of nature and more the principle of freedom which is decisive. The different approaches lead to profound questions, such as how human beings are conceptualized or whether and how body and mind should be differentiated. Closely linked to this is the understanding of nature as an essence. Originally, this understanding is based on the model of laws of nature and on the idea of something perpetually unchanging. In relation to a human being, this is reflected in the idea of a "true self." In contrast to this view, however, the "essence" of a living being can easily be regarded as being dynamic. If striving for improvement is regarded as the very core or essence of human existence, change even of fundamental characteristics

[6]See also the early remarks of David Hume: "'Nature' means something different when the concept is used as the opposite of 'miracle', 'what is unusual' or 'what is artificial'" (Hume 1739/1740, 475).

[7]In his famous and influential essay on "Nature" John Stuart Mill differentiated between two principal meanings of the word nature. In one sense, "Nature means the sum of all phenomena, together with the causes that produce them; including not only all that happens, but all that is capable of happening" (Mill 1874, 5). The other notion of nature refers to "not everything which happens, but only what takes place without the agency, or without the voluntary and intentional agency, of man." (Mill 1874, 8). The core of this essay is a sharp criticism of the employment of the word Nature as a term in ethics (Mill 1874, 9 ff.) or the "doctrines which make Nature a test of right and wrong, good and evil, or which in any mode or degree attach merit or approval to following, imitating, or obeying Nature" (Mill 1874, 13). However, this critique is based on the implicit assumption that human beings, in principle, are embedded in nature and that "man has no power to do anything else than follow nature" (Mill 1874, 64).

or abilities can be identified as part of human nature.[8] The idea of an essence is ambiguous on its own. In another traditional distinction, nature is understood as that which is not influenced by human beings and is contrasted with the results of human interventions. Somewhat differently and more narrowly described, this corresponds to the distinction between nature and technology. Anything which arises from itself and according to its own laws is natural. Because this is scarcely comprehensible and there is scarcely any such thing as completely untouched nature (Birnbacher 2006, 4 ff.; Oyama 2002), "naturalness" and "artificiality" are comparative concepts: Things are more or less natural (Birnbacher 2006, 4 ff.; Roughley 2011, 23). Furthermore, the concept of natural can refer to the way something came into existence, its genesis, or to its quality and appearance. Both might diverge: an artificial genesis can lead to a result that is judged to be natural on the basis of its appearance (Birnbacher 2006, 7 ff.). As with the understanding of normality, naturalness arguments thus require specification of the framework or the circumstances from which one is arguing. All of this shows that the concept of nature and the distinction between naturalness and artificiality are based on many assumptions and rich in implications (cf. also Bayertz 2003; Birnbacher 2006). References to images of humanity, technology assessments or descriptions of society quickly enter the picture. The large number of implications associated with the concept of human nature is reflected in the enhancement debate.

As a result, the differentiations between "therapy – enhancement," "normality – supranormality" and "naturalness – artificiality" are important for understanding enhancement. They are not sufficient in and of themselves, but rather assist understanding in combination. However, because they are rife with preliminary assumptions and implications, they do not lead to a clear definition of enhancement. We need a contextualization.

16.2.3 Contextualization

From the outset, enhancement is a complex concept because it describes relations and includes both descriptive and normative elements. A particular aspect (human traits, capacities, moods, life span) is changed in specific respects by specific means or instruments from a certain starting point to an end-state. Exact identification of the aspects and the respects in which these aspects are changed is important, because the chosen means regularly have several effects. The end point is only relatively defined and not fixed: "Optimization is a teleological approach while enhancement opens up an infinite step-by-step process during which criteria and direction of enhancement might change" (Grunwald 2013, 203). Changes specified

[8]See Birnbacher 2008, 101: "Modifying or transforming his own nature more directly and deliberately by means of technology does not constitute a radical change in human nature taken in its comprehensive sense but affirms this nature."

in this way are evaluated as improvements according to specific criteria or standards at a particular time in the assessment of individuals or institutions making a decision. This basal description already contains a multitude of elements.

So that the idea of enhancement does not incorporate everything encompassed by the basal description but becomes more powerful for describing problems and making differentiations, it must be contextualized. Contextualization means that assumptions, contextual conditions, reference points and constituent elements of a concept are worked out more clearly. Further statements are relativized in the sense that they only apply and hold true within the underlying framework. Contextualization leads neither to the consequence that nothing can be said at all nor to results being completely arbitrary, because assumptions and contextual conditions must, for their own part, be convincing. According to the respective object of knowledge and interest in knowledge, the necessary contextualization involves differing accentuation. Against this background, the concept of enhancement can appropriately be described as an umbrella term (cf., however, with different considerations Chadwick 2008, 26, 30; Coenen et al. 2009, 6).

The current enhancement debate which is being discussed here is closely related to new biotechniques as a means of enhancement (see also Hildt 2013, 3 f.). As technologies are converging (see, as a broad approach, the contributions in Roco and Bainbridge 2003), these new biotechniques comprise genetic engineering or pharmacological means as well as neurotechnologies or, if applied in a particular manner, nanotechnologies or information and communication technologies. If we take this as a basis, practices such as drinking coffee or wearing a pair of glasses do not fall within the definition of the topic.[9] The link to new biotechniques results from several characteristics which, however, need not all be simultaneously present in a particular measure: novelty, closeness to the human body, the notions of constructability and controllability of the human body, the in-principle potential for alteration which these technologies bring with them, the degree of technologization, the level of invasiveness, and the lack of reversibility. At this point we can see to what extent the concept of enhancement is a matter of risk perception and technology assessment, while at the same time raising fundamental questions about how the human can be understood. Noteworthy in this context is also the concept of human engineering (cf. Ramsay 1970: "Fabricated Man"), which grasps some aspects but is too closely linked to the concept of steering. Steering implies knowability of cause and effect relations or the existence of a central instance that steers human enhancement. Such notions are, as we can conclude from the description of the fields (Sect. 16.2.1), misleading.

The starting point from which it makes sense to speak of enhancement regularly is a state which can be referred to as "not sick". Although grey zones must be kept in view, the concept of enhancement builds on a negative distinction from "treatment" and thereby on the distinction between health and disease. Additionally,

---

[9]They may be part of a comparative, evaluative assessment of biotechnological enhancement compared with other common and accepted practices, but nevertheless, they are not very productive.

ideas of "normality" and the distinction between "normality" and "supranormality" play a central role in understanding enhancement as an issue of controversies. Notions of "normality" can in turn be linked with notions of "naturalness". "Nature" and the distinction between "naturalness" and "artificiality" provide an important framework as well. The desired end state of enhancement measures is an alteration of the earlier state which can vary from still falling within the "normal" spectrum all the way to change resulting in a supranormal state. At this point, the debate on enhancement sometimes distinguishes between moderate and radical enhancement (Agar 2010; Ach and Lüttenberg 2011, 240 f.) or between "doping" and "enhancement" (Grunwald 2013, 204 f.).[10] With regard to the end state or to the relationship between the point of departure and the end state, it is necessary to specify who actually evaluates the change as positive.[11] Is this decided by the person in question him- or herself (see, i.e., Menuz et al. 2013, 171 ff.)?[12] Or by which persons or institutions according to what criteria and in what way is this to be assessed? Breaking down the issues already shows that the concept of enhancement entails many assumptions and is extremely dynamic.

Furthermore, the profusion of implications of central concepts such as "health," "normality" and "nature" is reflected in the understanding of enhancement. Human enhancement is closely linked to understanding of the human being, with the construction of his or her boundaries or with questions of corporeality (for a distinction between body and corporeality see Böhme 2002) and the differentiation between body and mind. Technology-related assessments round this out: due to the range of possibilities and the intensity of interventions new biotechniques are identified as novel (cf. Lüthi 2013); in part means and methods, in part results are perceived as "artificial". Cultural perceptions of technology, though, change with technological developments (as well as, in turn, technologies are culturally embedded, see Grunwald 2002, with broader considerations on the concept of technology). Hence, descriptions of human boundaries based on the body as well as assumptions of the exceptional status of biotechnical enhancement measures, i.e., incorporating tools as part of the body in comparison to using them externally, are contested (Clark 2003; Harris 2007; Bostrom and Roache 2008). Human

[10]Differing moral evaluations or at least differing conclusions are based on these distinctions.

[11]See Merkel et al. 2007, 295: "the improvement implied by an enhancement is relative in at least two senses. First, what counts as an enhancement, i.e. improvement, depends on the standpoint from which the desired enhanced state is defined as advantageous, relative to certain values. [...] An enhancement in that particular, value-relative context may, therefore, not appear to be an enhancement for anybody else, or could even amount to a worsening or a disadvantage from the point of view of other people." Cf. also Holm and McNamee 2011, 291 ff.

[12]An approach which bases on the proposal that each individual determines for him- or herself whether the outcome of an intervention can be described as human enhancement or not allows an individualistic definition of enhancement which is independent from definitions of "disease" or "species-typical normal functioning" and tightly linked to (socially influenced) personal considerations. As a consequence, removing a limb is an enhancement if the person undergoing the intervention considers the removal to be an improvement. See Menuz et al. 2013, 171 ff.

enhancement is part of a development within which naturalistic self-descriptions are questioned and replaced by culturally and socially defined forms of description (cf. also Böhme 2002, 5 ff.; Bayertz 2003, 132; Heilinger 2010, 18).

But is it a problem that the concept of enhancement proves to be rife with presuppositions and implications and that it requires a contextualization? It would only be problematic and a lack of a sufficient definition if the aim were to link strict consequences with any form of biotechnical enhancement – in the sense of: Human enhancement is morally reprehensible or human enhancement must either be forbidden or allowed. In the initial debate, enhancement was in fact from the outset constructed as a concept which was labeled as either unethical or desirable. If such hasty decisions are avoided, there is no problem with constantly having to reassure oneself about the issue being addressed and that the issue is itself the subject of problematization. Precisely this necessity of permanent reflection has been made plain by the considerations up to now. Hence, the idea of "enhancement" is inherently reflexive. It stimulates debate, and it is sharpened as well as shifting within this dynamic field of debate.

16.3 Beyond the Human? Controversies on Enhancement

In the US heated controversies over the topic of enhancement have begun at an early stage. With its paper "Beyond Therapy" the President's Council on Bioethics triggered a wide-ranging social and scientific debate (President's Council on Bioethics 2003). In Europe, by contrast, the topic was initially only addressed occasionally. General discussions in Germany were triggered by the problems of stem cell and embryo research and focused primarily on determining the moral and legal status of the human embryo (cf. also Rothhaar 2014). People argued in a controversial and highly emotional way. In the meantime, however, human enhancement has also become an issue which is now being discussed throughout Europe (cf. Nationaler Ethikrat 2005; Coenen et al. 2009; or the contributions in Missa and Perbal 2009). How the debate originated provides an explanation for the way it has been conducted for a long time. We can notice typical characteristics of politicized debates where arguments have to be commensurable and stated in a form which allows decisions by balancing conflicting goods. Hence, both sides appear to be referring to the same values and rights. Arguments and counter-arguments seem to show a striking symmetry (cf. Heilinger and Crone 2013). But closer examination reveals numerous needs for differentiation and contextualization.

16.3.1 References to Human Nature

"Human nature" is one of the concepts to which arguments refer or which is implicitly reflected in some arguments. However, both the understanding of human nature and the question of whether and to what extent nature, as it is understood, can provide a normative standard is a matter of dispute. In the latter respect, the problem is less one of naturalness being regarded as a self-evident value or a fallacy that could be criticized, but rather that there is a scarcely resolvable interweaving of descriptive and normative elements in discussions of "human nature".

The rejection of enhancement measures refers to "human nature" rather frequently. To some extent, a naturalness bonus (Birnbacher 2006, 21 ff.; Ida 2009, 63 ff.; see also Bittner and Inthorn 2011) rooted in the lifeworld (German *Lebenswelt*) underlies the skeptical judgments. More closely worked out considerations also employ "human nature" as a topos which shall give reasons for objections to enhancement (cf., for instance the President's Council on Bioethics 2003). However, the concept of human nature usually remains a vague, ambiguous and compact concept. Sometimes human nature is described as what is not made but emerges of its own accord; a concept which is conceived as the opposite concept to the logic of domination (President's Council on Bioethics 2003, 287 ff.; Habermas 2003, 44 ff.; Sandel 2007, 85 ff.). In part, human nature is understood as a combination of essential characteristics that are typical of the human species, i.e., human consciousness, reason, emotions, sentience and sociability (in terms of a "Factor X," see Fukuyama 2003). In part, it is employed indirectly by describing preconditions for the respect for others as human beings which is indispensable in a rational society; keeping essential elements which arise organically as part of nature distinct from what has been manufactured is part of these preconditions (Habermas 2003). Sometimes it is deliberately used as a complex concept which cannot be described exactly, and directly contrasted with enhancement technologies in light of their characteristics and consequences (Fukuyama 2003, 172):

> What is it that we want to protect from any future advances in biotechnology? The answer is, we want to protect the full range of our complex, evolved natures against attempts at self-modification. We do not want to disrupt either the unity or the continuity of human nature, and thereby the human rights that are based on it.

That human nature is worthy of protection is justified by differing arguments. Respect for "the given" in the sense that certain processes should be barred from human intervention, has partly religious motivations. The concept of "playing God" is a popular watchword (cf. President's Council on Bioethics 2003, 285; for a discussion of the complexity of this watchword see Coady 2009). In secular terms, an "ethic of giftedness" (Sandel 2009, 79) is proposed. As people cannot be held responsible for natural traits and abilities, the acceptance of the given relieves them from an all-embracing responsibility which would be destructive under both psychological and social aspects (Sandel 2007, 85 ff.). Enhancement and genetic engineering represent (Sandel 2009, 78, 86)

> [...] a Promethean aspiration to remake nature, including human nature, to serve our purposes and satisfy our desires. [...] If bioengineering made the myth of the 'self-made man' come true, it would be difficult to view our talents as gifts for which we are indepted, rather than as achievements for which we are responsible. This would transform three key features of our moral landscape: humility, responsibility, and solidarity.

Furthermore, the danger of alienation (Müller 2010; Agar 2010, 179 ff.) and impairments of the development of the self or authenticity (cf., e.g., Gordjin 2008, 232 f.; Kipke 2011) are presented as consequences:

> With biotechnical interventions [. . .] we will be at a loss to attest whether the resulting conditions and activities of our bodies and our minds are, in the fullest sense, our own as human. (President's Council on Bioethics 2003, 285).

Naturalness – in the sense of a distinction between born and made – is conceived as a precondition of personal identity (Habermas 2003). Others emphasize that the valuation of human accomplishments is based on results achieved through natural potential and abilities, not by artificial means, because not so much the results as the acquisition processes are relevant (Tobey 2004, 123 ff., see also Kipke 2011). Enhancement

> [. . .] shifts our natural mode of valuing one another in the personal, familial sphere away from process-orientation and towards product-orientation (Tobey 2004, 126).

Last but not least, genetic manipulations could affect the preconditions of social interactions and the foundations of societal integration. This is not about "moralizing human nature" in the sense of a resacralization but a matter of (Habermas 2003, 25):

> [. . .] the assertion of an ethical self-understanding of the species which is crucial for our capacity to see ourselves as the authors of our own life histories, and to recognize one another as autonomous persons.

New biotechnologies are regarded as being exceptional in comparison with other technologies because they enable a new type of intervention (Habermas 2003, 12) and scenarios can be constructed that would change basic background assumptions about social life. Just as elementary is the depth of misgivings about predictability and controllability of the effects of interventions in complex functioning human systems which have until now developed through evolution (cf. Merkel et al. 2007, 347). Occasionally, the inherent limitations of science are emphasized: Science uses models and creates its own necessarily reductionist abstractions resulting in conceptual boundaries and specific blindness (cf. Kass 2002, 277 ff.). Therefore, "the nature and meaning of living, and of life altogether, will forever lie out of reach" (Kass 2002, 293). Precisely because human nature in its complexity cannot be exhaustively comprehended, it should be protected against potentially far-reaching biotechnical interference.

In case enhancement measures meet with approval in a more or less broad and more or less differentiated way, some lines of argument draw explicitly or implicitly upon the topos "human nature" as well. The central difference from skeptical judgments on enhancement lies primarily in the view that alterations are judged to be a part, if not the core of human nature. "Nature" or the "natural" is not understood as a complex whole and a dense web of harmonious interdependencies which would be endangered by any intervention. On the contrary, it is argued that the human organism is not characterized by extreme connectedness and a correspondent fragility but rather by limitations on connectedness due to features of organisms such as modularity, redundancy and canalization (Buchanan 2011b, 134 ff., 143 ff., 181 ff.). Furthermore, attention is drawn to the indivisible interweaving of primordial nature and culture. Hence, creating an antagonism between them is refused; nature is considered as always having been shaped by culture. Detached from ideas of immediacy and bodiliness, developing mental capabilities by employing artificial means and coupling them with "natural" brain functions is seen as a regular phenomenon in human thinking (see Merkel et al. 2007, 343). Stated radically, artificial extensions constitute the human mind. Thus, enhancement by means of technical devices or human-machine interfaces are characterized as an integral part of human nature (Clark 2003, 174):

> Human-machine symbiosis, I believe, is simply what comes naturally. It lies on a direct continuum with clothes, cooking ('external artificial digestion'), bricklaying, and writing. The capacity to creatively distribute labor between biology and designed environment is the very signature of our species.

Therefore, human nature

> [. . .] is dynamic, partially human-made, and improvable. (Bostrom 2011, 65).

That seeking enhancement merits protection is justified with underlying reasoning which is partly similar to anti-enhancement arguments but arrives at contrary results. Theological considerations describe human creativity as an ability given by God; human beings could be seen in some sense as being co-creators with God (see for the different theological positions the contributions in Cole-Turner 2011; Sagoff 2005, 82 ff.). Because nature and nature's gifts are not always precious, there is no reason to accept "the given" or species-specific natures (Bostrom 2011, 57). The factors determining the formation of personal identity or the conditions for social recognition would not be undermined without further ado. In cases of genetic enhancements this would hold all the more to the extent that genetic determinism is misleading (Buchanan 2011b). The use of drugs as a means of enhancement could even help people toward authentic conduct (Juth 2011); at least it would not have to interfere with people's ability to live authentic lives (Dees 2007, 386 ff.). Compared with other techniques, biotechnical measures of enhancement are merely a part of continuous development and nothing exceptional (see, among others, Buchanan 2011b, 39 ff.). Their outcomes are not regarded as generally too risky. Trusting in the results of evolution also includes risks, and these results are not necessarily preferable (Buchanan 2011a, 26 ff., 173):

> Nature or evolution is not like a master engineer. The natural – the biological status quo – is rarely optimal, and sometimes it's not even acceptable. To make a rational evaluation of the possibilities of biomedical enhancement, we have to rid ourselves of pre-Darwinian, romanticized, rosy assumptions about nature and our own biology. Human nature is a mixed bag, with plenty of room for improvement.

To sum up to this point, various arguments referring to human nature are closely intertwined, as are descriptive and normative elements, and contrary conclusions are drawn. Hence, appeals to human nature are sometimes judged to lead nowhere (see, for example, Buchanan 2009, 2011b, 136, 138 f.). At least, just as the enhancement concept itself is often inadequately defined and thus unclear, recourse to human nature in the broad debate leads to lack of precision.

16.3.2 Dignity, Autonomy, Equality

In addition to human nature, certain fundamental values and human rights have always played an important role in the enhancement debate. Human rights seem to reinforce specific concerns in a particular way. They provide them with legal weight and establish links to the human rights discourse which is well organized and widely recognized (cf. Fenton 2008, 2). This discourse is occasionally described as already being the accepted language of international ethics (Baker 2001, 249). Human dignity, autonomy and equality emerge as key concepts which, along with the idea of human nature, seem to offer a common language and also partially refer back to this idea.

The concept of human dignity and its inviolability has made a new career in view of biotechnology. An example is its impact on the worldwide ban on reproductive cloning. Human dignity implies through reference to the dignity of the *human being* and to the concept of *dignity* to notions of human. Furthermore, its inviolability is laid down as a legal consequence. It thus suggests itself as a central legal foundation of arguments regarding respect of human nature in the sense of "the given" or of a combination of essential characteristics which have to be protected, specifically against enhancement measures (cf., i.e., President's Council 2003; Annas 2005, 37 ff.). Such views are based on the fundamental assumption that human dignity has to be understood as a person's intrinsic value as a human being, which must never be violated. The definition of this intrinsic value is related to religious concepts of being human, to concepts of nature or to Kantian ideas that a human being should always be treated as an end and never as a mere means, and should neither be made an object nor instrumentalized. With respect to equality, another objection to the use of enhancement technologies is that they might give those who use them an unfair advantage and might provoke problems of distributive justice.

Opposing positions highlight the view that human dignity can be understood as the capabilities required for performing central human functions or as representing the conditions needed for a flourishing human life, which are not necessarily endangered but, rather, might be improved by enhancement technologies (cf. Fenton 2008, 5). As a moral status or as the quality of being worthy, dignity could be increased by some forms of enhancement and the concept of dignity could include enhanced humans or future post-humans (Bostrom 2009, 173 ff.; Bostrom 2011, 61 ff.). From the point of view of autonomy, it is stated that enhancement can extend the spectrum of a person's behavioral possibilities, e.g., cognitive abilities which are improved by attention- or memory-increasing drugs, and could lead to more rational and 'freer' decisions (cf. Heilinger and Crone 2013). Independently of empirical questions of effects, autonomy – defined as self-determination and the right of the individual to decide for him or herself – is enhanced by an increase in options of choice. Against the background that private freedom encompasses the use of new technologies for personal advancement, restrictions are regarded to be only justified if they involve interference with outweighing opposing public interests or the rights of other individuals. Beyond that, enhancement is even described as a moral duty on the basis of the assumption that there is no significant moral difference between acts and omissions (Harris 2007, 79 ff.). As to the fairness question, biotechnical enhancement measures might be a possibility to compensate for the "unfairness" of natural inequalities in strength or abilities and to advance distributive justice (cf. Buchanan 2011b).

Similar to the human nature argument, it can be observed that contrary conclusions are based on particular values or rights (see also Heilinger and Crone 2013 with regard to "freedom"). Closer examination of the ideas "dignity", "autonomy" and "equality", however, reveals implicit basic concepts, conditions and presuppositions that underlie each particular line of argument and involve quite different approaches. Any analysis has to take these diverse backgrounds into account.

16.3.3 Contextualization

For a long time, the enhancement debate was characterized by arguments divided into anti-enhancement and pro-enhancement views or anti-enhancement and anti-anti-enhancement positions (Buchanan 2011b, 13 ff.; see also Parens 2006 proposing the framing gratitude vs. creativity). But as we have seen, the issue of enhancement is very demanding. It cannot be described appropriately without a contextualization. Different contexts raise differing ethical questions (cf. also Juengst 1998, 43 ff.). Often, the initial scenarios being discussed are not the same. Evaluations depend on criteria and forms of description.

Consequently, the initial scenarios, the background and underlying assumptions of chosen approaches and the status of particular arguments must be described with sufficient clarity (see also Heilinger and Crone 2013). Contextualization does not weaken reasonings but makes them more transparent and comprehensible. Enhancement is a problem which reveals how varied and in need of contextualization concepts of "nature", "freedom", "dignity", "autonomy" or "equality" are. At the same time it provokes discussions of these concepts, their underlying assumptions and basic ideas. In the present context this leads to a linkage with the debate on human nature and human rights.

16.4 Interrelationships Between Human Rights and Human Nature

A popular description of human rights understands them as the rights of individuals who belong to the human species. They are rights which have to be respected and which all human beings enjoy equally, simply because they are human. On this basis, rights with specific contents are developed, whose holder is the individual human being. But how can the normativity of human rights be substantiated? What guarantees their universal and egalitarian validity? Are they exclusively rights of the individual human being or could other entities be entitled to them? What concrete contents do human rights have? Why should people regard themselves as under an obligation to respect or enforce the human rights of others (see the differentiation of Beitz 2009, 59 ff., between the demand and the supply side)?

Recourse to "human nature" has traditionally played a central role in the foundation and justification of human rights. At the same time, these conceptions, their background, their preconditions and their outcomes have always been as heterogeneous as the notion of "human nature" itself (see Sect. 16.2.2). Classical approaches developed comprehensive, theologically influenced and teleologically constructed ideas of "nature" and "natural order" as well as "natural rights" that are embedded in these all-embracing views. Modern natural right theories have abandoned teleological metaphysics and derived natural rights in terms of an assumed "state of nature" preceding civil society. They have arrived, however, at results which could not be more divergent (cf. Reinhardt 2014, 141 f.). More recent approaches diverge further, especially in terms of their conceptual starting positions. The functions of the recourse to human nature have also always been heterogeneous. This recourse may be ontological, i.e., center on describing the nature of human rights. It might seek to provide a foundation of human rights designed to be relatively independent of codifications or – to a considerable extent – of social or political conditions. It might aim only at explaining why particular interests or needs should be protected, at describing the range of values or goods to be protected, or at refining what follows from designating a value as a human right (see Beitz 2009, 48). The functions can be combined in a unitary justification context. However, after the collapse of the traditionally all-encompassing closed system of justification treating them separately and paying attention to a particular function is also widespread. The quite different approaches, the possibility of critical reconstruction and recombination of diverging strands and the diversity of the role of human nature in human rights thinking explain the continuing prominence of this paradigm. At the same time, every closer analysis must specify in what context and with what function which conceptualization of human nature is being employed.

If interrelationships between human nature and human rights are re-conceptualized nowadays, human nature is brought into play at different levels and in different respects. With regard to existing codification of human rights, recourses to human nature serve as, for example, a foundation, categorization or framework for interpretation or critical examination of codified rights. Particularly human rights are often conceived as being distinct from embodied positive rights and as rather emerging from moral standards or values that can be invoked as a basis for criticism of actually existing laws and social practices (Donnelly 1982, 400; cf. also Beitz 2009, 49; Reinhardt 2014, 149 ff.). Beside fundamental considerations, there have been several attempts to link the goods deserving protection or the contents of rights to human nature. This might refer to properties or natural interests which are attributed to human beings. Beyond that, the concept of human nature is taken in a broader sense, including not only human properties but also human potentialities (see Tepe 2014, 66 ff., with a view to the capabilities approach and the ontological anthropological approach; see also Donnelly 1982, 398 ff.). Other attempts aim at explaining the reasons why people respect the human rights of others – not only because they protect an interest that anyone might be expected to have but also because empathy can be regarded as a natural human faculty (see von Harbou 2014). These considerations strive to understand and refine the idea of "human nature" in the context of human rights in a sensible way; even in this concrete form, though, they quickly lead to different terminology because any concretization must necessarily go beyond mere recourse to human nature and be more precisely stated.

Abandoning both classical approaches based on comprehensive conceptions of a natural order and conceptions of a state of nature means, that the legal system itself provides the background and starting point for the foundations of human rights. The legal system is a social subsystem within contemporary, functionally differentiated society. It is characterized by its own, relatively independent communication and actors, structures, procedures and operations. As a social system, it is not the same as codifications of rights or international treaties which establish consolidated texts to which actors and communications can refer. The relative independence of the legal system by no means implies that relationships with its environment are irrelevant, but they are mediatized within the legal system. Against such a background it stands to reason that human rights cannot be based on a unitary foundation but are supported by various sources and concepts (cf. also Beitz 2009, on the basis of a "practical conception"). This means that the goal of establishing an essential foundation for human rights on the basis of human nature must be abandoned (cf. also Pollmann 2014) as well as the exclusivity of arguments referring to nature to develop or interpret concrete human rights (see also von Harbou 2014, 104 ff.). Nevertheless, human nature does not lose all significance.

All approaches which make use of human nature in arguments in connection with human rights, however, are processed and mediatized in the legal system. In the following, the differentiation between observations internal to the legal system and those external to it becomes relevant. Within the legal system there are differing recourses to the idea of human nature as a part of an argument for justifying and concretizing particular rights. To some extent, they contradict and compete with each other, such as approaches referring merely to properties or existing abilities and approaches which include human potentialities. Considering their self-understanding and their convincibility, these approaches must aim at internal consistency. From the perspective of each particular approach, diverging approaches are mutually irreconcilable. Switching to an external point of view, it is, however, possible for communications or actors in the legal system to sometimes be based on one approach, sometimes on another. Courts, for example, avoid making binding and final decisions favoring any particular theoretical approach. As long as consistency of judgments is not fundamentally laid open to question, in making their decisions they can work with the theory of innate human rights in one case, follow a non-naturalistic line of argument in the next, and then choose the capability approach. What may be contradictory from the internal perspective of the point of view of a particular approach can be completely compatible with a functioning legal system from an external perspective. Flexibility is even necessary for the required degree of complexity, the relative openness and the adaptability of the legal system.

Against this background the concept of human nature might be assigned a special role in human rights discourse. Within the legal system, the idea of human nature is proposed by several approaches. To be convincing, any of these approaches depends on not simply stopping at reference to human nature but on contextualizing it and making it more precise with the help of other terminology. From an internal point of view, the approaches prove to be different or even divergent. In observing the legal system from an external perspective, however, precisely the vagueness and ambiguity of the concept of human nature offers advantages. The concept seems to provide a uniform point of reference. In fact, because of its ambiguity and rich implications, constantly new and different concretizations can be carried out again and again. In addition, the concept has the potential to set processes of reflection in motion. All this supports the legal system as a sub-system in capturing the plurality of perspectives in modern society, in grasping problems in a multifaceted way and in deciding issues and at the same time remaining relatively open. Such a background – and not a one-dimensional notion of human nature or a preassigned interrelationship between human nature and human rights – provides an appropriate framework to handle the enhancement issue as a new challenge.

16.5 Enhancement as a Catalyst for Reflexivity

16.5.1 Enhancement, Human Nature and Human Rights: Interplay and Challenges

Human enhancement is an issue that enriches the discourse on human nature and human rights and, in turn, benefits from being part of such a discourse. We have worked out that the idea of "enhancement" must be contextualized and is inherently reflexive (Sect. 16.2). In addition, it sets off new debates with regard to other key concepts.

These involve on the one hand the concept of human nature. The meaning of nature and naturalness with its longstanding and richly diverse tradition has always been ambiguous. As a result of closer analyses (see Sect. 16.2.2) its function can be seen in offering a basis for self-formation processes or for ongoing specification and reflection of self-concepts or of descriptions how the human life-form can be understood (see, with different approaches, Müller 2008b, 31 ff.; Bittner and Inthorn 2011, 182 ff.). This function is intensified by the biotechnical possibilities for altering human beings. Natural limits to human and social change are vanishing to a greater and greater degree. Nature, which until now has been something observed and described as being in principle the way it is, is now being made the subject of responsibility and accountability (Sagoff 2005, 90). At a fundamental level, completely novel scenarios are imaginable: Conversion of human reproduction to cloning techniques, genetically altered human beings, human-machine beings, chimeras, artificial life. Human enhancement leads us to rethink the very questions of what it means to be human, how we are to construct human boundaries or the difference between human beings and their environment, or how we are to understand the differentiation between body and mind. Naturalistic self-descriptions have to be replaced by culturally and socially defined forms of description. Technical developments make this possible. They do not, however, provide an exclusive explanation. The emergence and application of techniques are always embedded in social developments; therefore, transformations must be more broadly understood as part of societal or cultural codes of communication. Enhancement, as we have described it, is part of and a result of modern society and of a combination of its characteristics (Elliot 2004; Müller 2008a, 194 f.; Coenen et al. 2009, 38 ff.; Grunwald 2013, 206 ff.): the performance-oriented or performance-enhancing society, the competitive society, the functionally differentiated society, the globalized society composed of countries with different cultures and divergent rules, the pluralistic society or the individualized society.

On the other hand, completely new discussions and reflections on the functions and contents of human rights are emerging. The background for the foundation of human rights is the legal system itself, but this does not mean that the acknowledgment of human rights would be an arbitrary formal construction. In describing interests deserving protection or protected goods the law must be open for implications provided by the environment and sufficiently receptive. As it is nevertheless relatively independent, it refers to and employs varying approaches. Human enhancement raises numerous novel questions at the fundamental level and in connection with concrete issues. What characterizes the "human being" upon which human rights are based? What is the relationship between human being and species? Can human rights be transferred to other entities, for example to chimeras, artificially constructed living beings or even androids (cf. Koops 2013a, 179 f.)? Do existing legal contents, e.g., human dignity or autonomy, have to be re-conceptualized? Will we have to solve problems of unequal distribution keeping enhancement resources in mind, and should we be concerned about an emerging gap between enhanced and non-enhanced human beings? Or is the problem more the opposite: that enhancement techniques are increasingly leading to a uniform 'ideal type' of human being which will preclude any form of abnormality (cf. Koops 2013b, 186)? And must we develop completely new kinds of liberties, rights and duties, for example, cognitive liberty (Bublitz 2013, 241 ff.), fundamental rights to mental integrity, to emotions or to forget or an obligation on governments to stimulate the pluralism of humankind (Koops 2013a, 174 ff.)? The concrete questions are also diverse. At the core is that the development of human beings will be the subject of decisions, and that who actually decides and how will become a problem.[13] Who makes the choices regarding human enhancement? What decisions are ultimately to be left up to individuals to decide for themselves? The traditional approach of assuming that all decisions having an impact only on oneself and not on others are to be considered "private" and left up to the individual falls short of providing a convincing solution. Assessing consequences is always dependent on the form of observation and the answer to the question of whether only the individual is affected or other parties as well varies accordingly. Can we still operate with the construction of balancing individual rights on the one hand and conflicting rights or public interests on the other? And if we have to deal with an elementary uncertainty and unknown issues, what about fundamental principles such as the acceptance of individual decisions as long as no damage can be proved? To what extent is protection against oneself acceptable? Is it permissible for individuals to make autonomous decisions which undermine the prerequisites for their own autonomy? How does the requirement of informed consent function if the person changes due to personality altering medical treatment or mind-altering drugs? How shall we judge surrogate decisions or decisions from parents which have an effect on children? Decisions such as these might be legitimate only when they are undoubtedly in the interest of these individuals, so that subsequent consent to them can be assumed to be a matter of course (cf. Bayertz 2003, 141). As to human enhancement, we have to take into account the vague and variegated criteria for classifying an alteration as an improvement.

The necessity for reflection which is contained in each of these key conceptualizations – enhancement, nature, human rights – inherently and with particular clarity in their interplay, shows that we have moved beyond a time when enhancement could be described in a general way and either sweepingly rejected or welcomed. The fields must be differentiated and scenarios more clearly described. The enhancement problem can be divided into different domains and discussed in a differentiated way. From certain points of view, however, enhancement is also a unified topic which revolves around common basic questions. It can be referred to as a catalyst for reflexivity because it sets off novel lines of thought.

In the following, I concentrate on two points. The controversies on enhancement reveal conflicts both over knowledge and also over values; however, the formerly clear-cut distinction between these types of conflicts becomes blurred. The second point centers on questions of human rights as those rights held by human beings as individuals and the role of the species.

[13]Early Glover 1984, 44, 45 ff.: "The set of problems which raises deeper issues centres round the question: if we adopt positive genetic engineering, who is in a position to decide what future people should be like?"

16.5.2 Knowledge Conflicts and Value Conflicts

Discussions of enhancement follow various lines of reasoning. Some considerations are based on empirical assumptions and prognoses, mainly with regard to the use or the consequences of enhancement techniques. Other arguments reflect values or normative standards. Although approaches and arguments are often insufficiently elucidated and lack necessary contextualization, the controversies surrounding them reveal knowledge conflicts as well as value conflicts. Conflicts over knowledge are disputes over claims concerning the empirical evidence of particular statements of facts or the validity of prognoses. These may arise because of differing knowledge or uncertainty about the existence of phenomena or the occurrence of consequences. The enhancement debate involves numerous contested predictions, for example, that social relationships would change if human beings were cloned or embryos genetically altered by intent, that parents would come to regard their children as manufactured objects, that cloned or genetically altered persons would no longer be able to see themselves as autonomous persons or as the authors of their own biographical histories, or that the self-understanding of the species would be undermined in a way which destroys the basis of a deliberative society. Conflicts over values are caused by divergent judgments regarding values or goods. They are also fairly obvious in the enhancement debate, especially as personal values and attitudes depend on factors such as whether people are committed to religious beliefs, the position they take on the idea of human nature or the extent to which they appreciate individual freedom of decision.

For a long time both types of conflict have been addressed separately. As to knowledge conflicts, knowledge was regarded as principally achievable. Uncertainty and a lack of knowledge were temporalized to merely "not yet knowing" which can be overcome by attempts to gain knowledge along methodological parameters. Conflicts over knowledge were to be resolved by assumptions of the primacy of scientific findings and guidance by the state-of-the-art of science and technology, by forms of proceduralization of risk assessment and decision-making and by institutionalizing participation procedures. In addition, duties to observe the effects of decisions and to rethink them, if necessary, as well as efforts to ensure their reversibility were established. Value conflicts have been dealt with by differentiating "private" and "public" issues which are subsequently subject to either individual or collective decisions, by an assumed consensus on fundamental values, by the establishment of procedures for reaching reasonable compromises or by majority rules.

Meanwhile, the preconditions of these patterns of resolving conflicts have changed. The conception of a definable body of acquired or at least achievable knowledge which people agree upon or accept as state-of-the-art of science and technology is vanishing more and more. A common basis of shared knowledge that has been taken for granted for a long time is no longer present. Knowledge has been diversifying to a greater and greater extent. Basal differentiations are made, for instance, between knowledge specific to social (sub-)systems and lifeworld knowledge (knowledge derived from the lifeworld – German *lebensweltliches Wissen*). It is becoming apparent that every form of knowledge, even scientific knowledge, is a specific construction that uses its own forms of observation and interpretation. Kinds of knowledge to be differentiated are explicit knowledge that can be articulated and tacit knowledge that is difficult to express. In addition, knowledge production may lead to more knowledge in some respects, in other respects, however, it leads to greater uncertainty and newly emerging unknown issues. The unknown is thus the constantly generated reverse of knowledge and an unavoidable result of acquiring knowledge. One can be aware of some areas of the unknown whereas other areas of the unknown are inaccessible and can only be identified retrospectively. The construction of knowledge, uncertainty and the unknown differs according not only to context but also to perspectives, for example, of decision-makers and those affected by decisions or of laypersons and experts. Among the effects of the diversification of knowledge, the experiences with uncertainty as well as the unknown and the pluralization of knowledge perspectives is, that knowledge conflicts are no longer addressed only cognitively, but also to an increasing extent by referring to normative standards which seem to offer orientation. But the conditions under which conflicts over values are dealt with are changing as well: the differentiation between issues which are regarded either as private or as public is no longer convincing; fundamental values are no longer shared; the majority rule is no longer widely accepted because decisions often prove to be irreversible. As a result, the distinction between knowledge conflicts on the one hand and value conflicts on the other is becoming blurred, and the traditional patterns for mediating knowledge or value conflicts fall short of resolving them.

The controversies on enhancement illustrate these developments quite clearly. The profundity of the topic and the broad discussions involve lifeworld knowledge as well as knowledge specific to social (sub-)systems, lay knowledge as well as expert knowledge and explicit knowledge as well as tacit knowledge. There are only limited ways to separate knowledge from the practices of generating knowledge and to find a common basis of shared knowledge. In addition, the continuing and inescapable existence of uncertainty and unknown issues is obvious: Often the risks of enhancement techniques are not investigated because their application was originally developed in a medical context for therapeutic purposes, and they have been studied only in this context. An illustrative example are unintended negative consequences or adverse effects of the "off-label" use of smart drugs which can not be foreseen but identified only after a period of prolonged use of this drug. Moreover, several safety issues cannot be sufficiently explored because of restrictions imposed on research conducted on living human beings. Just as well, the social and psychological consequences of some enhancement techniques are difficult, if not impossible to predict. This is true not least because the new biotechnical possibilities are unprecedented. But even if impacts and risks are assessed, science uses models and reductionist abstractions; therefore it has conceptual boundaries and inherent limitations. To a substantial extent, enhancement, its effects and its risks refer to an unknown future. The present construction of this future is necessarily characterized by normative assumptions. For all these reasons, knowledge conflicts in the enhancement debate are multifaceted and cannot be resolved by merely pointing to empirical research. Beyond that, the concept of enhancement itself is a complex concept which includes both descriptive and normative elements (see Sect. 16.2). Already the judgment whether an alteration can be described as an enhancement depends on the standards according to which something is regarded as being an improvement, and these standards derive from interplay among numerous individual and societal factors and are often remarkably unclear and contested (see Sect. 16.2.1). All in all, knowledge conflicts and value conflicts are quite interwoven. Additionally, differentiated to the fields of enhancement having in mind how manifold they are, the emerging conflicts might be hard to mediate and new forms of dealing with those conflicts will be needed.

16.5.3 Individual Rights and the Role of Species

Human rights are understood as being those fundamental rights held by human beings as individuals. To move the individual into the center of attention and to assign rights to him or her has been the revolutionary development set in motion by the idea of human rights as innate rights or as natural rights. The assignment of rights to the individual means that not only the interests of the individual but also his or her self-understanding of the protected freedom and his or her decisions how to make use of it are normatively relevant. However, human rights necessarily go beyond the individual and transcend individual rights. First, they are rights that have to be acknowledged and therefore inherently possess supra-individual values and at least relative consensus on their being worthy of protection. Secondly, legal measures limiting the scope of protection of human rights are allowed under certain circumstances. Limiting measures, though, are referred back to the individual rights: the interests protected must nevertheless be taken into account by respecting underlying prerequisites such as the principle of proportionality. Hence, the construction of human rights is already quite complex and enables interpreters to take varying perspectives into consideration. The tension found in the relationship between the interests and perspectives of the individual and those of society or other persons is kept in the construction of the norm and dissolved procedurally. This comes constantly into view, not only in case of disagreements about the normative preconditions of a limitation but also with regard to the understanding of the scope of protection of human rights. Already among the classical questions is if "freedom" or "autonomy" have to be understood as concepts which can be defined as the right of the individual to act by his or her own volition or as concepts which point to rationality as an underlying principle.

The tension increases with the advent of biotechniques, their potential for intervention and human enhancement in the sense we have described it (Sect. 16.2). Given the historical context of the emergence of human rights, the very question arises whether it is a part of the implied background of the conception and structuring of human rights that natural limits of human beings and humanity and the respective self-understanding of human species are relatively stable and not in themselves the subject of discussions and decisions. It is precisely this point that is reflected, from an ethical point of view, by species-ethical reasoning. The concept of species-ethics shall express the interest in preserving universal modern morality and guaranteeing equal respect for all humans by all humans (cf. Lohmann 2014, 167 ff., with view to the considerations of Habermas 2003). In legal contexts, there are meanwhile controversial discussions on the role of the "species" as well (see the contributions in Dabrock et al. 2010). Irrespective of difficulties to define the human "species" and to find appropriate criteria, controversy over the notion and fundamental premises of the concept of human rights is taking place again. Is there a concept of human species which provides underlying presuppositions for human rights without which human rights cannot be understood? Could species-related arguments be used for objecting against and limiting the rights of the individual or do they even affect the contents of human rights? At first sight, a decisive role of species-related arguments might sound quite plausible. However, human rights are to a certain extent transferred to organizations in case that the application of the particular human right makes sense. Additionally, they cannot be based on a unitary foundation, especially not be understood merely on the basis of human nature, but are supported by various sources and concepts (see Sect. 16.4). This at least relativizes the role that concepts of human species might play within the understanding of human rights. And the complexity of human rights (see Sect. 16.4) is also the basis for an appropriate discussion of human enhancement.

Nevertheless, these considerations point to the interplay between the problem of human enhancement on the one hand and human rights on the other. Human rights with their protection of "freedom", "dignity", "autonomy" or "equality" do not provide normative standards which readily offer answers to questions of the legitimacy of enhancement measures. On the contrary, the enhancement issue opens up new discussions of the concept, the underlying assumptions and basic ideas of human rights.

16.6 Conclusion

Human enhancement has proved to be a complex concept with many assumptions and both descriptive and normative elements. We have focused on the application and consequences of new biotechniques which have partly been realized already and are partly envisaged in future scenarios. Enhancement technologies range from plastic surgery, smart pills, genetic diagnostics and intervention, clones and chimeras to the production of cyborgs or the creation of artificial life. The standards according to which something is regarded as being an enhancement have to be concretized as well as the persons or institutions who decide on enhancement measures. Furthermore, the profuse array of implications of central concepts such as "health", "normality" and "nature" is embedded in the approaches. The idea of "enhancement" is not only inherently reflexive. It also sets off new debates with regard to other key concepts. This is true for the already multifaceted notion of "human nature". And among other discussions, new reflections on the functions and contents of human rights or on their role and meaning as those fundamental rights held by human beings as individuals are necessary. All in all, enhancement is an issue that enriches the discourse on human nature and human rights and, in turn, benefits from being part of such a discourse.

References

Ach, J. 2006. Komplizen der Schönheit? Anmerkungen zur Debatte über die ästhetische Chirurgie. In *No body is perfect*, ed. J. Ach and A. Pollmann, 187–206. Bielefeld: Transcript.

Ach, J., and B. Lüttenberg. 2011. Ungleich besser? Zwölf Thesen zur Diskussion über Neuro-Enhancement. In *Entgrenzung der Medizin. Von der Heilkunst zur Verbesserung des Menschen?* ed. P. Wehling and W. Viehöver, 231–250. Bielefeld: Transcript.

Agar, N. 2010. *Humanity's end: Why we should reject radical enhancement*. Cambridge, MA: MIT Press.

Allhoff, F., P. Lin, and J. Steinberg. 2011. Ethics of human enhancement: An executive summary. *Science and Engineering Ethics* 17: 201–212.

Annas, G. 2005. *American bioethics. Crossing human rights and health law boundaries*. Oxford: Oxford University Press.

Baker, R. 2001. Bioethics and human rights: A historical perspective. *Cambridge Quarterly of Healthcare Ethics* 10: 241–252.

Bayertz, K. 2003. Human nature: How normative might it be? *Journal of Medicine and Philosophy* 28(2): 131–150.

Beitz, C. 2009. *The idea of human rights*. Oxford: Oxford University Press.

Benford, G., and E. Malartre. 2007. *Beyond human: Living with Robots and Cyborgs*. New York: Tom Doherty Associates.

Birnbacher, D. 2006. *Natürlichkeit*. Berlin/New York: De Gruyter.

Birnbacher, D. 2008. Posthumanity, transhumanism and human nature. In *Medical enhancement and posthumanity*, ed. B. Gordjin and R. Chadwick, 95–106. Dordrecht/Heidelberg/London/New York: Springer.

Bittner, U., and J. Inthorn. 2011. Der normative Status der Natur im Kontext medialer Debatten zur ästhetischen Chirurgie. In *Im Dienste der Schönheit?* ed. B. Lüttenberg, A. Ferrari, and J.S. Ach, 167–188. Berlin: LIT Verlag.

Böhme, G. 2002. On human nature. In *On human nature. Anthropological, biological and philosophical foundations*, ed. A. Grunwald, M. Gutmann, and E. Neumann-Held, 3–14. Berlin/Heidelberg/New York: Springer.

Boldt, J., and G. Maio. 2009. Neuroenhancement. In *Das technisierte Gehirn. Neurotechnologien als Herausforderung für Ethik und Anthropologie*, ed. O. Müller, J. Clausen, and G. Maio, 383–397. Paderborn: Mentis.

Boorse, Ch. 1977. Health as theoretical concept. *Philosophy of Science* 44: 542–573.

Bostrom, N. 2005. A history of transhumanist thought. *Journal of Evolution & Technology* 14: 1–25.

Bostrom, N. 2009. Dignity and enhancement. In *Human dignity and bioethics*, ed. E.D. Pellegrino, A. Schulman, and Th.W. Merrill, 173–206. Notre Dame: University of Notre Dame Press.

Bostrom, N. 2011. In defense of Posthuman dignity. In *H +/− transhumanism and its critics*, ed. G.R. Hansell and W. Grassie, 55–66. Philadelphia: Metanexus.

Bostrom, N., and R. Roache. 2008. Ethical issues in human enhancement. In *New waves in applied ethics*, ed. J. Ryberg, Th. Petersen, and C. Wolf, 120–152. Basingstoke: Palgrave Macmillan.

Bostrom, N., and A. Sandberg. 2009. Cognitive enhancement: Methods, ethics, regulatory challenges. *Science and Engineering Ethics* 15: 311–341.
Bublitz, J.-C. 2013. My mind is mine!? Cognitive liberty as a legal concept. In *Cognitive enhancement. An interdisciplinary perspective*, ed. E. Hildt and A.G. Franke, 233–264. Dordrecht: Springer.
Buchanan, A. 2009. Human nature and enhancement. *Bioethics* 23: 141–150.
Buchanan, A. 2011a. *Better than human: The promise and perils of enhancing ourselves*. Oxford/New York: Oxford University Press.
Buchanan, A. 2011b. *Beyond humanity? The ethics of biomedical enhancement*. Oxford/New York: Oxford University Press.
Chadwick, R. 2008. Therapy, enhancement and improvement. In *Medical enhancement and posthumanity*, ed. B. Gordjin and R. Chadwick, 25–37. Dordrecht/Heidelberg/London/New York: Springer.
Clark, A. 2003. *Natural-born cyborgs*. Oxford: Oxford University Press.
Coady, C.A.J. 2009. Playing God. In *Human enhancement*, ed. J. Savulescu and N. Bostrom, 155–180. Oxford/New York: Oxford University Press.
Coenen, Ch., M. Schuijff, M. Smits, P. Klaassen, L. Hennen, M. Rader, and G. Wolbring. 2009. *Human enhancement. Study*. Brussels: European Parliament.
Cole-Turner, R. (ed.). 2011. *Transhumanism and transcendence: Christian hope in an age of technological enhancement*. Washington, DC: Georgetown University Press.
Cornips, L., and M. van Asselt. 2013. Human enhancement in futures explorations. In *Engineering the human. Human enhancement between fiction and fascination*, ed. B.-J. Koops, C.H. Lüthi, A. Nelis, C. Sieburgh, J.P.M. Jansen, and M.S. Schmid, 45–61. Berlin/Heidelberg: Springer.
Dabrock, P., R. Denkhaus, and S. Schaede. 2010. *Gattung Mensch*. Tübingen: Mohr Siebeck.
Damm, R. 2011. Medizinrechtliche Aspekte der Ästhetischen Chirurgie. Normstrukturen, Regelungsprobleme und Steuerungsebenen. In *Im Dienste der Schönheit?* ed. B. Lüttenberg, A. Ferrari, and J.S. Ach, 207–246. Berlin: LIT Verlag.
Daniels, N. 2000. Normal functioning and the treatment-enhancement distinction. *Cambridge Quarterly of Healthcare Ethics* 9: 309–322.
Davis, K. 2003. *Dubious equalities and embodied differences. Cultural studies on cosmetic surgery*. Lanham: Rowman & Littlefield.
Dees, R.H. 2007. Better brains, better selves? The ethics of neuroenhancements. *Kennedy Institute of Ethics Journal* 17: 371–395.
Devereaux, M. 2008. Cosmetic surgery. In *Medical enhancement and posthumanity*, ed. B. Gordjin and R. Chadwick, 159–174. Dordrecht/Heidelberg/London/New York: Springer.
Donnelly, J. 1982. Human rights as natural rights. *Human Rights Quarterly* 4: 391–405.
Elliot, C. 2004. *Better than well. American medicine meets the American dream*. New York/London: Norton.
Fenton, E. 2008. Genetic enhancement – A threat to human rights? *Bioethics* 22: 1–7.
Frazer, S. 2003. *Cosmetic surgery, gender and culture*. Basingstoke: Palgrave Macmillan.
Fukuyama, F. 2003. *Our Posthuman future. Consequences of the biotechnology revolution*. New York: Picador.
Gimlin, D. 2012. *Cosmetic surgery narratives*. Harmondsworth: Penguin.
Glover, J. 1984. *What sort of people should there be?* Basingstoke: Palgrave Macmillan.
Gordjin, B. 2008. Converging NBIC technologies for improving human performance. A critical assessment of the novelty and the prospects of the project. In *Medical enhancement and posthumanity*, ed. B. Gordjin and R. Chadwick, 225–235. Dordrecht/Heidelberg/London/New York: Springer.
Grunwald, A. 2002. Philosophy and the concept of technology – On the anthropological significance of technology. In *On human nature. Anthropological, biological and philosophical foundations*, ed. A. Grunwald, M. Gutmann, and E. Neumann-Held, 179–194. Berlin/Heidelberg/New York: Springer.

Grunwald, A. 2013. Are we heading towards an 'Enhancement Society'? In *Cognitive enhancement. An interdisciplinary perspective*, ed. E. Hildt and A.G. Franke, 203–216. Dordrecht: Springer.

Habermas, J. 2003. *The future of human nature*. Cambridge: Polity Press.

Harris, J. 2007. *Enhancing evolution: The ethical case for making better people*. Princeton: Princeton University Press.

Heilinger, J.-C. 2010. *Anthropologie und Ethik des Enhancements*. Berlin/New York: De Gruyter.

Heilinger, J.-C., and Crone, K. 2013. Human freedom and enhancement. *Medicine, Health Care and Philosophy.* Published online, 22 March.

Hildt, E. 2013. Cognitive enhancement – A critical look at the recent debate. In *Cognitive enhancement. An interdisciplinary perspective*, ed. E. Hildt and A.G. Franke, 1–14. Dordrecht: Springer.

Hofman, B. 2001. Complexity of the concept of disease as shown through rival theoretical frameworks. *Theoretical Medicine* 22: 211–236.

Holm, S., and M. McNamee. 2011. Physical enhancement: What baseline, whose judgment? In *Enhancing human capacities*, ed. J. Savulescu, R.H.J. ter Meulen, and G. Kahane, 291–303. Chichester: Wiley-Blackwell.

Hume, D. 1739/1740. *A treatise of human nature*, 2nd ed, ed. L. A. Selby-Bigge. Oxford: Clarendon, 1978.

Ida, R. 2009. Should we improve human nature? An interrogation from an Asian perspective. In *Human enhancement*, ed. J. Savulescu and N. Bostrom, 59–70. Oxford/New York: Oxford University Press.

Jones, G., and M. Whitaker. 2012. Transforming the human body. In *Beyond human. From animality to transhumanism*, ed. Ch. Blake, C. Molloy, and S. Shakespeare, 254–279. London/New York: Continuum.

Juengst, E.T. 1998. What does *Enhancement* mean? In *Enhancing human traits. Ethical and social implications*, ed. E. Parens, 29–47. Washington, DC: Georgetown University Press.

Juth, N. 2011. Enhancement, autonomy, and authenticity. In *Enhancing human capacities*, ed. J. Savulescu, R.H.J. ter Meulen and G. Kahane, 34–48. Chichester: Wiley-Blackwell.

Kass, L.R. 2002. *Life, liberty and the defense of dignity. The challenge for bioethics*. San Francisco: Encounter Books.

Kipke, R. 2011. *Besser werden. Eine ethische Untersuchung zu Selbstformung und Neuro-Enhancement*. Paderborn: Mentis.

Koops, B.-J. 2013a. Concerning 'Humans' and 'Human' rights. Human enhancement from the perspective of fundamental rights. In *Engineering the human. Human enhancement between fiction and fascination*, ed. B.-J. Koops, C.H. Lüthi, A. Nelis, C. Sieburgh, J.P.M. Jansen, and M.S. Schmid, 165–182. Berlin/Heidelberg: Springer.

Koops, B.-J. 2013b. Conclusion: The debate about human enhancement. In *Engineering the human. Human enhancement between fiction and fascination*, ed. B.-J. Koops, C.H. Lüthi, A. Nelis, C. Sieburgh, J.P.M. Jansen, and M.S. Schmid, 183–192. Berlin/Heidelberg: Springer.

Kurzweil, R. 2003. *Human body version 2.0*. www.kurzweilai.net/human-body-version-20.

Langer, M., and B. Wimmer-Puchinger. 2011. Der gemachte Körper – die genormte Frau. In *Im Dienste der Schönheit?* ed. B. Lüttenberg, A. Ferrari, and J.S. Ach, 71–89. Berlin: LIT Verlag.

Lanzerath, D. 2011. Professionsethische Aspekte aktueller Praktiken der Optimierung der menschlichen Natur. In *Entgrenzung der Medizin. Von der Heilkunst zur Verbesserung des Menschen?* ed. P. Wehling and W. Viehöver, 251–270. Bielefeld: Transcript.

Lenk, Ch. 2011. Enhancement vor dem Hintergrund verschiedener Konzepte von Gesundheit und Krankheit. In *Entgrenzung der Medizin. Von der Heilkunst zur Verbesserung des Menschen?* ed. P. Wehling and W. Viehöver, 67–87. Bielefeld: Transcript.

Little, M.O. 1998. Cosmetic surgery, suspect norms, and the ethics of complicity. In *Enhancing human traits. Ethical and social implications*, ed. E. Parens, 162–176. Washington, DC: Georgetown University Press.

Lohmann, G. 2014. How to protect ‘Human Nature’ – By human dignity, human rights or with ‘Species-Ethics’ argumentations? In *Human rights and human nature*, ed. M. Albers, Th. Hoffmann, and J. Reinhardt, 163–174. Berlin/Heidelberg/New York: Springer.

Lüthi, C.H. 2013. Historical and philosophical reflections on natural, enhanced and artificial men and women. In *Engineering the human. Human enhancement between fiction and fascination*, ed. B.-J. Koops, C.H. Lüthi, A. Nelis, C. Sieburgh, J.P.M. Jansen, and M.S. Schmid, 11–28. Berlin/Heidelberg: Springer.

Menuz, V., Th. Hurlimann, and B. Godard. 2013. Is human enhancement also a personal matter? *Science and Engineering Ethics* 19: 161–177.

Merkel, R., G. Boer, J. Fegert, T. Galert, D. Hartmann, B. Nuttin, and S. Rosahl. 2007. *Intervening in the brain. Changing psyche and society*. Berlin/Heidelberg/New York: Springer.

Mill, J.St. 1874. Nature. In *Three essays on religion*, 3rd ed, ed. J.St. Mill, 3–68. London: Longmans, Green, Reader, and Dyer.

Missa, J.-N., and L. Perbal (eds.). 2009. *Enhancement. Éthique et philosophie de la médecine d'amélioration*. Paris: Vrin.

Müller, O. 2008a. Der Mensch zwischen Selbstgestaltung und Selbstbescheidung. Zu den Möglichkeiten und Grenzen anthropologischer Argumente in der Debatte um das Neuroenhancement. In *Die "Natur des Menschen" in Neurowissenschaft und Neuroethik*, ed. J. Clausen, O. Müller, and G. Maio, 185–209. Würzburg: Königshausen & Neumann.

Müller, O. 2008b. Der Mensch und seine Stellung zu seiner eigenen Natur. Zum Status anthropologischer Argumente in der bioethischen Debatte. In *Mensch ohne Maß? Reichweite und Grenzen anthropologischer Argumente in der biomedizinischen Ethik*, ed. G. Maio, J. Clausen, and O. Müller, 15–57. Freiburg/München: Alber.

Müller, O. 2010. *Zwischen Mensch und Maschine. Vom Glück und Unglück des Homo faber*. Berlin: Suhrkamp.

Müller, O., J. Clausen, and G. Maio (eds.). 2009. *Das technisierte Gehirn. Neurotechnologien als Herausforderung für Ethik und Anthropologie*. Paderborn: Mentis.

Nationaler Ethikrat. 2005. *Forum Bioethik: Auf dem Weg zum besseren Menschen? Der Einsatz der Medizin zur Steigerung körperlicher und geistiger Fähigkeiten*. Berlin: Official Transcript.

Oyama, S. 2002. The nurturing of nature. In *On human nature. Anthropological, biological and philosophical foundations*, ed. A. Grunwald, M. Gutmann, and E. Neumann-Held, 163–170. Berlin/Heidelberg/New York: Springer.

Parens, E. 2006. Creativity, gratitude, and the enhancement debate. In *Neuroethics. Defining the issues in theory, practice, and policy*, ed. J. Illes, 75–86. Oxford/New York: Oxford University Press.

Pollmann, A. 2014. Human rights beyond naturalism. In *Human rights and human nature*, ed. M. Albers, Th. Hoffmann, and J. Reinhardt, 123–138. Berlin/Heidelberg/New York: Springer.

President's Council on Bioethics. 2003. *Beyond therapy. Biotechnology and the pursuit of happiness*. Washington, DC: President's Council on Bioethics.

Ramsay, P. 1970. *Fabricated man. The ethics of genetic control*. New Haven: Yale University Press.

Reinhardt, J. 2014. Human rights, human nature, and the feasibility issue. In *Human rights and human nature*, ed. M. Albers, Th. Hoffmann, and J. Reinhardt, 139–160. Berlin/Heidelberg/New York: Springer.

Roco, M.C., and W.S. Bainbridge (eds.). 2003. *Converging technologies for improving human performance. Nanotechnology, biotechnology, information technology and cognitive science*. Dordrecht: Kluwer Academic Publishers.

Rothhaar, M. 2014. Species, potentiality and their manipulation. In *Human rights and human nature*, ed. M. Albers, Th. Hoffmann, and J. Reinhardt, 175–185. Berlin/Heidelberg/New York: Springer.

Roughley, N. 2011. Human natures. In *Human nature and self design*, ed. S. Schleidgen, M. Jungert, R. Bauer, and V. Sandow, 11–33. Paderborn: Mentis.

Sagoff, M. 2005. Nature and human nature. In *Is human nature obsolete? Genetics, bioengineering, and the future of the human condition*, ed. H.W. Baillie and T.K. Casey. Cambridge, MA: MIT Press.

Sandel, M.J. 2007. *The case against perfection. Ethics in the age of genetic engineering.* Cambridge, MA: Harvard University Press.
Sandel, M.J. 2009. The case against perfection. What's wrong with designer children, bionic athletes, and genetic engineering. In *Human enhancement*, ed. J. Savulescu and N. Bostrom, 71–89. Oxford/New York: Oxford University Press.
Schramme, Th. 2007. A qualified defense of a naturalist theory of health. *Medicine, Health Care, and Philosophy* 10: 11–17.
Schramme, Th. 2011. The body as source of prudential value. In *Human nature and self design*, ed. S. Schleidgen, M. Jungert, R. Bauer, and V. Sandow, 67–81. Paderborn: Mentis.
Stock, G. 2002. *Redesigning humans. Our inevitable genetic future*. Boston/New York: Houghton Mifflin Company.
Stroop, B. 2011. Traurige Entlein und glückliche Schwäne? Glück in der Debatte um ästhetisch-chirurgische Eingriffe als Enhancement. In *Im Dienste der Schönheit?* ed. B. Lüttenberg, A. Ferrari, and J.S. Ach, 143–165. Berlin: LIT Verlag.
Synofzik, M., and Th. Schlaepfer. 2008. Stimulating personality. Ethical criteria for deep brain stimulation in psychiatric patients and for enhancement purposes. *Biotechnology Journal* 3: 1511–1529.
Tepe, H. 2014. Rethinking human nature as a basis for human rights. In *Human rights and human nature*, ed. M. Albers, Th. Hoffmann, and J. Reinhardt, 59–80. Berlin/Heidelberg/New York: Springer.
Tobey, D.L. 2004. What's really wrong with genetic enhancement: A second look at Our Posthuman Future. *Yale Journal of Law and Technology* 6: 54–160.
Toledano, P. 2011. *A new kind of beauty*. Stockport: Dewi Lewis Publishing.
von Harbou, F. 2014. The natural faculty of empathy as a basis for human rights. In *Human rights and human nature*, ed. M. Albers, Th. Hoffmann, and J. Reinhardt, 97–110. Berlin/Heidelberg/New York: Springer.
Wehling, P., and W. Viehöver. 2011. Entgrenzung der Medizin – Transformationen des medizinischen Feldes aus soziologischer Perspektive. In *Entgrenzung der Medizin. Von der Heilkunst zur Verbesserung des Menschen?* ed. P. Wehling and W. Viehöver, 7–47. Bielefeld: Transcript.
Witthøfft Nielsen, L. 2011. The concept of nature and the enhancement technologies debate. In *Enhancing human capacities*, ed. J. Savulescu, R.H.J. ter Meulen, and G. Kahane, 19–33. Chichester: Wiley-Blackwell.
World Health Organization. 1948. *Constitution of the World Health Organization*. http://www.who.int/gb/bd.

About the Authors

Marion Albers, Professor of Public Law, Information and Communication Law, Health Law and Theory of Law at Hamburg University. Main areas of research: Fundamental Rights, Information and Internet Law, Data Protection, Health Law and Biolaw, Police Law and Law of Intelligence Services, Theory and Sociology of Law. Selected Publications: "Risikoregulierung im Bio-, Gesundheits- und Medizinrecht", in: Albers (ed.), *Risikoregulierung im Bio-, Gesundheits- und Medizinrecht*, Nomos 2011, 9–33; "Höchstrichterliche Rechtsfindung und Auslegung gerichtlicher Entscheidungen", in: *Grundsatzfragen der Rechtsetzung und Rechtsfindung*, VVDStRL Bd. 71, de Gruyter, 2012, 257–295; "Privatheitsschutz als Grundrechtsproblem", in: S. Halft/H. Krah (eds.), *Privatheit. Strategien und Transformationen*, Stutz, 2013, 15–44; "Realizing the Complexity of Data Protection", in: S. Gutwirth/P. de Hert/R. Leenes (eds.), *Reloading Data Protection*, Springer, 2014, 213–235.

Bianka S. Dörr, Dr. iur., LL.M., Attorney at law, senior lecturer and researcher for private and biomedical law at various universities in Switzerland, co-leader of the SNSF-funded research project "Self-determination at the end of life in Swiss law" at the University of Lucerne, former Visiting Professor for Medical-, Health- and Private Law at the University of Augsburg. Selected Publications: "Research with Human Biological Material and Personal Data in Biobanks – Legal and Regulatory Framework in Switzerland", in: C. Lenk/N. Hoppe/K. Beier/C. Wiesemann (ed.), *Human Tissue Research: A European Perspective on the Ethical and Legal Challenges*, Oxford University Press 2011, 99–107; "Once Given – Forever in a Biobank? Legal Considerations on the Protection of Donors and the Handling of Human Body Materials in Biobanks from a Swiss Perspective", in: P. Darbrock/J. Ried/J. Taupitz (ed.), *Trust in Biobanking*, Springer 2012, 139–152.

Thomas Hoffmann, PD Dr. phil., Assistant Professor of Practical Philosophy and Member of the Center of Human Rights Studies at the Otto-von-Guericke-University in Magdeburg. Selected Publications: Th. Hoffmann, Das Gute (forthcoming 2014); "Die Moral der Strafe und die Grenzen staatlicher Bestrafung",

M. Albers et al. (eds.), *Human Rights and Human Nature*, Ius Gentium: Comparative Perspectives on Law and Justice 35, DOI 10.1007/978-94-017-8672-0,

in: Zeitschrift für Menschenrechte 1/2012; "Jeremy Bentham", "Richard Rorty", both in: G. Lohmann, A. Pollmann (ed.), *Menschenrechte: Ein interdisziplinäres Handbuch*, J.B. Metzler 2012; "Erste Natur, Zweite Natur und das Gute für den Menschen", in: T. Hoffmann, M. Reuter (ed.), *Natürlich gut. Aufsätze zur Philosophie von Philippa Foot*, Ontos 2010; "Menschenrechte, Demokratie und globale Gerechtigkeit", in: Zeitschrift für Menschenrechte 1/2009.

Bernd Ladwig, Prof. Dr. phil., Professor of Political Sciences at the Free University Berlin. Selected Publications: B. Ladwig, "Global Justice, Cosmopolitanism, and Moral Path Dependency", in: Philosophy and Social Criticism 2012 (forthcoming); *Gerechtigkeitstheorien zur Einführung*, Hamburg 2011; "Menschenrechte und Tierrechte", in: Zeitschrift für Menschenrechte 1/2010; "Menschenwürde als Grund der Menschenrechte? Eine Kritik an Kant und über Kant hinaus", in: Zeitschrift für Politische Theorie 1/2010; *Gerechtigkeit und Verantwortung. Liberale Gleichheit für autonome Personen*, Berlin 2000; Editor, with D. Jörke: *Politische Anthropologie. Geschichte, Gegenwart, Möglichkeiten*, Baden-Baden 2009.

Georg Lohmann, Prof. Dr. phil., Professor of Practical Philosophy and Member of the Center of Human Rights Studies at the Otto-von-Guericke University Magdeburg. Selected Publications: G. Lohmann, "Unterschiedliche Kulturen – warum universelle Menschenrechte?", in: A. Holderegger/S. Weichlein/S. Zurbuchen (ed.), *Humanismus. Sein kritisches Potential für Gegenwart und Zukunft*, Basel: Schwabe Verlag 2011, 217–232; "Die rechtsverbürgende Kraft der Menschenwürde. Zum menschenrechtlichen Würdeverständnis nach 1945", in: Zeitschrift für Menschenrechte, 1/2010, 46–63; "On the Relation between Moral, Juridical and Evaluative Justifications of Preimplantation Genetic Diagnosis (PGD)", in: Ethical Perspectives, Journal of the European Ethics Network, Vol. 10 (2003), Nr. 3–4, 196–203; "Unparteilichkeit in der Moral", in: K. Günther/L. Wingert (ed.), *Die Öffentlichkeit der Vernunft und die Vernunft der Öffentlichkeit. Festschrift für Jürgen Habermas*, Suhrkamp: Frankfurt a. M., 434–455; Editor with S. Gosepath, *Philosophie der Menschenrechte*, Suhrkamp: Frankfurt/M. 1998, 2. ed. 1999, 3. ed. 2002.

Corinna Mieth, Prof. Dr. phil., holds a Chair for Practical Philosophy (esp. Political Philosophy and Philosophy of Law) at the Institute for Philosophy I at Ruhr-Universität Bochum. Publications: C. Mieth, *Positive Pflichten. Über das Verhältnis von Hilfe und Gerechtigkeit in Bezug auf das Weltarmutsproblem*, Berlin/New York: de Gruyter 2012; "On Human Rights and the Strength of Corresponding Duties", in: G. Ernst and J.-Ch. Heilinger (ed.), *The Philosophy of Human Rights*, Berlin/New York: de Gruyter 2011; "Menschenwürde und soziale Gerechtigkeit: Überlegungen zu Kant, Rawls und Margalit", in: Ch. Thies (ed.), *Der Wert der Menschenwürde*, Paderborn: Schöningh 2009; "World Poverty as a Problem of Justice?", Ethical Theory and Moral Practice 2008, 11(1).

Arnd Pollmann, Dr. phil, Privatdozent for Practical Philosophy at the University of Magdeburg, Germany; co-founder of the interdisciplinary working group "Arbeitsstelle Menschenrechte" at the University of Magdeburg and co-editor of the German "Zeitschrift für Menschenrechte" (Journal for Human Rights). Main interests: ethical theory, moral philosophy, political philosophy, human rights, and human dignity. Selected Publications: A. Pollmann, *Philosophie der Menschenrechte. Zur Einführung*, Hamburg: Junius 2007 (together with C. Menke); *Unmoral*, München: C.H. Beck 2010; Editor, with G. Lohmann, *Menschenrechte. Ein interdisziplinäres Handbuch*, Stuttgart: Metzler 2012.

Jörn Reinhardt, Dr. phil., Research Associate at the University of Hamburg (Chair of Public Law, Information and Communication Law, Health Law, and Legal Theory). Fellow of the Young ZiF (Center for Interdisciplinary Research, Bielefeld). Selected Publications: J. Reinhardt, *Der Überschuss der Gerechtigkeit. Perspektiven der Kritik unter Bedingungen modernen Rechts*, Weilerswist: Velbrück 2009; "Reiterativer Universalismus. Sprache – Kultur – Menschenrechte", in: C. Bäcker/M. Klatt/S. Zucca-Soest, *Sprache – Recht – Gesellschaft*, Tübingen: Mohr Siebeck 2012, 291–308.

Markus Rothhaar, Dr. phil., Professor for Bioethics at the Catholic University of Eichstätt-Ingolstadt. Selected Publications: M. Rothhaar, "Menschenwürde qua Autonomie und Anerkennung: Kant und Fichte", in: J.C. Joerden/E: Hilgendorf/N. Petrillo/F. Thiele (ed.), *Handbuch Menschenwürde und moderne Medizin*, Berlin: Duncker & Humblot 2013, 73–98; "Philosophische und Klinische Ethik", in: A. Frewer/A. May/F. Bruns (ed.), *Ethikberatung in der Medizin*, Berlin/Heidelberg/New York: Springer 2011, 33–43; articles on "Reproduktionsmedizin", "Stammzellforschung", "Status des Embryo", in: R. Stoecker et al. (ed.): *Handbuch für Angewandte Ethik*, Stuttgart/Weimar: Metzler 2011, 424–432.

Tetsu Sakurai, Professor, Graduate School of Intercultural Studies, Kobe University, Japan. Selected Publications: Tetsu Sakurai and Usami Makoto (ed.), *Human Rights and Global Justice*, Archiv fuer Rechts- und Sozialphilosophie (ARSP) Beiheft, Franz Steiner Verlag, forthcoming; "Liberal Eugenics Faces the Challenge of Global Justice" (in Japanese), Intercultural Studies 2010 (Vol. 35); *Introduction to Legal Philosophy and Legal Ideas* (in Japanese), Minerva Publishing 2010 (co-edited with Takeshita Ken, Tsunoda Takeshi and Ichihara Yasuhisa); "Historical Origin and the Development of the Concept of Person" (in Japanese), in: Inoue Tatsuo (ed.), *Lectures on Contemporary Legal Philosophy*, Shinzansha 2009; *Liberal Eugenics and Justice* (in Japanese), Nakanishiya 2007.

Peter Schaber, Prof. Dr. phil., Professor of Applied Ethics at the University of Zurich. Selected Publications: P. Schaber, *Menschenwürde*, Reclam 2012; "Human Dignity, Self-Respect, and Dependency", in: P. Kaufmann et al. (eds): *Humiliation, Degradation, Dehumanization. Human Dignity Violated*, Springer 2011, 151–158; *Instrumentalisierung und Würde*, Mentis-Verlag 2010; "Are There Insolvable Moral Conflicts?", in: P. Baumann/M. Betzler (ed.): *Practical Conflicts. New Philosophical Essays*, Cambridge, Cambridge University Press 2004, 279–294; "Value Pluralism: Some Problems", in: Journal of Value Enquiry 33 (1999), 71–78.

Ingrid Schneider, PD Dr. phil., Reader (*Privatdozentin*) in the Department for Political Sciences, and Senior Researcher in the Research Group 'Medicine and Neurosciences' at the Research Centre for Biotechnology, Society and the Environment (BIOGUM), University of Hamburg. Recent publications: I. Schneider, "To Be or not IP? Exploring limits within patent law for the constitutionalization of intellectual property rights and the governance of synthetic biology in human health", in: Law and the Human Genome Review, No. 37 (2012), 193–233; *The European Patent System. Shifts in Governance through Parliaments and Civil Society* (Das Europäische Patent system. Wandel von Governance durch Parlamente und Zivilgesellschaft), Frankfurt/New York: Campus 2010.

Mateusz Stępień, Dr, LLD, Assistant Professor in the Department of Sociology of Law, Faculty of Law and Administration, Jagiellonian University in Krakow, Poland. Relevant publications: M. Stepien, "Konfucjański przewodnik samodojrzewania" (The Confucian guide of self-cultivation), in: J. Włodarski, K. Zaidler, M. Burdeelski (ed.), *Chiny w oczach Polaków* (China through the eyes of Poles), Danzig 2010; "Deklaracja Bangkocka – czyli azjatycki głos w sprawie praw człowieka" (Bangkok Declaration – The Asian voice on human rights), Annales UMCS, vol. XVII (2010); "The question of non-human primates morality", Studies in the Philosophy of Law 2010; "The three stages of judge's self-development", in: A. Amaya/Hock Lai Ho (ed.), *Law, Virtue and Justice*, Oxford: Hart Publishing 2013.

Harun Tepe, Professor of Philosophy at the Department of Philosophy and the Department of Human Rights and Director of Centre for Research and Application of the Philosophy of Human Rights at Hacettepe University in Ankara. Selected publications: "'Ethics of Journalism' or On Ethical Problems in the Mass Media," in: I. Kuçuradi (ed.), *Ethics of Professions: Medicine, Business, Media, Law,* Berlin/Heidelberg/New York: Springer 1999, 101–114; "Identity, Cultural Identities and Human Rights," in: I. Kuçuradi (ed.), *Human Rights in Turkey and the World in the Light of Fifty-year Experience*, Ankara: Hacettepe University 2002, 76–85; "Epistemological Dilemmas of Contemporary Ethics," in: H. Tepe, S. Voss (eds.), *Ethics, The Proceedings of the Twenty-first World Congress of Philosophy Volume 1*, Ankara: Philosophical Society of Turkey 2007, 25–30; "The Place of Human Rights in Public Policy," in: *Social Transformation, Bioethics and Public Policy*, Ankara: Turkish National Commission for UNESCO 2012, 43–55.

Frederik von Harbou, Doctoral candidate at the University of Zurich and legal trainee in the district of the Superior Court of Justice of Berlin. Selected Publications: F. von Harbou, "A Remedy called Empathy: The Neglected Element of Human Rights Theory", in: Archiv für Rechts- und Sozialphilosophie, No. 2/2013 (forthcoming); "Anspruch und Anthropologie. Unparteilichkeit und Universalismus als rechtsethische Herausforderungen", in: *Gleichheit und Universalität*, Archiv für Rechts- und Sozialphilosophie 2012, Beiheft 128, 151–164.

Index

M. Albers et al. (eds.), *Human Rights and Human Nature*, Ius Gentium: Comparative Perspectives on Law and Justice 35, DOI 10.1007/978-94-017-8672-0,

MIX
Papier aus verantwortungsvollen Quellen
Paper from responsible sources
FSC® C105338

If you have any concerns about our products,
you can contact us on
ProductSafety@springernature.com

In case Publisher is established outside the EU,
the EU authorized representative is:
Springer Nature Customer Service Center GmbH
Europaplatz 3, 69115 Heidelberg, Germany

Printed by Libri Plureos GmbH
in Hamburg, Germany